The Retail Jeweller's Guide

The Retail Jeweller's Guide

Sixth edition

Kenneth Blakemore
Formerly Managing Editor
Watchmaker, Jeweller and Silversmith

Revised and expanded by Eddie Stanley

OXFORD AUCKLAND BOSTON JOHANNESBURG MELBOURNE NEW DELHI

Butterworth-Heinemann
Linacre House, Jordan Hill, Oxford OX2 8DP
225 Wildwood Avenue, Woburn, MA 01801-2041
A division of Reed Educational and Professional Publishing Ltd

A member of the Reed Elsevier plc group

First published by Iliffe Books Ltd 1969
Second edition 1973
Third edition 1976
Fourth edition 1983
Fifth edition 1988
First published as a paperback edition 1994
Reprinted 1997
Sixth edition 2000

British Library Cataloguing in Publication Data
A catalogue record for this book is available from the British Library

ISBN 0 7506 4650 0

Typeset by Avocet Typeset, Brill, Aylesbury, Bucks
Printed in Great Britain by MPG Books Ltd, Bodmin, Cornwall

Contents

Foreword

We have had a very long and happy association with Eddie Stanley who has acted as our Training Consultant, conducting many seminars for us over the years and acting as a constant source of information and guide to many of our employees. I was therefore delighted and honoured to have been asked by Eddie to write the Foreword for the sixth edition of the *Retail Jeweller's Guide* and I wish to take this opportunity of thanking him for all his support, not only of our Company but of the jewellery industry as a whole. Eddie is well respected and highly qualified and the work that he has done in contributing to this new edition of Kenneth Blakemore's *Retail Jeweller's Guide* is absolutely relevant to the needs of the Retail Sales Professional.

As a Company that has won the Retail Jeweller's Training Award many times, we consider the Jewellery Education Training programme to be of vital importance to the development of our people and believe that it ties in absolutely with our own in-house training programme.

The changes to this edition include a section devoted to the valuation of jewellery, an up-date to the Hallmarking Act which was amended on 1 January 1999, and the inclusion of a section on gemmological instruments and their uses. This book is therefore of vital importance to all students on JET 1 and JET 2 as well as those students taking the JET Gemmological Course. It is also useful for people employed within the manufacturing and pawnbrokingtrades.

In these times of increasing change it is more important than ever that we ensure that everyone within our unique trade becomes even more knowledgeable about jewellery to enable us to offer that essential point of difference and allow the customer to enjoy the experience of buying from a 'real jeweller'.

Mark Adlestone FGA
Joint Managing Director, Beaverbrooks the Jewellers
April 2000

Preface

This, the sixth edition of the *Retail Jeweller's Guide*, has again been expanded to include new sections. Eddie Stanley has contributed an invaluable chapter on gemmological instruments and their use, as well as a further one on valuing jewellery.

Since 1988, when the previous edition was published, there have been a number of developments which needed to be incorporated in this new edition. Gemmology, for instance, is a continually developing science and companies in Europe, America and in the Far East have developed an increasing variety of synthetic gems and synthetic simulants in recent years. There has also been a great increase in gem enhancement over the same period. For instance, it has been stated that today there is hardly any sapphire on the market whose colour has not been improved by heat treatment. It has therefore been necessary to incorporate additional material in this edition to take account of these developments. There have also been changes in the hallmarking legislation which anyone selling jewellery or silverware or carrying out appraisals needs to be aware of.

I have also taken the opportunity which a new edition provided to correct a number of those errors which seem inevitably to creep in to such a wide-ranging book and also to amend the text here and there in the light of what I have learned during the intervening years since the last edition was published. This edition also benefits from the revision carried out by Eddie Stanley, who brought to the task a wide-ranging experience in all aspects of the trade. Therefore, this new edition contains many minor amendments to the text designed to make the book both more accurate and more comprehensive.

I am indebted to a great number of people who over the years have taken the time to talk to me about various aspects of the trade. Without their help I could not have produced this book. I would also be remiss if I did not express my appreciation for all the help and encouragement I received from the late Gordon Andrews, the one-time secretary of the NAG whose extensive knowledge of the trade made him an invaluable editor during the production of the earlier editions of this book. Gordon Andrews' contribution to trade education, which he pioneered in the immediate post-war years, sometimes seems to be forgotten today and it was his desire to provide students studying for the Retail Jewellers examinations with an up-to-date text book which led him to persuade me to produce the *Retail Jeweller's Guide*.

Kenneth Blakemore

Acknowledgements

My contribution to this sixth edition has only been made possible by the kind assistance given by the following:

Mr D.W. Evans of the London Assay Office and the British Hallmarking Council, for assistance and guidance with regard to the changes in hallmarking on 1 January, 1999.

Mr Michael Alchin, Birmingham Assay Office Master, and his staff, namely Mr Clive Broadfield and Phyllis Benedikz, the Assay Historian, for their assistance in the updating of Chapter 8 on the techniques of assaying.

For the Gemmological section, thanks are due to Mr John Harris of Carlisle for permission to use his slides of gemstone absorption spectra.

Mr Michael Marshall for the proof reading of the chapters on the valuation of jewellery and on gemmological instruments.

Mr John Bodenham of Bodenham & Shorthouse, Manufacturing Jewellers of Birmingham, for permission given to use illustrations from their collection.

Mr Matt Moyse of Rotary Watches Limited, for permission to use their flow diagram of the quartz watch, in Chapter 10.

Mr J.G.N. Nicholls, Managing Director of Harvey Thompson, for his assistance during the research of pawnbroking in the UK.

Eddie Stanley

Chapter 1

Precious metals

Gold

Craftsmen have been making jewellery from gold for more than 6000 years, and it is not difficult to understand why the craftsmen of so many different civilizations have worked in this metal. Gold has a beautiful yellow colour, and it is a wonderful metal with which to work, being the most malleable of all metals. A troy ounce of it can be beaten out into a sheet 100 ft square. This sheet will be 1/280 000 in. thick, so thin as to be transparent. Gold is also very ductile. That same troy ounce could have been rolled down between grooved rollers and then drawn, and drawn again, through ever finer dies, to make 50 miles of gold wire.

When the craftsman produces a piece of jewellery from this sympathetic metal, the jewellery has a satisfying weight about it. Gold is a very heavy metal, considerably heavier than lead. Lead has a specific gravity of only 11.4 compared with a specific gravity for gold of 19.3. But gold has much more to offer than beauty and weight. The physical characteristic which has probably had most to do with its importance in the history of mankind is its great nobility. It is almost certainly because gold does not tarnish like copper, or rust away like iron, that great men down the ages have used it to display their wealth to the world, loading their tables with gold plates and gold vessels, and decorating their women with gold ornaments, the beauty of which they knew would never diminish.

It was also, in part, the relative indestructibility of gold that persuaded so many nations in the past to mint their most valuable coins from it. Hoards of gold coins, unearthed after centuries in the ground, have come to light looking as bright as newly-minted pennies. People would have had confidence in such coins, the glory of which they knew time would not decay, and confidence has always been of prime importance where the coinage is concerned.

In the language of the chemist gold is a 'relatively inert substance'; it does not combine with the oxygen in the air. That is why oxide does not form on it, as iron oxide collects on the surface of iron to form what is known as rust. Even strong acids do not attack gold, with the exception of that potent mixture of hydrochloric and nitric acids that has been given the name *aqua regia* – royal water – because it can dissolve the noblest of metals, gold.

Although a rare substance, gold occurs in at least small quantities in most countries. It is carried along in minute quantities in the currents of many rivers, including the Rhine and the Danube, but these specks are not worth trying to trap. Gold can be found even in Britain. There is a gold mine on a hillside between Barmouth and Dolgelly in North Wales, which was originally worked by local village people as a co-operative. Later the royal family came to have first call on the gold found there, and every royal bride in this century has been married with rings made from gold from this Welsh mine. The Welsh gold is in the form of grains and nuggets; the largest nugget found in this mine weighed about four ounces. Incidentally, Welsh gold has a characteristic colour, a reddish-yellow with a green tinge.

Gold has been mined in many countries at one time or another. Today, the main sources are South Africa, with a production of some 20 million ounces a year, the USSR, Canada, the USA, and Australia.

Alluvial gold

In some parts of the world there was a time when gold could be had merely for the effort of picking it up. If a man happened to be in the right place at the right time, he had merely to pan the sands and pebbles of a river bed or, at worst, sink a shallow shaft into the ground to find a fortune in gold.

The gold found in this way was alluvial gold, sometimes called 'placer' gold. This alluvial gold had originally been what we call 'native' gold, that is, it existed in crystalline form in a matrix, usually of quartz. The gold and its attendant matrix originated as a result of changes in the Earth's crust. It was formed by chemical reaction when hot magma from deep within the Earth was forced up to the surface through fissures in the Earth's crust many millions of years ago. Over the centuries many of the quartz reefs that contained the crystalline gold were weathered away by rain and frost and the debris from this weathering, which contained the gold, was gradually washed down the hillsides away from what prospectors call the mother lode.

Alluvial gold is sometimes found as 'dust' or sometimes in the form of nuggets. A nugget is a lump of a quarter of an ounce or more, resulting from tiny grains of gold and sand compounding together. Some really big nuggets have been discovered. Perhaps the most famous is the Welcome Stranger, found in a cart rut in New South Wales in the middle of the nineteenth century. The Welcome Stranger was two feet long.

Gold rushes

It was the finding of placer gold that started the famous gold rushes of the nineteenth century; the first was the California rush of 1849. This was started as a result of a landowner in the Sacramento Valley, the Swiss adventurer John Augustus Sutter, sending out a party of men to harness a stream on the lower slopes of the Sierra Nevada to drive a sawmill. James Marshall, the foreman entrusted with the job, found yellow specks in the muddy bed of the mill-race that he was building. Word soon spread abroad. The first effect of Marshall's find was that the price of shovels rose from $1 to $10. Then some 50000 men crossed the United States on foot in search of a fortune. Another 50000 booked passages on ships. These men, the 'forty-niners', came pouring into the Sacramento Valley like a swarm of locusts destroying as they came. Ironically they ruined the man responsible for it all, John Augustus Sutter. The colony he had founded in the fertile valley was laid waste and he died in poverty.

Most of those who came to California for the gold were content to pan the river gravels, watching the gold, which is much heavier than the gravel containing it, sink to the bottom of their pans. But there were some who realized that there was an even greater prize for anyone who could find the mother lode from which all this alluvial gold had originally come.

In 1859 two Irishmen, Peter O'Riley and Pat McLaughlin found the mother lode high up in the Sierras. The gold could be seen in an outcrop of quartz, the outward and visible sign of the richest reef that had ever been found. The reef was only a small one, and in ten years the Big Bonanza mine was worked out. But in that time it has produced £130 million worth of gold and the silver found in association with this gold was worth another £170 million.

The most famous and the best documented of the gold rushes was that to the Yukon. This began in 1896. A certain George Washington Cormack, a sailor who had deserted from an American ship, married a Siwash Indian girl and settled down to live with her family. One day he was in a creek of the Yukon river with his Indian 'brothers' looking for suitable stands of trees to cut for lumber, when he noticed a glinting in the sand at the water's edge. Siwash George, as he was called, had found the first gold in the Yukon.

Word of Siwash George's find soon spread and prospectors from far and wide began to arrive with their packs and pans at Rabbit Creek on the Klondyke, where gold was there for all to see. The people from the little settlement of Forty Mile nearby poured out of the town to pan for gold. Rumours of the Yukon strike gradually filtered through to the outside world and in June 1896, two ships, the *Excelsior* and the *Portland*, arrived in Seattle harbour carrying the first of the gold from the Yukon. There was a ton of gold on board the *Portland* alone. Then the rush began in earnest.

Men, many with no knowledge of gold or how to live in the wilderness, poured into the north-west corner of Canada heading for the Klondyke river. A town began to spring up on the banks of the Klondyke – the famous Dawson City. In 1896 its population was 500. A further 1000 people arrived in 1897, and by 1898 there were over 30 000 inhabitants. In that year £3½ million worth of gold was taken out of the creeks, and for another 20 years men continued to find gold along the sandy banks of the Klondyke and its tributaries.

Australia, too, had her gold rushes in the nineteenth century. Gold was first found there in 1839. As the century progressed nuggets were picked up here and there on that continent, but the great discoveries came after Edward Hammond Hargreaves, an unsuccessful prospector, returned from California to his native country in 1850, determined to discover gold there.

The following year he struck lucky at Ophir in New South Wales. Soon people were flocking to Australia from all over the the world to dig for gold. Obscure places, such as Ballarat and Bendigo Creek, acquired a world-wide fame overnight and the population of Australia increased from 400 000 to 1 260 000 in the next decade.

Birth of the rand

The great find of the century was, however, made neither in Australia nor in North America, but in the Transvaal. There are a number of versions describing how the great gold-rich quartz reef at Witwatersrand came to be found in 1886, but it is now generally accepted that it was discovered by a prospector called George Harrison, who had the bad luck to be killed by a lion before he could benefit from his good fortune. It is doubtful whether Harrison had even an inkling of the vast potential of the reef. What he found was a small outcrop of a reef which stretched for nearly 300 miles from the Far Eastern Transvaal into the Orange Free State. In places this reef went 10 000 feet below the surface, by far the biggest gold-bearing reef that had ever been found, and not only is it still yielding gold in prodigious quantities, but even today new mines are being sunk into still untapped areas along its length.

The Witwatersrand reef was unlike the other gold-bearing reefs. The thinly sprinkled gold, which is alluvial and not native, is contained in a conglomerate. The crystalline quartz of the original reef had been crushed and then reconstituted under enormous pressure. Such a reef could not be worked by individual prospectors. It called for mining and crushing operations that required both capital and industrial organization. Men who had made fortunes in the diamond fields of South Africa, Cecil Rhodes and Barney Barnato among them, became interested in gold. The mines were dug, the machinery brought in, and a well-organized tent-town grew up to house the miners. This tent-town was to develop into the great city of Johannesburg and the Rand gold-mining industry was born. Concentration became inevitable if the area was to be efficiently exploited. Once there were hundreds of mines in the Rand. Today there are only 46 mines owned by seven large companies, but those 46 mines supply the world with a large proportion of its gold.

By the end of the nineteenth century alluvial gold was growing scarce. It was becoming increasingly unlikely that a man with nothing but a pan, a shovel, a grubstake and a good constitution would pick up a fortune somewhere in the wilderness. Gold mining had become an industry where rich men grew richer and poor men had little chance to strike it rich.

Down the mine

Some 300 000 black workers and about a tenth of that number of white are employed at the mines of the Rand. Most of the black workers labour below ground, going down in the cages suspended from the pithead gear, down perhaps a mile or so to the tunnels, known as drives, that lead to the slopes where mining actually takes place. Here they drill holes in the rock face, and in the holes place cartridges of explosive. In a year some 27 000 tons of explosive are detonated in the Rand, to loosen 50 million tons of gold-bearing rock.

Extraction

Every ton of rock mined yields, on average, only about half an ounce of gold. The specks of gold in the quartz conglomerate are usually too small to be seen with the naked eye, and the extraction is a complicated business. First the rock goes to the stamping mills, where it is crushed in a series of great mechanical pestles and mortars. The crushed rock then goes to the tube mills, or the ball mills, where it is tumbled with large pieces of rock, or steel balls the size of cannon balls, to reduce it to a sludge. From this sludge the gold is then chemically extracted.

The capital required to put a gold mine into production today is enormous, but the rewards, of course, are considerable. The Western Deep Levels in the Transvaal (Figure 1.1), which came into full production in 1970, had already cost the Anglo-American Corporation £15 million by 1960. But this mine is expected to go on producing for at least 60 years, and to yield gold worth £800 million. The cost of bringing such a mine into production today would be about four or five times that figure.

The concentration of the mines of the Witwatersrand reef was not solely the result of a need for more capital than the average prospector could provide. A contributory cause was a sudden and dramatic drop in yield in the year 1889, when many miners thought that the reef was worked out and sold their claims for what they could get. But the fault lay not in the reef, which was as rich in gold as ever, but in the amalgamation process for extracting the gold from the sludge. This consisted of putting the

Figure 1.1 Western Deep Levels mine on the Witwatersrand Reef (by Courtesy of De Beers)

sludge containing the gold into barrels with mercury, which has a strange affinity for gold. The gold adhered to the mercury, and the mixture was then transferred to a retort and heated. The mercury, which has a much lower vaporization point than gold, was driven off as a gas, leaving the gold behind. This method worked adequately for the gold-bearing ore from near the surface, but below 500 ft the character of the ore changed and the miners found that they were getting less and less gold from their ore.

By 1890 many of the small operators had already given up gold mining and gone to search for diamonds around Kimberley, or had returned to their farms. Three Scotsmen, two brothers called Forrest and a man named MacArthur, came to the rescue of the miners of the Rand. With others, they had been working for some time in Glasgow on an alternative method of gold recovery. In 1890, they arrived in Johannesburg to demonstrate the cyanide process that they had evolved. In this process the sludge containing the gold is put into tanks with cyanide of potassium which dissolves the gold. The liquid containing the dissolved gold is next filtered off and clarified. Zinc dust is then added to precipitate the gold, which is washed in sulphuric acid to get rid of excess zinc and then placed in an electric oven and roasted, or calcined. The calcine produced is only about 60 per cent pure at this stage. It is then mixed with a flux of sand, borax and manganese, and smelted in an electric furnace. During the smelting the impurities combine with the flux to form a slag, and the gold, which is much heavier, sinks to the bottom of the crucible. After the slag has been removed, the gold is poured into iron moulds to produce ingots each weighing about 1000 ounces.

The mercury amalgamation process is still retained at some mines as a first process, because it is a relatively cheap way of recovering about 40 to 50 per cent of the gold in a rich ore. But at most mines, the processed ore goes straight to the great tanks of green cyanide which are a feature of modern gold mines.

Refiners

The gold poured from the smelting furnace after the recovery process is still far from the 99.98 to 99.99 per cent purity needed for making carat golds for the jewellery trade. Therefore it must be further refined. This degree of fineness is necessary, not merely because pure gold makes it easier to arrive at the exact proportions of gold and base metals in carat gold, but also because the wrong impurities can affect the colour or the working qualities of a carat gold.

When the gold reaches the refinery it contains some silver, some base metal including copper, and also, perhaps, platinum and osmiridium. In the refinery, the gold is first subjected to the chlorination process, which exploits the relative inertia of gold. The gold is heated in a furnace to a molten state. Chlorine gas is then passed down a tube into the furnace. This does not affect the gold, but it turns the silver and the base metals present into chlorides, which float to the top of the furnace where they can be ladled off. The gold is now 99.6 to 99.7 per cent pure; but still not pure enough. Therefore it has to go through one further process.

To get it into the right state for this process, the molten gold is poured from the furnace into a circular pit filled with water on the floor of the refinery. The rapid cooling solidifies the gold into granules which are placed in a large stoneware pan containing *aqua regia*. The gold and the combined acids are boiled together and the gold dissolves in the *aqua regia*. This gold-rich solution is allowed to cool for 6 to 8 hours, during which time any silver chloride left in the gold will have sunk to the bottom of the pan because silver chloride is an insoluble salt. The liquid containing the gold is then siphoned off leaving the silver chloride. An agent must be added to this liquid which will throw down the gold, but not the copper, platinum or any other metal dissolved with it in the *aqua regia*. Iron chloride is the agent most often used to precipitate the gold. Following precipitation the liquid is siphoned off, and the gold, now in a sandy form, is washed with hydrochloric acid to get rid of any traces of iron chloride. The platinum, copper and other metals left in the *aqua regia* are subsequently extracted because the refiner obviously does not waste any of the valuable by-products of the refining process.

In recent years a method of refining gold electrolytically, similar to the Betts process used for separating silver from lead ores, has been devised and this is much quicker than the traditional method.

Carat golds

The refiner now has almost pure gold ready for alloying. Gold has to be alloyed because, in its pure state, it is unsuitable for working because it is both heavier and softer than lead.

The refiner has to achieve three things when he alloys his gold with other metals for use by the jewellery trade: working properties, the necessary degree of purity, and the right colour. He aims at providing the jeweller with a metal suitable for any job that it may be required to do. He must also provide gold which will pass rigid tests at the assay office. He will further have to be prepared to offer his gold to the jewellery manufacturer in various colours, ranging from a natural yellow to purple.

Let us first deal with quality. In Britain there are four legal standards for gold: 22 carat, 18 carat, 14 carat and 9 carat. In some countries 8 carat gold is legal, in others 15 carat. The carat system is merely a method of expressing the proportions of gold to other metals in a particular alloy. Pure gold is reckoned as 24 carat, 22 carat gold contains 22 parts of gold to 2 parts of other metals, while 9 carat gold contains 9 parts of gold to 15 parts of other metals.

The law requires that 22 carat gold shall contain not less than 22 parts of gold, or 14 carat not less than 14 parts of gold, and so on. There is nothing to prevent a jeweller from selling a gold which assayed at 17 carats, but because the assay office would have marked such gold at 14 carat there would be no point in doing so.

Copper and silver are the principal metals used to alloy gold, though nickel, zinc, cadmium, iron and aluminium are also used. The refiner's problem is to reach a satisfactory compromise between working qualities and colour for each different carat quality. Copper produces a hard alloy, with good wearing qualities and a red colour; silver gives a more malleable gold of a yellow colour. The refiner has to balance one consideration against the other. If hard gold is needed for pins and snaps, he will include as much copper as possible, using only enough silver to give him the colour required. If malleability is called for, such as for deep drawing in a press, then he will use more silver than copper.

Taste in various markets, and fashion at different periods, influence the demand for the various colours of gold which the refiner can produce. On the Continent, for instance, people have tended in the past to like their gold redder than have people in Britain. Sometimes there has been a swing towards white gold, both in Britain and abroad, and every now and again multicoloured gold jewellery comes into fashion.

Red and yellow gold, as has been shown, are the result of varying the proportions of copper and silver in the alloy. A low-carat white gold can be produced by alloying with a high proportion of silver, but for the higher carat white golds nickel or palladium must be introduced. A blue gold is produced by adding 6 parts of iron to 18 parts of gold. Purple gold is the result of adding 6 parts of aluminium to 18 parts of gold. An alloy of three-quarters gold and one-quarter zinc produces a lilac colour. Silver and cadmium are used to produce green gold.

When the gold has been alloyed to produce the various carats, colours and grades, including special grades for spinning and stamping, casting, enamelling and making chain, the refiners then process these golds to produce the materials needed by the jewellery trade. They turn the golds into sheet of various gauges, plain and fancy wires, tube and solders. Even the solder used by the jewellery trade must pass an assay (Figure 1.2). They make findings, too, such as snaps and catches and settings, earring wires and bolt-rings and even complete eternity ring mounts and wedding ring blanks.

Gold coins

The first gold coins were minted by King Croesus of Lydia in 500 BC. Since then many countries have at some time had a gold coinage, and gold coins were in daily use in Britain over many centuries. Henry VII gave us the sovereign at the end of the fifteenth century, it was the equivalent of the 20 silver shillings, up to then the basis of UK monetary systems. The guinea came later, so named because it was made from gold from the Guinea coast of West Africa. At first the guinea, too, was the equivalent of 20 shillings, but after the Restozration at the end of the seventeenth century silver was in such short supply that silversmiths began clipping the silver coinage to obtain enough raw material to fulfil the flood of commissions from the new court.

Eventually, things deteriorated so that no one would exchange a good golden guinea for 20 of these decimated pieces of silver. People began to demand an extra

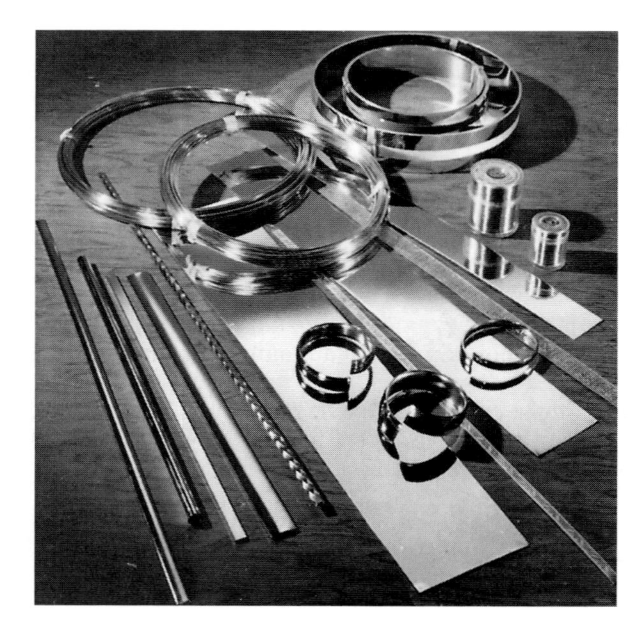

Figure 1.2 Precious metal, sheet, wire, tube and findings supplied to the manufacturing jewellery trade by the refiners

shilling to make up for the bits that had been clipped off. In 1717 the guinea was officially proclaimed to be worth 21 shillings. and so it remained for the next 100 years, after which it ceased to be minted.

By the early nineteenth century it was becoming widely recognized that a 21 shilling coin was unwieldy and was an unsuitable unit for the complicated calculations of an aspiring commercial nation. In 1817, therefore, the old sovereign, worth 20 shillings, was brought back again. The new sovereign was made of 22 carat gold and weighed 123.27447 grains. Sovereigns were last used as currency in 1917.[1]

Gold and economics

The gold used for jewellery represents only a fraction of the world output. A considerable part of the gold mined every year is put to economic uses. This means that it goes back into the ground again into great vaults such as those at Fort Knox in the USA. Gold remains the currency of international trade and lying in a country's vaults it serves to prop up the economy of that country.

Britain was on the Gold Standard until 1931. This meant, in simple terms, that the national as well as its international trading was based on gold. The currency was closely linked to gold, which was maintained at an official standard price of 84 shillings an ounce from 1717 to 1931, except for the ten years between 1915 and 1925. While Britain was on the Gold Standard, anyone owning a piece of gold could, in theory at least, take it along to the Mint and exchange it for sovereigns at the standard rate. Also, it was laid down by the Bank Charter Act of 1844, that for every pound note issued, the equivalent in gold must exist in the vaults of the Bank.

World War I upset this established order. Between 1915 and 1920 the price of gold fluctuated wildly, soaring to 120 shillings an ounce. Authority had lost control, and the law of supply and demand had taken over.

In 1925, the authorities reasserted control, and Britain returned to the Gold Standard, but only until the slump in 1931. In that year paper currency and gold finally parted company. Since then gold has had no fixed price; like that of any other commodity its price has fluctuated according to supply and demand.

The price of gold

The price of gold rose less during six centuries than it has risen during the past 30 years. In 1344, gold was worth 22s. 9½d. an ounce. By the sixteenth century the price had risen to 45s. an ounce. During the seventeenth century it rose as high as 86s. an ounce, and then came the long period when the price remained stable at 84s. The quoted price of gold is normally for fine gold, which is of a minimum fineness of 996 parts per 1000.

When Britain left the Gold Standard the price of gold rose rapidly to 130s. in 1932, and 140s. by 1935 – the period known to retail jewellers in those days as 'the gold rush'. The then Chancellor of the Exchequer appealed to the patriotism of the people to sell their sovereigns to the banks for the face value of 20 shillings. Avarice prevailed over patriotism and hoards of sovereigns were unearthed and brought into jewellers' shops to sell for from 25 to 27 shillings each. And it was not only their sovereigns that people brought to the jeweller – they brought their old watches, their Victorian jew-

1. Under the 1936 Coinage Offenses Act it is an offence to impair, diminish or lighten coins or a coin so that it would resemble a UK or foreign coin.

ellery, their old spectacle frames, and disused dentures. Many jewellers in those days were too busy buying gold to have time to sell.

The price of fine gold has continued to rise. By 1945 it was 172 shillings, by 1949 it had risen to 248 shillings, and in 1982 it fluctuated around £175 per ounce. The 1987 price was £291 per troy ounce. Fine gold is now £160 oz troy (September 1999).

Industrial uses

In addition to its use by the jewellery trade and its economic role gold has many industrial applications today. It is used, for instance, for electrical contacts and instrument springs. Its resistance to corrosion makes it especially applicable in the chemical engineering and the textile industries and for use in space satellites.

Substitutes for gold

Ever since rich men began to use gold to boast their wealth to the world, there must have been a market for jewellery and hollowares that had the appearance of being made of gold but were within the means of the less wealthy. The bronze jewellery of early civilizations was the costume jewellery of the day. Joan Evans, when writing of the post-Roman period in her *History of Jewellery*, differentiates between the sophisticated jewellery of the Mediterranean culture reserved for the governing classes using 'materials that were inherently precious' and the 'barbarian technique of gilt bronze of the governed'.

Pinchbeck

By the eighteenth century the use of copper and zinc alloys to imitate gold was an established practice, and the most famous of these alloys was pinchbeck. Its inventor was a watchmaker who had a workshop in London's Fleet Street. Christopher Pinchbeck (1670–1732) evolved his alloy of 83 parts copper to 17 parts zinc, with the idea of marketing an expensive-looking watch case at an attractive price. In the years that followed, however, many jewellers worked in pinchbeck, and made more use of it than did the watch trade. In the eighteenth century there was a considerable vogue for pinchbeck chatelaines, for instance, and these were manufactured in their thousands.

Gilding-metal

Pinchbeck would today be called a gilding-metal. The jeweller now has a range of copper and zinc alloys available: 90 copper to 10 zinc, 85 to 15 and 80 to 20 gilding-metal. These are strong, corrosion-resistant alloys of good colour, with slightly different working characteristics. Those with more zinc are the more malleable. A still softer grade, normally a 75 to 25 alloy is used for wire. For casting, which, since the development of centrifugal casting, has replaced stamping as the principal fabrication technique of the costume jewellery industry, lead and tin replace some of the zinc to provide better flow.

Though gilding-metal has a golden colour it is now seldom used in its naked state. The customer of today expects imitations of gold to look as golden as the real thing, and so the manufacturer puts a veneer of gold on his gilding-metal.

Rolled gold

Until the recent development of the hard gold plating technique, the most satisfactory method of putting a veneer of gold on to gilding-metal was to make rolled gold. Rolled gold was a British invention, dating from 1817. It is made in the same way as Old Sheffield Plate, and was a logical development of this process which Thomas Bolsover invented in 1743. Under heat and pressure a block of gold is fused on to a larger block of base metal to make an open sandwich which is then rolled out repeatedly between highly polished rollers. The sandwich becomes increasingly thin, until sheet of the required gauge is achieved. It was this rolling process which gave rolled gold its name.

In Pforzheim, which since the twentieth century has been the rolled gold capital of Europe, rolled gold with gold on one side only is called 'doublé'. Rolled gold with gold on both sides is known as 'treble'. The situation is confused, however, by the fact that 'doublé' is also used as a description of quality.

Rolled gold wire is made by producing a gold seamless tube. This is done by first deep-drawing sheet gold to form a cup. The end of this cup is sawn off to produce a short fat tube. This has a base metal core inserted in it. It is then drawn down in the same way as gold wire. Great care has to be taken in the production of rolled gold sheet, tube and wire, because only a limited amount of polishing can be done, otherwise the gold veneer will be penetrated and the metal beneath brought to light. Dust is as much the enemy of the producer of rolled gold as it is of the watchmaker. Any dust or grit that gets on to the tools between which rolled gold jewellery is formed will mark the surface and it will be impossible to polish out such imperfections.

Silver is sometimes used in place of gilding-metal as the basis for rolled gold, and so are bronze (an alloy of copper and tin) and nickel silver (an alloy of copper nickel and zinc). When silver is used a two-colour decorative effect is sometimes achieved by cutting through the gold to show the silver beneath. Modern textured finishes achieved by diamond milling also result in the penetration of the gold surface. Articles decorated in this way are subsequently plated, and are sometimes marked 'RGPl'. However, rolled gold is not really a suitable raw material from which to make this type of jewellery. It is preferable to make it in a base metal and then gold plate it with a deposit comparable to the gold layer on rolled gold.

Various qualities of gold are used for rolled gold. A British Standard published in 1960 advocated a minimum standard of 9 carat. Johnson Matthey described 9, 10, 12 and 14 carat rolled gold as 'the more usual qualities'. All these qualities were supplied in a choice of three colours: red, yellow or white.

The thickness of the gold veneer is usually measured in microns (a thousandth of a millimetre). The established standard for a good quality rolled gold watch case is 20 to 30 microns. This should mean that no part of the case is covered with less than a thickness of 0.02 millimetre of gold. To test these cases, samples are taken from every batch and the base metal backing dissolved away with acid, leaving a case-shaped skin of gold which is then measured all over with a micrometer. It is generally reckoned that a 20 micron case will give 20 years of wear – a year for every micron.

Standards based on the micron were, in the past, generally reserved for articles such as cigarette cases, cuff links and watch cases, where it is essential for the gold to be evenly distributed because their edges are subjected to considerable wear. When this even spread of gold was less important, the gold content was often expressed in milliemes. This merely indicated the proportion of gold to base metal, and said nothing about its distribution over the surface of the article. Jewellery, which is described as of 10 millieme quality, will have 10 grams of fine gold to every kilogram. The well-known doublé, advertised by the Pforzheim manufacturers, was a 10 to 12 millieme rolled gold. Charnier had 25 grams to the kilo, and Union 50 grams to the kilo. Today the micron unit of measurement is being increasingly used for jewellery, whether produced from rolled gold or as a result of gold plating.

Gilding

The idea of covering base metal with a skin of gold goes back centuries before Bolsover's time. No one knows who invented gilt, or when, but it is known that from very early in history craftsmen were applying gold to the surface of other metals using a bonding agent similar to niello. Later, mercurial gilding was invented. This process is the reverse of the gold recovery technique originally used in the Witwatersrand gold mines. In the recovery process gold is amalgamated with mercury to separate the gold from the rock. In mercurial gilding the amalgamation was done with the idea of attaching the gold to another material. The combined gold and mercury were applied to, say, a base metal chalice, and then the mercury, which has a low vaporization point, was driven off by heating, leaving the gold fused to the chalice. This process, which is still used occasionally, is very effective. Pieces of silver made 400 or more years ago that still have the original gilt clinging to them exist. It is, however, a very dangerous process because mercury fumes are lethal.

Electrogilding

After G.R. Elkington took out a patent in 1840, the electro-deposition of metals began on a commercial scale. Before many years had passed the plating bath had become a familiar piece of equipment in the jeweller's as well as the silversmith's workshop.

The plating bath is filled with a solution containing gold potassium cyanide. The plater hangs the jewellery to be plated in the bath, so that this jewellery becomes the cathode. If the bath is thought of as a battery, the jewellery acts as the negative pole. The plater switches on the current, and the gold atoms in the solution all move along and attach themselves to the cathode. The solution does, of course, eventually become exhausted of its gold. Sometimes this is overcome by making the positive pole of the bath, the anode, from a sheet of gold. When the current is switched on, gold atoms are detached from the anode, go into the solution, travel across the bath and attach themselves to the cathode. Electrodeposition then becomes a continuous process until the anode is stripped of its gold. Nowadays, however, it is usual to use platinum, or platinum-plated anodes that are permanent, and either to replace the solution when the gold content drops below a certain level or to top it up with a replenisher. This method allows much closer control of the plating solutions.

In the past electrogilt had some disadvantages when compared with rolled gold. Because 24 carat gold was normally deposited in the plating bath, the gilt was very soft. Also gilding usually gave only the yellow colour associated with pure gold. With rolled gold it was easy to produce any of the colours in which carat golds are available. Further, the technique of electrogilding allowed the deposition of infinitely thin coatings of gold, and offered a temptation that was not always resisted. There is at least some basis of fact to the stories circulating in the trade about parsimonious electroplaters. A great deal of gilt jewellery has come on to the market which had only a very thin plating of soft gold and would not stand up to much wear. Moreover, plating was sometimes done in primitive conditions, often in a bucket in a backroom. Standards of cleanliness were not high, and unless all the grease is removed from a piece of jewellery before plating, the gilding will simply peel off.

As a result of recent developments, some of the drawbacks to electrogilding have disappeared. Today, new processes allow carat golds of various shades to be deposited by the addition of copper, silver, and other metallic salts to the bath. Also it is now possible to deposit gold which is a great deal harder than the electrogilt of the past by varying the current density.

Hard gold plating

The relatively new technique of depositing a plating of hard gold is now widely used both in the watch case industry and the jewellery industry.

The details given by Johnson Matthey in their electroplating data will suffice to explain what the process achieves. The salts and anodes that they supply give a deposit of 98 per cent gold and 2 per cent silver, with a Vickers hardness of between 115 and 120 as compared with a hardness of between 65 and 70 for normal electrogilt. The plating 'can be produced to any normal thickness required'. In other words, by using this technique a coating of a known micron thickness can be applied that will have better wearing qualities not only than gilt, but better even than rolled gold. It has been said that a 20 micron hard gold plated case will wear as well as a 30 micron rolled gold case.

Hard gold plating is carried out in laboratory conditions (Figure 1.3). Success depends on very careful control of pre-cleaning and of the contents of the plating bath. Care must also be taken to control temperature, current density and the rate of agitation of the pieces being plated. Plating with a gold alloy is more exacting than plating with pure gold. After plating there are more cleaning operations, and then the thicker platings are heat-treated.

This process, like any other, can of course be debased. The production of cheap hard gold plated jewellery, just flashed with gold and produced without the necessary strict control, could make the public as suspicious of this as they are of gilt and rolled gold. Efforts are being made within the trade to establish minimum standards for hard gold plate to prevent this loss of reputation, but there are still no legally imposed standards.

Since the Trade Descriptions Act was passed jewellers have become increasingly aware of the necessity for differentiating between a rolled gold article and one which has been gold plated.

Gold core and gold shell

In the past a number of descriptive names were given by the trade to gold-covered articles, which are now not permitted by the assay offices. One of these was gold core. This was a 9 carat rolled gold, a fifth of which was gold when it was despatched from the refiners. Some of the gold was, of course, bound to be polished away in the process of turning the sheet metal into jewellery.

Gold shell was a name sometimes given to an electroplated article having gold of sufficient substance deposited on it to stand up by itself, if the base metal were etched away.

Figure 1.3 A computer-controlled plating bath. The timing of the various stages of the process is programmed; all the operator needs to do is load the jigs

Gold back and front

This is best explained by describing a gold back and front heart-shaped locket. This consists of a hollow heart-shaped frame of gilding-metal to which two heart-shaped pieces stamped out of gold, usually 9 carat, have been attached. When the locket was closed, all that could be seen was made of gold. The manufacturer was thus able to offer what appeared to be a gold locket at a considerably lower price than would have been possible had the whole been made of gold. Gold back and front articles have been declared illegal by the assay offices.

Lacquer

Lacquer is a synthetic resin suspended in a volatile solvent which evaporates, leaving the resin behind as a film over an article. Lacquer can be either clear or coloured. Gold-coloured lacquer is used to simulate gold on very cheap jewellery. Coloured lacquers are used to simulate enamel. Colourless lacquer is used to protect other finishes. It is not unusual for gilt jewellery and compacts to be given a coating of colourless lacquer. Provided that it is not subjected to too much wear, lacquer can last for quite a long time.

Silver

It is not known when men first started to turn silver into ornaments and vessels, but silver was used by both the ancient Egyptians and the Sumerians. Athens was founded on silver, just as in the nineteenth century Johannesburg was founded on gold. When the mines that had made the Athenians rich were eventually worked out, the silversmiths of the classical period, including the Romans who were great lovers of silver, got their raw material from the mines of Thrace and of Spain.

The European silversmiths of the Middle Ages got their silver, from which they wrought plate for the great new Gothic cathedrals, from Germany and from Bohemia. In the sixteenth century the Spanish conquerors of Mexico discovered seemingly inexhaustible lodes, and started to send galleons laden with bullion back to Spain. They enriched not only their mother country, but England as well. English privateers flourished for over a century on the Spanish Main, bringing into Plymouth Sound many a bright cargo pillaged from the King of Spain.

Silver from Mexico and from South and Central America satisfied the needs of silversmiths and coiners of Europe until the nineteenth century, when the great finds and silver rushes in the USA began. Harry Comstock's famous lode was discovered in 1859, and the first shaft of the great Anaconda mine was sunk in 1894. And there were many more smaller finds. In Australia, too, there were silver rushes as well as gold rushes. The most famous find there was Broken Hill, which the miners gradually reduced to rubble, so that today there is no sign of a hill.

Recovery today

The leading silver-producing nation is still Mexico which produces 35 per cent of all silver mined this side of the Iron Curtain. The USA comes next, contributing 25 per cent, followed by South and Central America, where 20 per cent of the total is mined. Canada contributes 10 per cent and Australia 5 per cent. The total world production, excluding that mined in Russia, China and the other Communist countries, which the US Bureau of Mines estimates to be about 10 million ounces a year, was over 220 million ounces in 1966. The current figure is not very different from this. Silver is the commonest, and therefore the cheapest, of the precious metals. However, as a result of the attempt by Bunker Hunt in 1980 to corner the silver market the price of silver did rise to £22 an ounce. In 1987 the price of silver was £4.67 per oz troy. Fine silver is now around £3.20 per oz troy.

Silver as a by-product

Apart from its widespread occurrence, there is another reason why silver is relatively cheap. Most of what is recovered is a by-product of the mining for other metals. Between 80 and 85 per cent of all the silver recovered today is found in lead, zinc, copper, nickel, tin or gold ores. The most important of these sources is lead ore. Where lead is found, so almost invariably is silver. About 45 per cent of all silver comes from lead mines.

Silver in Britain

At one time or another silver has been mined in most countries of the world. In Britain silver has been recovered in small quantities over a long period. In the olden days it was a by-product of the Cornish tin and lead mines. More recently the lead mine at Millclose in Derbyshire produced a few thousand ounces a year, but this closed down in 1961. Recent statistics do not include Britain as a source of silver, but the British Isles may soon be included again, because lead mines have been opened up in Eire and Scotland which yield ore with a silver content.

The properties of silver

The craftsman finds silver a particularly sympathetic metal in which to work. It has many properties in common with gold, and for some purposes the smith finds it preferable.

Like gold, silver is a malleable and ductile metal. It cannot be beaten out quite as thin or drawn quite as fine as gold, but it has excellent working qualities none the less. It can be rolled to make sheet 0.0005 in. thick, or a troy ounce can he turned into a mile of wire. It bends easily without a tendency to crack, so that it is a good metal from which to make boxes. It can be tempered to make blades for fruit and butter knives. Its malleability means that it can be raised under the hammer or deep drawn in the press to make hollowares.

Silver is a noble metal, second only to gold in nobility. It does not combine readily with the oxygen in the atmosphere, and is resistant to the attack of many corrosive substances. It is unaffected by the acetic acid in fruit juices, or by hydrochloric or weak nitric acid. However, it is attacked by strong nitric acid and by sulphuric acid, neither of which attacks gold.

The advantages of silver over gold include its greater tensile strength, its relative lightness and its greater reflecting power. Its strength, and the fact that fine silver has a specific gravity of only 10.5 compared with 19.32 for fine gold, means that it is a better metal than gold for making into domestic hollowares. A pint tankard made of gold would be quite unwieldy and less strong than one made of silver.

Silver is often thought of as having a soft white beauty, but in fact silver is the brightest of all the metals when newly polished. It reflects 95 per cent of the incident light that falls on it, compared with 92 per cent for gold, 82 per cent for copper, and 62 per cent for rhodium, the metal which is used invariably to plate white gold to improve the colour.

Silver, unlike chromium and steel, is not unattractive when scratched. Because it is a soft metal the scratches that it acquires do not have such sharp edges as the scratches taken by harder metals. Also the scratches on silver become blurred with wear and with polishing, and it is this process of scratching and blurring which, over the years, imparts to silver that unique patina which antique collectors set such store by.

A disadvantage of silver is its affinity for sulphur. Silver egg-spoons are particularly vulnerable. The sulphur in eggs attacks silver turning the surface a nasty black-brown colour. This tarnish was not easy to remove employing the cleaning methods of the past.

Unhappily, the air around us is full of sulphur, in the form of sulphur dioxide, a gas which pours out of domestic and industrial chimneys. This is the main cause of tarnish on silver, and was a contributory factor in the decline in silver sales in the years after the last war. Fortunately, the chemist has done much to solve the tarnish problem and to make silver acceptable in a servantless age, by inventing various forms of tarnish protection. This subject will be discussed later in this chapter.

A further drawback of silver for the craftsman is its susceptibility in the alloyed form in which it is used in the trade, to firestain. Here again, however, the scientists have helped, and the problem has been minimized.

The silver used by the trade is alloyed with copper. During annealing and soldering, when the alloy is heated, the copper oxidizes and a black stain appears on the surface of the metal. As operation succeeds operation the stain tends to penetrate the metal. If this happens the firestain can be extremely difficult to remove. Surface stains can be removed by rubbing with a stone or by pickling in acid, but if the stain is deep-seated the silver has to be electroplated to hide it. Today, however, the refiners anneal their sheet and tube in an oxygen-free atmosphere, so that the material at least reaches the manufacturer unimpaired by firestain. Such stain as the silver now acquires in the workshop is usually easily removed, and the old practice of plating to hide firestain is today used far less widely. Plating, incidentally, imparts slightly greater tarnish resistance to an article, because pure silver resists tarnish a little better than sterling silver.

Refining

Although most silver today is a by-product, there are lodes, particularly in Mexico, rich enough to be mined for their silver alone.

These silver ores are crushed and the silver is recovered by the cyanide process. The crushed silver ore is dissolved in a cyanide solution, usually a calcium cyanide solution, and the silver is then precipitated by the addition of zinc. The resulting precipitate is between 75 and 90 per cent pure. After being washed with acid to remove surplus zinc, this impure silver is melted in a stream of air, which results in the oxidizing of any base metal present. The oxides are then skimmed off the surface of the silver which will now be about 98 per cent pure, and ready for despatch to the refinery.

When the silver is a by-product, other methods of recovery are used. The most important source of silver is lead ore. Two processes are used to separate the silver from the lead. These are known as the Parkes process and the Betts process.

The Parkes process uses the fact that zinc has a greater affinity for silver than lead, and so any copper, tin and other impurities are first separated out, then the lead containing the silver is melted and zinc added. The zinc combines with the silver and is removed from it by distillation.

The Betts process is electrolytic. The lead containing the silver is cast into sheets which serve as anodes in a bath containing lead fluosilicate. The passage of an electric current through the bath etches away the lead from the anode leaving a silver slime which can be scraped off and refined. This and the Parkes process both give a silver which is about 90 per cent pure.

The 90 per cent pure silver has to be further purified at the refinery before the various grades of alloyed silver, with the required working properties, can be produced. This purification can be achieved by further cupulation, that is, oxidation of the base metals present by heating the silver in a stream of air. Nowadays, though, most of the refining of silver is done electrolytically. The silver is cast into anodes, and these are placed in refining cells. The passage of an electric current through these cells results in crystalline silver of a purity of between 99.95 and 99.99 per cent being deposited on the cathodes.

The refiners produce silver of three qualities. They produce sterling silver, consisting of 92.5 per cent silver and 7.5 per cent base metal, mostly copper. Small quantities of Britannia standard silver, which contains 958.4 parts in 1000 silver, and which was introduced in the seventeenth century to discourage the clipping of the coinage by the silversmiths, are still used for reproduction pieces. Then there is 800 silver, which, as will be gathered from its name, is only 800 parts in 1000 silver, or 80 per cent pure. This standard is still widely used on the Continent. However, it is illegal to offer wares of this standard for sale in Britain unless they are not described as silver.

In addition to these different qualities the refiners supply the manufacturing silversmith with different types of silver for different jobs. Cadmium is often added to silver when sheet suitable for spinning is to be produced. High melting point alloys, free of zinc and cadmium, are made for enamelling.

Nowadays the manufacturer can buy a vast range of materials from the refiners. There is the sheet silver of various gauges which can be turned into coffee pots, cruets, napkin rings, sports trophies and spoons. Sheet silver gauges are expressed in BMG (Birmingham metal gauge). This is an irregular scale of measurement. The difference between 40 BMG at the top of the scale and 39 BMG is a hundredth of an inch. The difference between 1 BMG and 2 BMG is only a thousandth of an inch. Besides sheet, the refiners supply the trade with tube of various cross-sections, and with plain and fancy wires produced by stamping or rolling. They also supply rod, flakes, grain or shot used for casting, as well as findings, and the various solders with different melting points. Just how these materials are turned into hollowares, flatwares and smallwares will be described in a later chapter.

Silver coins

Silver had great importance in the past as a coin metal. Most countries had a silver coinage. In Britain this dates back hundreds of years. The term 'sterling' is said by some to have derived from the name 'starling', or 'sterling', given to the silver pennies used in Britain 700 years ago. They are said to have got the name because they were made by Easterlings, coiners from Hansa in North Germany. Other derivations suggested for the term are that it was the name of a twelfth century silver coin bearing a star, or that it came from the starlings on the coat of arms of Edward the Confessor which appeared on coins minted in his reign. It is not known which, if any, of these legends is true.

Silver coinage ceased in Britain in 1947, though deterioration had started 25 years earlier. Until 1922, British silver coins had been made of 925 silver, the same standard as the legal standard for wrought plate. In 1922, coins which were only 50 per cent silver began to be minted. During the next quarter of a century various alloys containing 50 per cent silver were used at the mint, and then finally in 1947 silver disappeared altogether as a result of the coinage act of that year. Since then British 'silver' coins have been made of an alloy called cupro-nickel, which consists of 75 per cent copper and 25 per cent nickel.

If, however, there is no longer a demand for silver from the mint, this has been more than compensated for by the growing demand from industry. The great popularity of photography resulted in an enormous demand for silver nitrate, the light-sensitive chemical used in the preparation of photographic emulsions. Silver is also important to the electrical and electronics industries because besides being the best conductor of heat of any metal, it is also the best conductor of electricity. Its nobility recommends it to the food and pharmaceutical industries.

Protection against tarnish

From time to time experiments have been carried out in an effort to produce a silver alloy that would not tarnish, but so far without success. If tarnish is to be avoided, therefore, silverwares must be protected in some way from the sulphur dioxide in the atmosphere. Until recently this could only be done by plating the silver with a metal that did not tarnish, or by coating the silver with lacquer.

Silverwares have been plated with gold down the centuries, but presumably in the early days when the atmosphere was less polluted with sulphur dioxide, the object of this was to enrich rather than protect the silver. Anyway, as any museum curator will confirm, mercurial gilt is not particularly tarnish resistant. Electrogilding offered better protection, but it had two drawbacks. It was very soft, and therefore did not stand up to wear very well. Also, it changed the silver's appearance, and not everyone liked their silver to look golden. Hard gold plating overcomes the first of these objections but does not overcome the second.

Lacquer is a much cheaper way of protecting silver without greatly altering its appearance. Modern transparent lacquers are a quite satisfactory way of protecting decorative silverwares for a number of years. Lacquer, of course, cannot be used for articles likely to be subjected to hard wear like flatware, or for anything that will be subjected to heat, such as an entrée dish or an ash tray. Even the lacquer on decorative silver will eventually wear off, and if silver treated with lacquer is accidentally cleaned with an abrasive, this will scratch the lacquer and cause peeling. The unprotected areas of the silver will then tarnish.

Recently, the scientists have evolved new and better ways of protecting silver. One technique is to plate the silver with a rhodium-based alloy. Rhodium, one of the platinum group of metals, provides very good protection. It is both a hard and a noble metal, and, when alloyed, its colour, if not acceptable to antique collectors, is sufficiently like silver to persuade less exacting owners to rid themselves of the burden of regular cleaning. It is not a cheap way of protecting silver because the initial cost is fairly high, but if properly applied a plating of rhodium alloy should give years of protection even to pieces having to stand up to a good deal of rough usage.

Another of the new methods of protection is to give the silver a coating of silicone. This is cheaper than rhodium plating, but the protection does not last nearly so long. On the other hand, the silicone coating alters the appearance of the silver only very slightly. Treated pieces tend to yellow with age, however, and the length of time which silicone protection lasts depends upon the amount of wear the silver receives. It could be months, it could be a year or more.

Besides the silicone treatment which is carried out in a workshop, there are domestic cleaners on the market that leave a silicone coating behind. This prevents tarnish for anything from a couple of weeks to months, depending on the amount of wear and also on the amount of sulphur dioxide in the atmosphere.

Silver not in daily use can be protected simply by storing it in the special papers or bags available. These all work in much the same way. The paper, or the fabric, from which the bags are made is impregnated with a substance (colloidal silver is sometimes used) which combines readily with sulphur dioxide. Thus the paper or bag acts as a filter extracting the sulphur dioxide from the air before it reaches the silver.

Cleaners and polishes

There is some confusion, which is understandable because there are so many products on the market today, between the silver cleaners and the silver polishes. The early cleaners worked on the principle of scouring the tarnish off the silver. They were abrasives hard enough to remove the tarnish but not to cut the silver. The rubbing involved also polished the silver; they were combined cleaners and polishes. Since the war a number of 'dips' have appeared on the market. These liquids, in which the silver is placed, dissolve the tarnish away and take the labour out of cleaning, making it possible to live with decorative silver in a servantless age. But for silver to look really nice it still has to be polished, though this involves no more trouble than polishing any other household possession. More recently, foam polishes, similar to those used for cars, have made their appearance, and these clean and polish at the same time. The domestic tarnish preventative liquids on the market also clean and polish the silver in one operation.

A solution came on to the market some years ago which, when applied to silver or silver-plated wares, actually deposits silver on the surface. A cleaner based on this solution has also been produced.

Substitutes for silver

Plated wares

The most convincing imitations of silverwares are base metal articles plated in some way with a coating of silver. An early technique for doing this was called 'close plating'. The article to be plated was first dipped in molten tin, and pieces of silver foil were applied and fixed in position by burnishing or hammering. A bond between the foil and the tin was effected by passing a soldering iron over the surface which melted the low melting point tin. A variation on this technique was known as 'French plating'. In this process the tin was omitted and the silver foil heated to a high temperature and burnished directly on to the base metal article.

Old Sheffield plate

Sheffield plate, invented by Thomas Bolsover in 1743, was a far more satisfactory technique. Bolsover's invention came at a propitious time. In the early years of the eighteenth century the middle classes were assuming an importance they had never had before. The merchants were growing rich, some of them very rich indeed, and they were beginning to ape their betters, the aristocracy, by setting fine tables laden with plate. They bought candlesticks and epergnes, tea-pots and coffee-pots, salts and dredgers from the great Huguenot silversmiths and from the native English smiths. Bolsover's invention made it possible for the slightly less successful members of the middle class to put on a show. Old Sheffield plate, as we call it, looked like silver and cost a great deal less. It was soon in great demand. From the time when the invention began to be fully exploited in the years following 1765, Sheffield became a boom town.

Old Sheffield plate was made, like rolled gold, by fusing a block of noble metal, in this case silver, on to a block of base metal, usually copper, and then rolling out the block to form sheet of various gauges.

Much of the Sheffield plate was made in the neo-classical style, usually known as the Adam style, and new techniques had to be developed to impart to it the necessary decoration. The traditional techniques of chasing and engraving could not be used because they would have cut into the silver surface and brought to light the copper underneath. Therefore the decoration had to be stamped or rolled on to the metal. If engraving was called for shields of silver were soldered into the pieces.

Old Sheffield plate is now attracting collectors in growing numbers. Unfortunately many of the old pieces became worn in use and were restored by electroplating when the copper began to show. This, of course, reduces the value of the piece from a collector's point of view.

Electroplating

Old Sheffield plate was eventually superseded by George Elkington's exploitation of electroplating in the 1830s and 1840s. In this process the silver is deposited on either copper or nickel silver. Nickel silver made its appearance in Europe in the 1820s, but it was not new. The Chinese had been using a similar alloy for centuries. The nickel-copper alloy which they used was known to European travellers to the Orient as white bronze.

In the early years of the nineteenth century experiments to discover a satisfactory alloy were carried out in Germany, which is why nickel silver is sometimes called German silver. The Germans evolved an alloy consisting of nickel, copper and zinc. The nickel provided hardness and tarnish resistance, the zinc gave a low melting point and improved the ductility. This alloy was originally used unplated as a silver substitute, and is still used in this way to produce very cheap flatware.

A similar metal was evolved in Sheffield in the eighteenth century. This metal, which consisted mostly of tin with some antimony and copper, was called white metal or Vickers metal after James Vickers who first used it to make holloware and flatware in the 1780s. It was later called Britannia metal.

Nickel silver began to be important when Elkington started to use it as the base metal foundation of e.p.n.s., electroplated nickel silver. The principles of electroplating have already been described in connection with gold plating. The only difference is that, for silver plating, silver salts are used in the bath and a silver anode is used to replenish the solution. The description nickel silver may well prove to be unacceptable under the Trade Descriptions Act 1968, but this has still not been tested in the courts.

Plating standards

There are unfortunately no legal standards for silver plating. The term A.1 was widely used by most firms for their top quality, but as a *Which?* report published a few years ago revealed, A.1 means very different things to different firms. A good A.1 quality would have a plating 0.001 in thick, about the thickness of tissue paper. Some firms expressed this as so many pennyweights per dozen tablespoons, or so many pennyweights per square foot. A plating of 16 pennyweights per square foot will give a lifetime's wear. The pennyweight was abolished on February 1, 1969 under the Weights and Measures Act, 1963 – but old usages die hard.

Recently, hard silver plating has been introduced. This promises even better wearing properties than normal silver plating, but at the present time only one company, a German one, has offered hard silver plated cutlery and flatware.

A British Standard Specification (BSS 5577:1984) has been produced for table cutlery and although this does not have the force of law a number of manufacturers have now adopted it. This Specification is very complicated: it covers both the over materials used and their application and lays down testing procedures. It specifies, for instance, a complicated formula for nickel silver. Probably the most important section of this standard so far as the retail jeweller is concerned, is that relating to the deposits on silver plated flatware. Two different standards are laid down, 'standard' and 'special'. For place settings the silver deposit for 'standard' is not less than 20 μm and for serving pieces not less than 12 μm. For the special category a thickness of not less than 33 μm is specified and for serving pieces not less than 19 μm.

Platinum metals

In 1731, a Spaniard, Don Antonio de Ulloa, wrote *A Narrative of a Journey to South America*, which contains a reference to 'platina', the diminutive of the Spanish word for silver. This 'little silver' was sometimes used by the Indians to make beads, but for the most part the Indians considered platinum to be a nuisance. Its high melting point meant that the native craftsmen found it unworkable, and the platinum that was found in the gold and silver mines in the Choco district of Colombia was dumped with other refuse in the Bogota river to await the miners of the nineteenth century.

The first platinum known to have reached England was sent here by a Jamaican assayer, Charles Wood, in 1741. Scientific work on the metal was first carried out by William Watson, who described its interesting properties in the Philosophical Transactions of the Royal Society in 1750. Then, in 1789, came the first recorded use of platinum for a ceremonial vessel. In that year Charles III presented Pope Pius VI with a platinum chalice.

The problem of working platinum continued to limit its use until the invention of the oxy-hydrogen blowpipe by the American scientist, Robert Hare, in 1847. Meanwhile, it had been discovered that platinum was not a single metal, but a family of six metals. In 1800, William Woolaston and Smithson Termant had established a partnership to try to produce a malleable platinum for commercial use. This partnership was to prove fruitful in a way neither man could have anticipated. In 1802, as a result of his study of platinum, Woolaston isolated palladium, naming it after the planet Pallas, which had also just been discovered. Next, he isolated another of the platinum metals, rhodium. Its name derives from the Greek word for rose, because its salts are rose coloured.

Smithson Termant's contribution was his discovery of osmium and iridium. Then a Russian scientist named Claus discovered ruthenium, Ruthina being the Latin name for Russia. The platinum family was now complete.

Until 1916 the biggest source of supply of platinum metals, which are almost always found together, was the alluvial mines of South America, including the tailings of old gold and silver mines and the bed of the Bogota river. Another source was the Ural mountains in Russia, where platinum had been discovered in 1823. Today, some platinum metals come from Canada as a by-product of the nickel and copper mines in Ontario province. However, the main source of these metals is Rustenburg in South Africa. The refiners refuse to give output figures, for reasons best known to themselves, but the world output of platinum must be in excess of 500 000 ounces.

The recovery and refining of platinum differ only in detail from the recovery and refining of gold already described. An important factor in determining refining techniques is that platinum like gold can be dissolved in *aqua regia*.

Platinum recommends itself to the manufacturing jeweller for a number of reasons. Once the problem of melting it had been overcome it proved to have good working properties. It is almost as malleable and ductile as gold. It does not oxidize during working. Its strength and colour make it ideal for mounts for diamond jewellery. Further recommendations are its nobility (it resists the attack of almost everything except *aqua regia*) and its reflective power which, though not as good as that of silver and gold, is more than acceptable.

Platinum has one disadvantage as a metal for making jewellery. It is very heavy. It has a specific gravity of 21.43, and a large platinum brooch will weigh down a light fabric. To wear sizeable platinum earrings is a real labour of love.

The lower specific gravity of palladium, only 12.16, was one of the arguments urged for its adoption by the jewellery trade. Another argument was its relative cheapness. It offers many of the same physical properties and much the same appearance at about a third of the price of platinum. For some reason, however, palladium has never become popular except for settings for diamond rings.

The only other platinum metal of much importance in the jewellery trade is rhodium, which is harder and whiter than platinum, and the most reflective of all the platinum metals. It is widely used for plating. Silver jewellery is often rhodium plated, and it is a common trade practice to plate platinum and white gold jewellery with rhodium.

Like gold and silver, platinum has to be mixed with other metals to improve its working qualities. Palladium, iridium and copper are the metals most commonly used for this purpose. Platinum has been assayed and hallmarked in Britain since January 1975, following the Hallmarking Act of 1973. There are now four standards for platinum:

850 parts per 1000 pure
900 parts per 1000 pure
950 parts per 1000 pure
990 parts per 1000 pure.

Tests carried out by one of the refiners in the 1950s revealed that some manufacturers were at that time using platinum considerably below the now accepted standard.

There has been a considerable renewal of interest in platinum among jewellery manufacturers recently. Platinum is now considerably more expensive than gold and has been so for many years. Also, because it is not the subject of international speculation, the price of platinum tends to be more stable. Supplies, too, are now more readily available than they have been in recent years because the mines in the Marensky Reef at Rustenburg in South Africa have come into full production.

Like the other noble metals, the platinum metals have many industrial uses today. To cite just a few, it is used for vessels for the chemical industry, for electrodes, for the decoration of china and book covers and for automobile exhaust systems designed to reduce lead pollution.

Base metals and the jeweller

Apart from the noble metals and their imitators, a metal which is often found in jewellers' shops nowadays, is stainless steel. This was discovered in 1913 by a Sheffield man, Harry Brearley, almost by accident, while he was experimenting to find a non-rusting steel that could be used for rifle components.

At first, no one was much impressed by Brearley's discovery that an alloy containing 13 per cent chromium, 3 per cent carbon, and 84 per cent iron was resistant to rust. One reason for this cool reception was that the new steel proved difficult to work. Also, the experimental knife blades made from it were insufficiently hardened and proved to be poor tools. These early stainless steel knives, made before the working of this metal was fully understood, gave rise to the myth that stainless steel blades do not take as good an edge as the old carbon steel blades. In fact, recent tests carried out in butchers' shops have shown that not only are stainless steel blades sharper but they also keep their edge four times as long as carbon steel blades.

Until the 1950s the most important use of stainless steel connected with the jewellery trade was for the making of knife blades. Even before the war, however, a handful of firms had been far-sighted enough to realize that this new metal had other uses and they began to produce stainless steel hollowares and flatwares.

We tend to talk about stainless steel as though it were one alloy only, but in fact there are a number of stainless steels. They all contain a proportion of chromium. It is the presence of this metal that gives stainless steel its resistance to rust, but the proportions of the constituent metals are varied to provide the different working qualities demanded by various industries. The stainless steel most commonly used for flatwares and hollowares is called 18/8. This contains 18 parts of chromium and 8 parts of nickel. This alloy has good working properties and a high resistance to rust.

The terms 'austenitic' and 'martensitic' are sometimes seen in descriptions of stainless steel. Austenitic steel, named after the Assayer to the Mint, Sir W. Roberts Austen, is the 18/8 carbon-free stainless steel already described. Martenistic steel is stainless steel with a small carbon content, about a half per cent, and is used for making knife blades.

The term stainless steel is not really accurate. Rustproof would really be a better name for it. Certain foodstuffs, salt and vinegar for instance, do stain stainless steel, but most stains can be easily removed simply by washing in soapy water. Solid detergents if they are not completely dissolved can, however, cause pitting. More permanent stains also sometimes appear on knife blades.

One of the drawbacks to stainless steel as a metal for making domestic hollowares and flatwares is that it scratches. Because the metal is hard these scratches are unsightly and permanent and do not polish out like the scratches on silver. Thus highly polished stainless steel does, after a time, look the worse for wear, and this has led to a great increase in the popularity of the satin-finished stainless steel to which there was some initial resistance. Scratches do not mar its appearance to anything like the same extent as they do the bright-finished wares.

Pewter

Pewter is a tin alloy. Many people believe that pewter contains lead, but most modern pewter is entirely lead-free. In fact the Board of Trade restricts the lead content in pewter measures and tankards to not more than 10 per cent, in the interests of public health.

Our ancestors were not so mollycoddled. The tankards they used often contained as much as 30 per cent lead. They were sometimes made of that villainous looking 'black metal', that was 50 per cent lead and 50 per cent tin, and there is really no evidence to show that anyone's health suffered from drinking out of these leaden vessels.

Even in the old days, however, other metals were used to alloy tin to make pewter. 'Common pewter' consisted of 83 per cent tin and 17 per cent antimony, while 'fine pewter' was an alloy of 81 per cent tin, and 19 per cent copper.

Until the eighteenth century, pewter was a very important metal, everybody ate off it and drank from it. Then the increased production of china, and the invention of Sheffield plate, resulted in a decline in the craft of the pewterer, and by the middle of the nineteenth century it had almost died out altogether.

The marks to be found on pewter vessels are as interesting in their own way as those on silver. In 1503 it became obligatory for every master pewterer to stamp his wares with his touch mark, and this marking continued to be obligatory until 1824. The early marks consisted of heraldic devices embodying the maker's initials. The touch marks were all stamped on a plate kept by the Pewterers Guild. Unfortunately, the first plate was destroyed in 1666 during the Great Fire of London. Later plates, however, have been preserved.

After 1690, the master pewterers were allowed a little more latitude in the form that their touch mark could take, and some decided that it would be better advertising to use their whole name instead of just their initials. Some also added the word 'London' to give their wares a metropolitan cachet. Others went too far. They began to use marks that bore too close a resemblance to the hallmarks stamped on silver at Goldsmiths Hall, and the Goldsmiths Company had to take out an injunction restraining the pewterers from attempting to deceive the public in this way.

Countless pewter vessels and pewter plates must have been made in the old days when every household had its garnish of pewter, but surprisingly little remains for the collectors of today. When pewter lost its popularity in the eighteenth and nineteenth centuries, thousands of unfashionable plates and tankards and jugs must simply have been melted down for scrap; the owners could hardly have been expected to foresee that there would be a pewter revival in the second half of the twentieth century. The best pewterwares are still produced by the traditional technique of casting in iron moulds, but less expensive pewterwares are produced by spinning.

Bronze

Bronze, the oldest alloy known to man, has been highly regarded at many periods in history. It had many uses for spears, armour, hammers and cannon. The craftsmen of ancient China cast great bronze incense burners and bells; the sculptors of ancient Greece cast their graceful statues in this brown metal, a mixture of copper and tin, with usually some zinc. The Phoenician traders of the Bronze Age came all the way to Cornwall to obtain the tin which their craftsmen at home, in what is now Syria, needed to make the bronze for fashioning into jewellery and weapons. The manufacturers of the nineteenth century also regarded bronze highly, and made clocks and mantelpiece statuary for Victorian drawing rooms from it. Recently, there has been a revival of interest in small bronze figurines. Many of these are, however, not true bronzes but what are known as cold cast bronzes and are far less valuable.

Chromium

Chromium is normally met only in the form of chromium plate, which, like so many other things, has a bad name simply from being badly used. Many have suffered from the shoddy chromium plating too often supplied by motor car manufacturers. As a result 'chrome' is regarded with suspicion and is expected to pit and peel and rust. Properly applied, chromium plate can provide a hard, tarnish-resistant, bright finish that will stand a great deal of wear. It has, however, a rather bluish look compared with silver, and many people, perhaps because of its associations, think of it as cheap looking. Chromium plate is met with in the jewellery trade only on cheap watch cases, cheap powder compacts, spoons and on very cheap costume jewellery. Experiments some years ago resulted in the production of coloured chromium plate, but this has proved to have little commercial importance.

Chapter 2

Diamond and coloured gemstones

Fashion and economic considerations have brought great changes to the world of gemstones over recent years. There is a great demand these days for real jewellery in the medium price range, and to meet this demand the jewellery manufacturers have had to employ a wide range of relatively inexpensive gem material in place of rubies and emeralds, which have become increasingly expensive. If these stones are used today they tend to be very small compared with those used some decades ago. The reason for this rise in the price of fine facetable gem material is that such material is becoming scarcer at a time when world demand is increasing. More and more people throughout the world are now buying jewellery. The Japanese, for example, who only a few years ago showed little interest in gems, are now buying them in great numbers. The rise in prices due to increasing demand and diminishing supply applies not only to fine emeralds, rubies and sapphires, but to all fine facetable material. Between 1970 and 1981, for instance, the price of good-quality aquamarine rose from £60 per carat to £500 per carat.

There is every likelihood that the trend will continue. Julius Petsch in Idar Oberstein told a party of British jewellers when he talked to them in 1974 that fine facetable stones were going to become scarcer, and that they would in the future often have to persuade their customers to accept jewellery set with either poorer quality stones or cabochons. Though the jewellers in question expressed doubts that their customers would in fact find cabochons acceptable, it must be remembered that only a century ago the cabochon enjoyed a great vogue and some of the most attractive Victorian jewellery featured cabochon garnets and amethysts. Since Julius Petsch made his prediction many more cabochons have been seen on the continent. We have also seen a revival of interest in the less well-known gemstones, while beads and polished slices of a variety of gem materials have been used to enrich jewellery, notably jade, coral, dyed agate, rhodonite, rhodochrosite and various quartz minerals.

So whereas jewellers a quarter of a century ago dealt with little else in the course of trade except the 'big four' (diamond, emerald, ruby and sapphire), in future they are likely to deal with a host of gems. They will have to be more open-minded about what constitutes a gem, more receptive to the new and unfamiliar and, of course, more knowledgeable about gem minerals generally. And not only will they be dealing with gemstones set in jewellery, they will probably also be selling more gem minerals in the form of *objets d'art*. In recent years more and more people are using hardstone carvings, boxes and bowls as decorations in their homes.

With the scarcity of major stones of fine quality, it obviously behoves jewellers to educate their customers to accept alternative attractive stones such as blue topaz, fine opal and aquamarine. Two new and very attractive alternative stones have made their appearance in recent years. These are the mauve form of zoisite, found in Tanzania and known as tanzanite, and the gem-quality grossular garnet of fine emerald green colour called tsavorite, found in both Tanzania and Kenya. It is hoped that supplies of both stones will be increasingly available in the future and that they will become increasingly acceptable to the public as they become more familiar. There is, however, some doubt about the supply of these and many other stones found in the developing countries. The governments of these countries are inclined, understandably perhaps, to nationalize their natural resources. Having done so, however, often they fail to exploit these resources and (as in the case of tanzanite) the supply tends to dry up or to become sporadic and unpredictable.

Besides needing to be familiar with the whole range of gem minerals, the jeweller in the future will also have to be very much on guard where gems are concerned. New ways of altering gem minerals and new and ever more plausible reconstituted and synthetic gems are an ever-present problem. Altered stones or synthetics can result in the jewellers losing their money or jeopardizing their reputations, or both. Even plastic simulants of gem materials such as amber can deceive the unwary.

Altered gem minerals are nothing new. The vast majority of aquamarines have for years been heat-treated to turn them from the more commonly found green colour to a more desirable water blue. Commercial citrine and cairngorm have been produced for a long time by heat-treating inferior amethyst, while the heat-treating of zircons (turning them from brown to blue or colourless crystals) goes back centuries. Agate has also been dyed in Idar Oberstein for a century or more, while materials such as jade and lapis-lazuli are often stained or, in the case of turquoise, improved by rubbing wax into the surface. The results of heat-treatment are more or less permanent, and jewellers, if they have been aware of it, have come to accept the practice as a necessary and useful way of improving on nature.

In the past few years there has been a good deal of experimentation and some commercial colour alteration by means of X-rays, neutron bombardment and so on. While it is generally felt that such methods of improving stones should be just as acceptable as heat-treatment if the results were permanent, in practice it seems that this is not always the case. In 1974 a number of very experienced Idar Oberstein dealers paid high prices for some parcels of topaz that exhibited a beautiful deep aquamarine blue coloration. Unfortunately after a matter of days the colour faded away altogether. The best known example of the use of these new techniques is the coloration of diamonds. Diamonds bombarded in cyclotrons and atomic piles have been coloured green, blue, brown and yellow. Also, X-ray bombardment has been used to turn kunzite green and to deepen the yellow of yellow sapphires, albeit only temporarily.

The new and more sophisticated synthetic gemstones are a problem to jewellers, because they could be deceived by them. It must be borne in mind, however, that those who produce them do not do so with intent to deceive the jewellery-buying public. These synthetics are seen by their producers as less expensive substitutes for natural stones, which are rare and expensive. Pierre Gilson and Carroll Chatham, who were the most successful pioneers in this field, were probably motivated to simulate natural minerals as closely as possible by the scientific challenge this presented.

A question posed by the existence of synthetic gems almost indistinguishable from real ones is whether the jewellery trade should accept such stones as part of its stock-in-trade. The reactions of individual jewellers to this question vary considerably. Many would prefer to continue to sell only natural gems, others are less certain of their position. The fact is that synthetic gems are already being used fairly extensively in jewellery in the United States, and to a lesser extent in Britain. It must also be borne in mind that the cultured pearl, over which there was once a similar controversy, has now gained universal acceptance, and natural pearls have almost ceased to exist as a commercial proposition. Today cultured pearls are used side-by-side with diamonds, and few people see anything wrong with this. It could be argued, of course, that cultured pearls, unlike gems produced in a laboratory, are at least partly natural.

In discussing the various stones that are most likely to come the jewellers' way in their day-to-day business it is convenient to consider them under four headings. First the 'big-four' are described, then the rest of the gemstones and decorative minerals. The third section is given over to a description of those gems and decorative materials that derive from living substances, the so-called 'organic gems' such as pearl, jet and amber. Lastly the various synthetic stones are reviewed. The list of gemstones and organic substances included in this book is not exhaustive. At one time or another all kinds of minerals and natural substances have been mounted in jewellery or used to decorate *objets d'art*, but many of these are met with very infrequently. For those who wish to widen their knowledge of gemstones to take account of all the various substances, a number of specialist books exist, notably Robert Webster's comprehensive book *Gems, their sources, descriptions and identification* (5th edn, Butterworth) or B. W. Anderson's *Gemstones for Everyone*.

Diamond

Diamond is crystallized carbon, and chemically speaking, is the simplest of all the gem minerals. It is not particularly rare. Since the discovery of the diamond pipes in South Africa in the nineteenth century, diamonds have been recovered in much greater numbers than have emeralds, rubies and sapphires. But if the diamond is neither a rare nor a complicated mineral, it is nevertheless a unique and a wonderful one. It is not the result of some caprice of fate that diamond has assumed its paramount position among gemstones. Its position is a tribute to its unique physical properties.

Hardness

The best known of these properties is hardness. Diamond is the hardest of all natural substances. To say that it is immeasurably hard is not just to use a figure of speech. It is impossible to measure the hardness of a substance that is harder than anything else. A Vickers hardness of between 10 000 and 50 000 has been suggested as a possible figure for diamond, and just how hard this is can be appreciated when it is realized that the Vickers hardness of steel is around 500. It is this characteristic, the great hardness of diamond, which makes it so valuable to industry today. For every carat of diamond sold to the jewellery trade, four carats are sold to industry for use as a cutting and grinding medium in almost all forms of high-precision engineering.

The hardness of diamond is the result of its closely packed atomic structure; the atoms are so tightly packed that it is virtually impossible to pack them any tighter. It is interesting to compare diamond with that other well-known form of crystalline carbon, graphite, used in the manufacture of pencil leads. Graphite is as soft as diamond is hard, and this is because its atomic structure is very loose.

The name diamond derives from the Latine *adamantem*, and the Greek *adamantinus*, both meaning unconquerable. But diamond is far from unconquerable. It can be conquered by flame (in air at about 800°C), and whereas it was once thought that diamond would resist any blow, this belief also proved to be false. A diamond can be shattered into pieces by a blow, and it can be cleaved, that is, split with a steel blade along one of the cleavage planes. These cleavage planes run parallel to the four faces of the octahedron, the classic, though by no means, the only, form in which diamond crystals are found.

That diamond can be cleaved in this way is most useful to the diamond cutter, for by cleaving a large crystal, one blow will accomplish what might take days to do with a saw. But the diamond cutter has to remember that a blow delivered in the wrong place can break a diamond into pieces. The story of the cleaving of the 3106 carat Cullinan is one of the legends of the trade. The cleaving of this stone was carried out in 1908. The task was entrusted to Joseph Asscher of the famous house of Asscher, who studied the stone for months, fully aware that the slightest error could reduce this great stone to a heap of fragments. The day came, and he stood over the stone and struck the cleaving blade. The blade broke. He tried again. This time he was successful. From the cleaved pieces a number of fine stones were cut including the 530 carat Cullinan Great Star of Africa, now in the Royal Sceptre, and the Cullinan II of 317 carats, now in the Imperial State Crown.

Optical properties

Diamond has unusual optical properties which are responsible for its adamantine lustre and its celebrated fire. Because diamond is so hard it take a very high polish. As a result of this the table facets of a cut stone reflect 18 per cent of the light falling on them, compared with the 4 per cent reflected by the table facets of a glass imitation.

Diamond is also highly refractive. Refraction occurs when light passes from one substance to another. If a stick is partly immersed in a pool of clear water, the stick will appear to be bent. This is because the light rays travelling to the eye from that part of the stick below water level are bent when they pass from the water to the air above it. The angle of bend which a substance imparts may be expressed by a figure called an index of refraction. The refractive index of diamond is 2.419. That of quartz is only between 1.53 and 1.54, while that of the corundum gems, ruby and sapphire, is between 1.765 and 1.773.

Both the high reflectivity and the high refractivity of diamond are exploited by the brilliant-cut with its 33 facets above the girdle and 25 below. These facets are angled so that the light entering the top of the stone is refracted on to the pavilion facets at such an angle that these facets will reflect a high proportion of this light back through the table.

Diamond is singly refractive. Some minerals are doubly refractive, that is they not only refract a light ray entering them but also split it in two. The mineral which demonstrates double refraction most clearly is Iceland spar (calcite). Another mineral which is doubly refractive is zircon, which sometimes poses as diamond.

Fire

Minerals refract different coloured light to different degrees, so that the various colours that make up white light, red, orange, yellow, green, blue, indigo and violet, are separated one from another. This colour dispersion is especially high in diamond. That is why this usually more-or-less colourless stone appears scattered with bright colours. This flashing of colours is known as the ‘fire’ of a diamond.

Though diamonds are often thought of as colourless gemstones, most are, in fact, at least faintly coloured. Many of those sold to industry are of an unattractive shade of brown, green or grey, but a tinge of yellow or green is far from uncommon in gem diamonds. The blue–white diamond is a stone much talked about but seldom seen.

Diamonds of many colours have been found, and those which have a distinct and pleasing coloration are often valuable, more valuable sometimes than colourless stones of the same size. Most common among the coloured diamonds are the lightly tinted stones, the canary yellows, the pinks, the pale greens and the topaz browns. Rare are the strongly coloured stones, the deep green of the Dresden Green, the blue of the Hope and rich reds, as strong as, but more brilliant than, the red of a fine ruby.

Valuation

The value of a diamond depends on four factors, what De Beers call the four Cs – carat weight, colour, clarity and cut. The assessment of these qualities is not easy, however. As the late Norman Harper, the well-known authority, wrote in an article, 'Putting a price on diamonds' (*Watchmaker, Jeweller and Silversmith*): 'Many jewellers with years of experience behind them are extremely cautious of such matters, others with more nerve and less enlightenment throw caution to the winds and sometimes bring discredit to their trade.'

For instance, it might seem to be a simple matter to arrive at the carat weight of a stone. So it is if the stone is unmounted, because it is only necessary to weigh it to the second decimal place on an accurate balance. But, as Norman Harper pointed out: 'Diamonds usually have to be valued when they are mounted, so that an assessment of the weight has to be made and these assessments are frequently wildly inaccurate. In many cases as much as 25 per cent out. If one considers a stone said to weigh 2 carats of I colour and SI2 clarity grade, which is in reality 1.50 carats, and assumes the value to be £2800 per carat, the error is in the order of £1400. "It spreads 2 carats" is a phrase often heard, and one must know that a stone spreading 2 carats with a diameter of 7.4 millimetres and a depth of 5.4 millimetres can weigh 2 carats, but another with the same diameter and a depth of 3.4 millimetres will probably only weigh 1.25 carats. This would increase the error mentioned above to £2096.' These figures seem very low today because diamond prices have shot up in the meantime making the errors even more serious.

Errors of judgement about the cut, or the 'make' as it is called, can lead to equally devastatingly wrong assessments. There are many common faults of make, such as a stone with a too thick or a too thin table, or a pavilion that is too deep or too shallow, a too large culet, incorrect height of the crown and so on, and such faults can affect the price of a diamond by as much as 50 per cent.

Clarity again has a big influence on value, and is not an easy thing to assess either. Most diamonds have inclusions or flaws, which need an expert to evaluate them. A diamond is generally accepted as flawless if it shows no imperfections when examined under a glass which magnifies 10 times, but the job of deciding whether a stone showing flaws is more or less valuable than another, calls for considerable experience. Scales of clarity do exist but they are no substitute for experience.

Olivine and garnet are two of the important inclusions in diamond. Others are ilmenite, bronzite, chrome-diopside, chrome-enstatite, diopside and graphite. Some cavities in diamond are coated with graphite and occasionally an octahedron of diamond is seen within a diamond. The term 'carbon spots' is loosely used in the trade to describe what may well be other included materials.

Some dealers use the terms 'very slightly imperfect' (v.s.i.) or 'slightly imperfect' (s.i). The word 'imperfect' is a derogatory term and the use of 'very slight inclusions' (v.s.i.) or 'slight inclusions' (s.i.) is more sensible when the presence of inclusions is being referred to. It might be very desirable if the whole trade adopted the CIBJO scale.

Colour is the most fugitive of the four Cs. It is a relative thing, and is affected by the light in which a stone is examined, the setting in which the stone is mounted and any other stones surrounding it. A diamond surrounded by blue sapphires will be flattered by the blue light reflected by them. A diamond in a gold setting may well look more yellow that it really is.

To quote Norman Harper again: 'When considering colour it has to be appreciated that colour is not a thing in itself; it is a sensation conveyed by the eye to the brain by vibrations of varying wavelengths and velocities. Eyes vary and so do brains, and just as some people have a well-educated sense of tonal differences so some others have a highly developed sense of colour. No matter how developed a sense may be it can always be improved with practice.'

A classification by colour widely used in the past was river, top wesselton, wesselton, commercial white, fine or top silver cape, silver cape, light cape, cape, dark cape, fine light brown, light brown, brown and dark brown. The use of the term 'cape' implies a yellow tinge. To relate the colour of particular stones to such a scale requires, besides skill and aptitude, a known source of light and a set of standard stones to use for comparison. By some dealers 'commercial white' is regarded as meaningless.

In the USA the Gemological Institute of America in the past used a grading system based upon the alphabet, D being the top grade and R the lowest. The American Gem Society uses a numerical form of grading, from 0 to 10. Both systems are open to objection because they tend to downgrade desirable stones in the eyes of the customers – grade 2, or grade B, sounds much worse than it is. The Scandinavian countries use the terms river, top wesselton, wesselton, top crystal, crystal, top cape, cape, light yellow, and yellow or the colour grades which correspond to the American numerals and letters for stones of half a carat and above. For diamonds under this weight the terms rarest white, white, tinted white, yellowish, and yellow are used. Some argue that these terms are to be preferred. In the UK there is no uniformity of practice.

The Diamond Committee of the European Organization CIBJO (Comité International des Bijoutiers, Joailliers et Orfèvres) has introduced standards for colour and clarity and has established authorized laboratories in a number of countries which it supplies with standard sets of stones. The CIBJO categories for colour are:

Exceptional white
Rare white
White
Slightly tinted white
Tinted white
Yellowish

The categories for clarity are:

Internally flawless
VVS (very very small inclusions)
VS (very small inclusions)

SI (small inclusions)
Piqué I
Piqué II
Piqué III

The clarity standards are dependent on 'the examination by an experienced professional under ten-power magnification in normal light by means of an achromatic aplanar lens'.

Acceptance of the CIBJO standards is recommended by the trade organizations for use in the UK. Obviously many advantages would accrue to the universal adoption of such categories, but it must be borne in mind that experienced professionals can differ with one another, particularly over borderline stones. A totally objective judgement of subtle variations in colour and clarity is just not possible.

Source

The first diamonds were, as far as we know, found in Southern India, probably among the sand and pebbles on the banks of the Krishna river, and it was from India that many of the great historic stones, such as the Koh-i-noor, the Orlov, the Sancy and the Hope, came.

By the early seventeenth century when the French jeweller Tavernier visited India on his travels on behalf of Louis XIV, the mining of alluvial diamonds on the banks of the rivers of Southern India was on a formidable scale. 'The first time I was at the mine,' he wrote, 'there were about 60 000 persons at work, men, women and children, the men being employed to dig, the women and children to carry the earth.' At Golconda he found a market established where the diamonds changed hands. Most of them were bought by the emissaries of Indian Princes to decorate their masters' thrones and their masters' persons.

In the eighteenth century diamonds were discovered in the gold mines of Brazil, and were subsequently found in the alluvial gravels along the rivers. Brazil gradually replaced India as the principal source of diamonds.

The most important chapter in the history of the diamond was, however, written in the nineteenth century in South Africa. The first diamond was found there in 1866, by the son of a Boer farmer. The boy was playing on the banks of the Orange River when an unusual pebble attracted his attention. He picked it up and kept it. This bright pebble of 21¼ carats, later to be known as the Eureka, was obviously something unusual, and was passed from hand to hand. But no one set much store by it until Dr Guybon Atherstone, an amateur gemmologist, identified it as a diamond. It was then sold to Sir Phillip Wodehouse, the Governor of the Cape, for £500, and subsequently put on show at the Paris exhibition.

No one seemed at first to appreciate the importance of the discovery of the Eureka, but another and bigger diamond was found elsewhere on the Orange River three years later, this time by a shepherd boy on the Zendfontein farm. The boy was given five hundred sheep, ten oxen and a horse for his stone. A few days later that same stone, the Star of South Africa, was sold for £10 000. Its discovery triggered off the great South African diamond rush.

The prospectors who flocked there found very little to begin with. In 1870, at the Jagersfontein farm in Orange River Colony, the overseer found a diamond in a dry working. Then there was a similar find on the Dutoitspan farm, another on the Bulfontein farm, and another on the De Beers farm. These farms, which families of

Boer immigrants had won from the wilderness, were soon to be world famous as the names of diamond mines.

On these farms, large areas of yellow ground were discovered which proved to be rich in diamonds. Miners poured into the country, and the bonanza was on. Then suddenly, about 50 ft below the surface, the yellow ground ran out. Disappointed, many miners sold their claims for what they could get for them and moved away. But some people stayed and bought. Among these were a music hall entertainer, Barney Barnato, and an Oxford graduate, Cecil Rhodes.

One day someone discovered that the ground below the yellow ground was even richer in diamonds. It was realized that the yellow ground was weathered material at the top of volcanic pipes going down into the depths of the Earth which were packed with a grey-blue volcanic rock. This was to become famous as the 'blue ground'. The pioneers had laboriously uncovered the Aladdin's caves of South Africa.

Barnato and Rhodes took stock of their investment, and they were not over-happy about what they saw happening in the mines. There were still too many miners, and too many claims. The tops of the pipes were shrouded in a spider's web of cables, carrying each man's little daily collection of blue ground to the surface. To men like Rhodes and Barnato such inefficiency must have seemed absurd. But that was not the worst of it. Men undermined their neighbours' claims. There were earth falls and men died. The pipes began to flood, and work had to stop until the water could be got rid of. Rhodes and Barnato decided that the time had come for all the little claims to be consolidated, and they began to buy them up. But soon it became apparent that the Kimberley diamond fields were not big enough to hold two men with a thirst for power. The battle was bitter but brief, and in the end it was Cecil Rhodes who won.

Cecil Rhodes was the chairman of a company called De Beers. In 1888 he purchased control of the great Kimberley mine for £5 338 650. 'We had to choose between the ruin of the diamond industry and control of the Kimberley mine', Rhodes said after the battle had been won and he had emerged as the virtual dictator of diamond mining in South Africa. An outcome of the consolidation of the South African mining interests was the establishment in 1890 of the Diamond Syndicate, whereby the entire diamond production was sold on a percentage basis to certain buyers. From these basic origins Sir Ernest Oppenheimer, who had become chairman of De Beers, formed the Diamond Corporation Ltd in 1930 when the great depression had forced the closing of virtually all the diamond mines. The idea behind the forming of this body was the realization that some at least of the diamond's attraction lay in its being not just a thing of beauty, but a saleable asset. It became obvious, therefore, that it was desirable to see the value of diamonds controlled. The company was formed to control the price at which diamonds were sold. Contract buying was also started, and the miners had the guarantee that their diamonds would be bought at a fixed price even when there was a slump. This policy of price control through centralized selling is still pursued. Today not only South African diamonds, but those found in many other parts of Africa, Australia and even the USSR come to the Central Selling Organization in London to be sorted and sold at the 'sights'. These sights, which are held four times a year are attended by leading dealers and cutters from all over the world, who take the parcels of rough that they have bought back to their respective countries to be cut.

Today, African diamonds are recovered from the blue ground in Botswana as well as South Africa by mining, from alluvial deposits in South West Africa, the Congo, Angola, and Sierra Leone, by dredging the sea-bed at the mouth of the Orange River for diamonds that have been carried there by the current (Figure 2.1), and by sifting the marine terraces where beach mining techniques are used. Some idea of the cost of beach mining can be judged from the fact that, on average, 25 tons of sand and gravel have to be moved and processed for every carat of rough recovered.

Besides Africa, Brazil remains an important source of diamonds. An increasing number of diamonds are nowadays coming out of Russia where diamond pipes have been found, but figures of Russian production are not available. Recently, too, it seems that diamond pipes in China are being worked on a considerable scale and so are the pipes which have been found in Australia which yield vast numbers of small and, for the most part, inferior stones.

Diamond simulants

The physical characteristics of diamond are so distinctive that it might be thought unlikely that anyone could succeed in passing off any other colourless and transparent stone as a diamond. However, synthetic simulants of ever-increasing plausibility have come on to the market and even an experienced jeweller has to be continually on guard.

The natural stone that has been most often passed of as diamond is zircon. The colourless zircon produced by the heat-treating of brown zircon can easily be mistaken in poor light for diamond. Zircon certainly looks more like diamond than any other natural stone, though compared with a good diamond a zircon really appears rather lifeless. There is a simple test that jewellers can apply if there is any reason to suppose that a stone might be a zircon. All they have to do is to look down through the table of the suspected stone using a magnifying glass. Because zircon is doubly refractive the edges of the facets at the base of the stone will usually appear to be doubled. The edges of the back facets of a diamond, because it is singly refractive, will appear as single lines. The jeweller must, however, be careful to rotate the suspected stone when carrying out this test because if a zircon is viewed in one particular direction this doubling of the facet edges is not seen.

Two other colourless stones which could conceivably be mistaken for diamond, though there is really little excuse for doing so, are rock crystal, the colourless form of quartz, and colourless topaz. Rock crystal has comparatively little fire, while topaz, which has more fire, has a characteristic slippery feel to it. Both these stones are also much softer than diamond.

In addition to natural stones a number of man-made stones have been passed off as diamonds. These synthetics are described in the third section of this chapter. However, among these man-made simulants is paste. Paste stones are made from a very soft lead glass. Admittedly they often have considerable fire when new, but their softness and the occasional presence of distinctive bubbles make it difficult to understand how anyone in the trade could mistake paste for diamond.

The most recent diamond simulant is moissanite. Natural silicon carbide was discovered by Dr Henri Moissan while analysing part of the Diablo meteorite in 1893. Moissanite was named after Dr Henri Moissan in 1905. Natural moissanite is rare, however it has been created in a laboratory producing an excellent diamond simulant. Further coverage can be found under the latter part of this chapter under Synthetic gemstones.

Diamond cutting

The aim of the diamond cutter is to make the most of the diamond's high reflectivity and refractivity. Cutters too have often to make the most of an imperfect rough stone given to them for faceting. In his *The Art of the Diamond Cutter*, A. Monnickendam

Figure 2.1 Stripping the overburden from the marine terraces at the mouth of the Orange River to recover alluvial diamonds eroded from the blue-ground and carried down the river (by courtesy of De Beers)

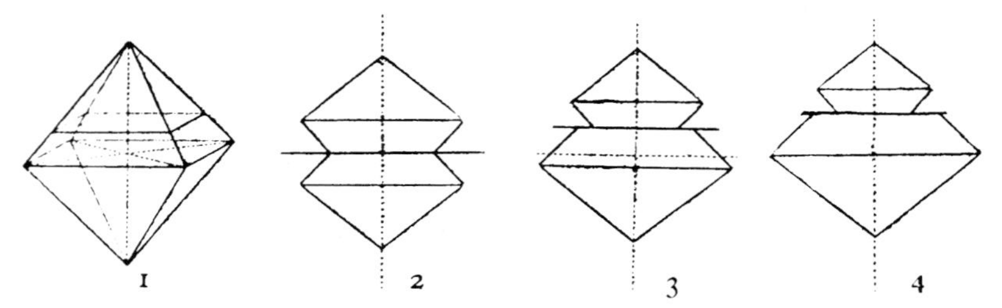

Figure 2.2 When inspecting an octahedral crystal the designer will be influenced by any faults or inclusions in the stone when deciding whether to get one big stone and a small one out of the rough, or two smaller stones. If they decide to get two equal stones out of the octahedron (2) 60 per cent of the rough will be lost, If the stone is cut to obtain the two cut stones outlined in (4) only 51 per cent of the valuable material will be lost

wrote: 'The master cutter is constantly on the alert for ways of eliminating marks and imperfections of all kinds, and at the same time aims at the best possible weight consistent with correct proportions.'

It will be appreciated from this that the diamond cutter has first of all to study the stone. This is usually the job of the designer, who is often the head of the firm. After the stone has been studied, the designer marks it with Indian ink as a guide to the cutters who will do the actual work on it.

Suppose that the rough diamond to be cut is an octahedral crystal, the classic shape, and not one of the irregularly shaped stones that make up a considerable proportion of all the diamonds recovered. An octahedron is like two pyramids joined at their bases. To obtain the profile of a brilliant-cut stone the top of one of the pyramids must be removed (see Figure 2.2). One technique of shaping a rough diamond is by splitting it along one of the cleavage planes with a steel blade, but the removal of the point of an octahedral crystal is more usually done by sawing than by cleaving (Figures 2.3 and 2.4). Twinned crystals which are difficult to saw are being cut increasingly by laser. In a diamond-cutting factory there are batteries of saws, and the operative in charge of these mounts the rough stone in cement and fits it between two holders. The stone is then held, by the pressure of a counterweight, against the edge of the saw. The saw, consisting of a paper-thin disc of bronze impregnated with a paste made from powdered diamond and oil, rotates at 6000 revolutions a minute. Surely, but very slowly, diamond cuts diamond. It takes about a day to saw through a one-carat stone.

The next operation is to round off the stone to create the girdle dividing the top, or the table as it is called, from the base, or pavilion. This rounding operation is known as 'bruting', and is done by rubbing one diamond against another. One stone is held in a lathe and rotated; the other is cemented into a handle so it can be held against the stone in the lathe with the necessary force.

Once the stone has been rounded the faceting begins. This is done on a turntable called a 'scaife', which is impregnated with a mixture of diamond dust and olive oil. The diamond is held in a 'dop', a small metal cup with adjustable claws. People who visit diamond factories are often disappointed by the faceting department; the rows of scaifes look much like rows of gramophones playing silent music.

In the first faceting operations, which are carried out by the 'cross-worker', the flat table is polished on the top of the stone and four facets are polished above, and four below, the girdle. The angles of these, and indeed those of all the facets, are critical.

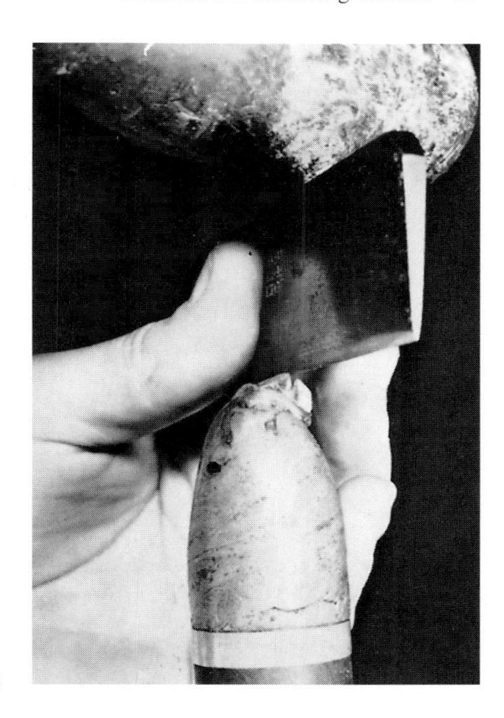

Figure 2.3 A cutter dividing a diamond by cleaving

Figure 2.4 A diamond being divided by sawing

Figure 2.5 The process of polishing a diamond on a cast iron scaife. All the cutter has to guide him is his eye and his experience. The cutter is shown inspecting the progress of cutting a facet with the aid of a loupe

Unless they are correct they will not fully reflect the light falling on them through the table, and the stone will lack brilliance.

Next, the stone passes to the brillianteer (or brilliandeer, or finisher) who puts on the other 48 facets to make up a total of 57. When the brillianteer has finished, the sides of the stone will be broken up into a pattern of stars and kites.

Apart from the round brilliant-cut, there are oval brilliant-cut stones, pear-shaped brilliant-cut stones, sometimes called drops or pendaloques, and there are the nowadays very fashionable boat-shaped or marquise-shaped brilliant-cut stones (Figure 2.6). These fancy shapes are often produced because they are the most economical way to cut particular rough crystals. The cutting of these fancy shapes calls for more skill and knowledge than the cutting of round brilliant-cuts, and this is reflected in the relatively high price asked for them. The cutting of emerald-cuts, sometimes called square-cuts or trap-cuts, is also very exacting. An emerald-cut stone has 16 traps above and 24 below the girdle. A diamond much used in modern jewellery is the long rectangular baguette. This is a simplified step-cut.

Occasionally jewellers will also come across diamonds in other styles. In older

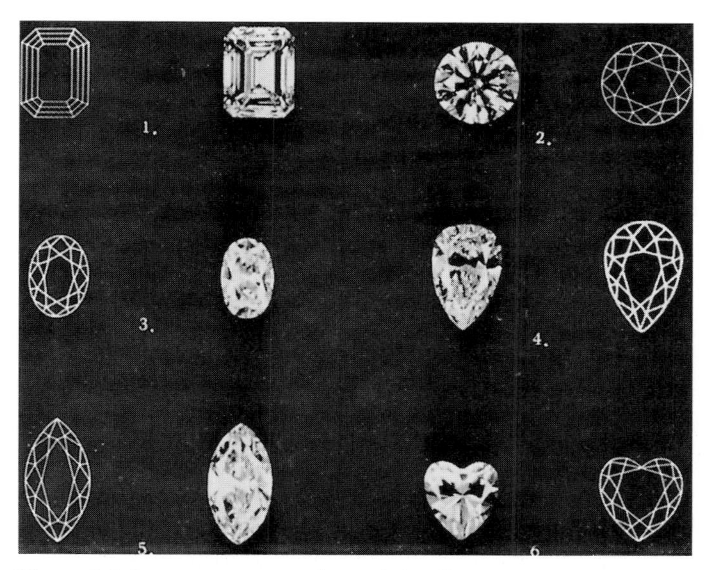

Figure 2.6 Some cuts commonly used for diamonds: 1 – step-cut, or emerald-cut; 2 – brilliant-cut; 3–6 – variations on the brilliant-cut; 5 – the fashionable marquise

pieces there are often many rose-cut stones, which have flat bases and a pointed top on which from 3 to 24 triangular facets have been cut. This is an old Indian style of cutting which was introduced into Europe during the seventeenth century. It was very popular until it was superseded by the modern brilliant-cut (Figure 2.7). Even today, rose-cut and old-fashioned brilliants are produced for use in reproduction pieces and to replace stones lost from antique mounts.

Then, too, there are simplified versions of the brilliant-cut which are used for small stones. Among these is the eight-cut which, as its name suggests, has eight facets on the table and a further eight on the pavilion.

Diamonds have been cut in many other and sometimes very complicated ways, such as the Jubilee-cut in celebration of Queen Victoria's Jubilee. This has a pointed top like a rose-cut diamond, but is also faceted below the girdle. Then there is the so called Royal-cut, which has no fewer than 144 facets. But stones cut in this way are seldom met with. Recently a new cut has been developed for the flat triangular crystals, known as macles, which imparts considerable brilliance to them. A square cut brilliant called a radiant cut has become popular of late.

Ruby and sapphire

Ruby and sapphire are both corundum gemstones. They consist of crystalline aluminium oxide; the only difference between them is their coloration, which is caused by trace elements of alien metallic oxides. Ruby is the red variety of corundum. All the other different coloured corundums, the green, brown, orange, yellow, pink, violet and purple, as well as the blue corundum, are known as sapphires.

The red of ruby is caused by the presence of chromic oxide. The amount of chromic

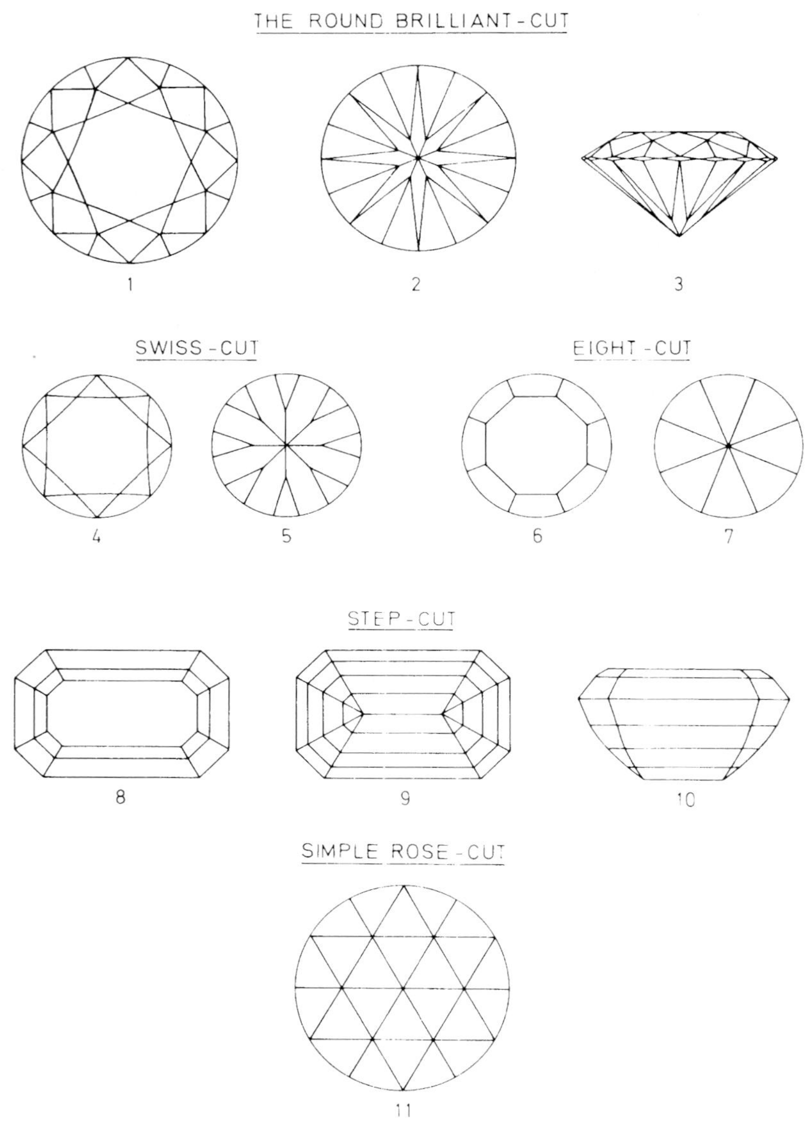

Figure 2.7 Diagrams of cuts used for diamonds: 1, 4, 6, 8, 11 – crown; 2, 5, 7, 9 – base; 3, 10 – side. The rose-cut was very popular in the eighteenth and nineteenth centuries

oxide in a crystal determines whether it is pale red in colour, or the much more desirable rich crimson. The blue of blue sapphire comes from titanium and iron oxide in the stone, and the colour varies according to how much of these oxides is present. Blue sapphires which have no more than a pale watery tinge of blue in them are found. Others, like many of the sapphires found in Australia, are almost black. Some people think that the deeper the colour the more valuable the blue sapphire, whereas it is in fact the middle blues, the cornflower blue and the French blue stones, that fetch the highest prices.

The traditional source of the best rubies is the district around Mogok in Upper Burma. Stone-age tools have been found there, testifying that rubies have been mined in the alluvial gravels of the Mogok region for thousands of years. Dr Edward Gulbelin, the eminent Swiss gemmologist, made a film there a few years ago which revealed that the mining methods used were still extremely primitive. The women, dressed in their colourful national costume, panned the streams with flat wicker baskets, while the men dug pits, called twinlons, 20 to 40 ft into the gravel. They worked away in these confined twinlons digging out the gravel and sending it up to the surface in baskets suspended from primitive bamboo cranes to be sorted. Nowadays mining in the Mogok region seems to have ceased entirely and Burma rubies are becoming increasingly rare.

The rubies found in Thailand, near Cambodia, the next most important source, are usually of a rather brownish-red colour, or sometimes violet-red. Rubies are also found in Sri Lanka, and more recently there have been worthwhile finds in Tanzania. An important ruby find has also been made in Kenya by a geologist looking for chrome garnets. Rubies of facetable quality from this source have been said to compare in colour with those from Burma, although the majority of the stones recovered seem to be suitable only for cabochons. Cabochons from this area, first exhibited at the Basle Fair in 1975, tend to be of a sugar-pink colour and to have many inclusions, but there is a strong possibility that the best material is not being released on the market and that this find is indeed an important one. The find is interesting because it illustrates the fact that major finds of gemstones are still being made. The mining of coloured stones is still very much in the hands of freelance prospectors who work in remote areas with primitive equipment. Systematic geological surveys are the exception rather than the rule in gem prospecting.

Ruby is found in very few places because chromic oxide is a comparatively rare substance, and, because the ruby-bearing areas of Burma, Thailand and Sri Lanka have been systematically mined for longer than anyone can remember, good rubies are becoming very scarce. It is hardly surprising, therefore, that a fine ruby today is extremely expensive.

Sapphires are much more widely distributed over the Earth's crust than are rubies. They are found in Burma, Thailand and Sri Lanka. Sapphires of a fine cornflower blue colour have been found in Kashmir, in a valley high in the Himalayas, though very few stones are coming out of this region today. Sapphires of many colours are found in Australia, which has recently become an important source of the blue sapphire so popular today as an engagement ring stone, though much of the material found there is very dark. Australia also produces some very attractive pastel-coloured sapphires. These are by no means the only sources of sapphires. They are also found in the USA, Africa and the USSR.

Heat treatment of corundum

Though the heat treatment of stones such as zircon, aquamarine and tourmaline to change or improve their colour goes back a long way, the heat treatment of corundum is a recent development. This treatment can turn the palest of sapphires, stones that would once have been discarded as valueless, into sapphires of a colour and quality reminiscent of fine Kashmir stones. Heat treatment can also remove the clouds from a cloudy ruby and render it transparent. It can remove the purple tinge present in some rubies and even change purple sapphires into more desirable rubies. Further, by heat treatment and the addition of titanium it has proved possible to 'imprint' stars in cabochon rubies and sapphires.

So far as is known all these changes are permanent and it is therefore currently not beholden on any European laboratory to report that a stone has been treated in this way any more than in the past a laboratory was expected to report that the colour in an aquamarine had been altered in a furnace. However, the GIA in the USA has decided that all enhanced stones must be declared.

The laboratories have evolved techniques for identifying these altered stones and in some instances their job has been made easier by the carelessness of the lapidaries entrusted with the necessary repolishing of these stones after treatment. Quite often these lapidaries have failed to repolish one or more facets and these facets display the characteristic shallow pitting caused by the process. On other stones this pitting can be seen in the area of the girdle. Besides heat-treated blue sapphires, fine orange-yellow sapphires have been identified as having had their colour improved by heat treatment.

Another form of heat treatment has also resulted in the production of the very deceptive colour-diffused sapphires that have appeared in the market. These were originally pale or colourless corundums, which have been coated with powdered chemicals and then heated to very high temperatures and cooled slowly. This has had the effect of covering the stone with a blue skin. Robert Crowningshield, the eminent American gemmologist, in the course of a lecture he gave in London in 1981, described a magnificent necklace set with cabochon sapphires which he had been called upon to test. The suspicions of the dealer who had sent the necklace to the New York laboratory had been raised by the fact that the stones were, he felt, too good to have been cut as cabochons. They would have produced faceted stones of far greater value. When the laboratory checked the stones they found that they were in fact colour-diffused sapphires.

Other worrying products of the heat-treatment furnace that have come to light have been Verneuil synthetics which have not only had the characteristic banding removed by the process but have also had seemingly natural 'fingerprints' induced in them. The absence of banding and the natural-looking inclusions could well lead to these stones being assumed to be natural.

Ruby and sapphire simulants

Besides synthetic rubies and sapphires, which are dealt with in the section on synthetics, the jeweller has to watch for other natural stones posing as rubies and sapphires. Red spinel could be mistaken for ruby by the unwary. Red spinel used to be called 'balas ruby', and the most famous specimen is the so-called Black Prince's ruby in the Royal Regalia. A good red spinel is, incidentally, a rare and valuable stone in its own right. A much less valuable stone sometimes mistaken for ruby is the almandine garnet. There are also red tourmalines, often called 'rubelites' because of their resemblance to rubies. Natural and synthetic blue spinels, blue tourmalines, deep blue aquamarines, the rare deep blue benitoite, and even blue paste stones, could all possibly be mistaken for blue sapphire.

Besides these natural stones the jeweller should also keep a lookout for doublets. A careful study of a suspected stone with a lens will reveal the joint between the natural crown, often garnet, and the fake pavilion. Doublets are obviously more difficult to detect if they are set in rub-over settings.

Cuts, cabochons and stars

The mixed-cut is used for rubies and sapphires and for most transparent coloured stones. It is a mixture between the brilliant-cut and the emerald-cut, brilliant-cut on the table and emerald-cut on the pavilion. Many rubies and sapphires are cut cabochon, because a large proportion of the stones recovered are not of facetable quality. Cabochon cutting consists of polishing the crystal to produce a smooth dome with a flat or slightly curved base. Such cutting is always used for stones showing the optical effect known as asterism. The asteria, or star, seen in these stones is caused by bunches of needle-like crystals, which lie at 120 degrees to one another, catching the light.

Star-stones are very highly prized in the USA, where synthetic star-stones are produced in considerable numbers. The value of a natural star-stone depends on the colour, how clearly defined the star is and how central it is in the polished gem.

Emerald and the other beryls

Many people think that a fine emerald, which is a rich velvety green in colour, is the most beautiful of all the coloured stones. Today, fine emeralds are extremely rare and expensive.

Emerald is a beryl, a silicate of aluminium and beryllium, a metal which has many specialized industrial uses. Emerald owes its colour to traces of chromic oxide, the rare substance that is also responsible for the red of ruby and the green of demantoid garnet.

The colour of emerald varies from a deep green to a pale and watery green, and colour has a big effect on its price; so has the clarity of the stone. Emeralds are very often flawed and usually contain many inclusions. A really clean stone is a rarity. Because most emerald crystals are more or less flawed, it is a common practice to protect them during cutting by soaking them in oil, which fills up any fissures and acts as a bonding agent. Often stones are oiled again after cutting to hide surface flaws. In the East, green oil is sometimes used for this purpose and to improve the colour at the same time. If such a stone is subsequently ultrasonically cleaned it is possible that this oil may be leached out and that the appearance of the stone will suffer considerably in consequence. Despite the fact that many people in the trade believe that oiling is an unacceptable form of enhancement, attempts to outlaw the practice have come to nothing.

Emeralds were mined in the Cleopatra emerald mines near the Red Sea in Egypt, as early as 2000 BC. These mines have long been worked out, and the main sources of fine emeralds today are mines in Brazil and in the Andes in Colombia. The Andean mines were worked by the Incas and by their Spanish conquerers after them. Although emeralds are still being recovered in fair quantities, after so many years of working there is not much good-quality emerald left. A number of new mines have, however, been opened recently in Brazil and Tanzania and run-of-the-mill emeralds have gone down in price as a result.

Russia has some emerald mines in the Urals. The stone was first discovered there in 1830, when a peasant noticed a green stone among the roots of a tree that had been toppled by lightning. Emeralds have also been found in Australia, Africa, Pakistan and India, and there have been small finds in both Norway and Austria.

Other beryls

Emerald is not the only beryl that has been cut as a gemstone. Almost as famous is aquamarine, a sea-blue or sea-green beryl. Unlike emerald, aquamarine is often remarkably free from inclusions, and clear and clean crystals of very considerable size have been found. Aquamarine has been found in Brazil, Russia, Malagasy, the USA, Burma, India and in a number of other places. Enormous crystals have been found recently in Mozambique. Most aquamarine is green rather than blue when found, and is subsequently changed to the favoured blue by heat-treatment. Once a relatively inexpensive stone, a fine aquamarine can nowadays be very expensive.

Another famous beryl is the pink morganite, named after the American financier, J. P. Morgan, who collected pink beryls. Some of Morgan's stones came from the USA, others from Brazil and Malagasy. Yellow beryls and brown beryls exist as do green beryls which the gemmologists distinguish from emerald because their coloration is not due to the presence of chromic oxide.

The emerald-cut, known sometimes as the trap-cut or the square-cut, got its name because it is the most common cut and that which most flatters beryl gemstones. To produce it 16 rectangular cuts are placed round the table, and another 32 on the pavilion.

Beryl simulants

Emerald simulants were made in the past by cementing a piece of green gelatine or a sintered copper compound, between a table and a pavilion of rock crystal or two pieces of colourless synthetic spinel. A new simulant has since appeared in which a layer of an unrevealed green substance is placed between a crown and a pavilion of colourless beryl. Doublets have been made from a top of almandine garnet and a pavilion of green glass. Also, rock crystal has been stained green to imitate emerald. Glass imitations have been passed off as aquamarines and so have pale-blue synthetic spinels.

Synthetic emeralds are discussed in the section on synthetic gemstones.

Other coloured stones

While ruby, sapphire and emerald are by far the most important coloured stones, an increasing interest shown in other coloured stones makes it imperative for the jeweller to know about so-called semi-precious stones. This is an unfortunate description, for to call a ruby precious and, say, a topaz semi-precious can confuse the public about relative values. With gemstones, quality can be more important than a name. A fine topaz could have 10 or even 20 times the value of a poor ruby.

What a great many people, even in the jewellery trade, do not fully appreciate is that fine coloured stones have always been exceptional. When a prospector discovers that ruby or sapphire, say, is present in a particular area, he still does not know whether it will be worth investing in their recovery. The gems may be of poor clarity, translucent or heavily included, and so unsuitable for use in jewellery and therefore not worth the cost of extracting them. Only a small proportion of finds prove to be economically viable to mine.

Blue john

Blue john would be scarcely worth mentioning were it not Britain's most famous native gem material. Blue john is the massive form of fluorspar. It is an attractive stone which usually has a transparent, or translucent, colourless ground through which striations meander, usually blue in colour. A green massive fluorspar is also sometimes found.

There is also a cubic crystalline form of fluorspar found in England and elsewhere in Europe, but the bold pastel-shaded crystals of this mineral, though very attractive, are seldom cut as gemstones because they are too soft to withstand wear. Blue john, on the other hand, has been used for jewellery since the Roman occupation. For many years this jewellery has been sold to tourists visiting the little town of Castleton, in Derbyshire, outside which, on one of the tors, the Blue John Mine is situated. The mine is almost worked out now, but in the past some very large pieces of blue john have been recovered from it.

Chrysoberyl

Like beryl, chrysoberyl is a beryllium gemstone, an oxide of beryllium and aluminium. It is usually a rather yellowish-green in colour. There are two types of chrysoberyl that show curious optical effects. They are the beautiful chrysoberyl cat's-eye, which contains needle-like inclusions that produce the characteristic chatoyant streak, and the very rare alexandrite, that looks green in daylight and red in artificial light. The split personality of the alexandrite is due to the fact that the stone absorbs both red and green light with equal facility. Therefore the colour which the stone shows depends on the colour of the light in which it is seen. Unfortunately, a natural alexandrite is seldom seen, though there have been some important finds in South America recently. Even in Russia, once the main source of the stone, synthetic imitations are sometimes sold as alexandrites to the tourists.

Dioptase

The copper silicate dioptase has been used in jewellery in recent years, mainly in the form of uncut crystals. It is occasionally faceted or cabochoned, though with a hardness of only 5 on Mohs' scale it is not really hard enough to use as a jewellery stone. The attraction of this mineral for jewellery designers is the beautiful emerald-green colour of the trigonal crystals, often contrasted with the white of the translucent massive rock-crystal matrix with which it is frequently associated. Dioptase is found in many mineral-rich areas, including Russia, the Congo, Chile and the USA.

Feldspar

Feldspars are a complicated and a geologically interesting group of minerals, but only a few members of the family are of interest to the jeweller. These are the mysterious moonstone, sunstone, amazonstone and labradorite.

The most important of these, moonstone, is an orthoclase feldspar, a potassium aluminium silicate. The pure form of orthoclase is adularia named after the Adular mountains in Switzerland where clear crystals are found. Moonstones, which are invariably cut as cabochons to bring out their beauty, have a blue shiller in them. This is caused by reflection of light from layers of albite, another feldspar, contained in the orthoclase. The most important source of moonstone is Sri Lanka.

Sunstone, which, as its name suggests, has a yellow glow in it, is an orthoclase containing the iron mineral hematite. Amazonstone is a green microcline feldspar, another potassium aluminium silicate, but slightly harder than orthoclase. It is named after the Amazon, but although it is found in Brazil it has never been found anywhere near the Amazon itself. The main source of this stone is, in fact, India.

Labradorite is a plagioclase feldspar, a soda-lime mineral. It is noted for the iridescent play of colour shown when viewed from the right angle; a play of colour not much different from that displayed by gem opal. Labradorite is basically a rather drab grey stone, and the blue and copper colour it shows is due to interference effects, caused by its laminated structure, when light falls on the stone. Labrador is the main source of the mineral, but it is also found in Finland. The Finnish material is called spectrolite.

Garnet

To try to sum up the garnet family in a few paragraphs is very difficult, simply because it has so many members. Of the six distinct types of garnet five are polished and used in jewellery.

The garnets most often seen by the jewellers are the deep-red almandine and the blood-red pyrope. At one time the source of most of the pyropes was the area once known as Bohemia. Archaeological finds suggest that a jewellery industry existed in this part of Europe as early as the fifth century AD and that the jewellers set the local garnets in gold. The garnets were polished as shaped slabs and set *cloisonné* to produce a stained-glass-window effect. In the nineteenth century, when garnets were immensely popular, a thriving industry grew up again in Bohemia, and today the factories in the Czech Republic turn out quantities of garnet jewellery, mostly in the nineteenth century style, set with a profusion of rose-cut stones.

Pyrope garnet is a magnesium aluminium silicate. Almandine garnet, which has been found in Australia, India and Sri Lanka, is an iron aluminium silicate. When there is a lot of iron present in almandine garnet it tends to be brown rather than red in colour.

The rarest and the most beautiful of all the garnets is the green demantoid, a species of the andradite garnet. This is a calcium iron silicate, and the grass-green colour is due to the presence of chromium. The best demantoid garnets come from Russia, but few are found there nowadays. Some have also been found in Italy, Switzerland and the Congo. There are also yellow and black andradite garnets. The black garnets have occasionally been used for mourning jewellery.

The grossular garnet, which is a calcium aluminium silicate, derives its name from the botanical name for gooseberry – *grossularia*. This is rather confusing because most of the grossular garnets seen by the jeweller are orange or yellow in colour. The green grossular garnets are not usually sufficiently transparent to be cut as gemstones. There are also red grossular garnets, but they are seldom of gem quality.

Some transparent green grossulars have been found in Pakistan, and there have been very important finds recently of facetable green grossular garnets coloured by chromium in both Kenya and Tanzania. These stones have been christened tsavorites by Henry Platt of Tiffany's, who also christened tanzanite when this came on the market. Platt first saw the tsavorites in the tanzanite mines in Tanzania, but was unable to buy them there. He became convinced, however, that the vein containing these beautiful emerald-green stones crossed the Kenyan border. Eventually he was proved to be right, for they were found in considerable quantities in the area of the Tsavor

National Park, hence their name. Supplies of these stones are now coming out of both Kenya and Tanzania, and they make very attractive and relatively inexpensive centre stones for cluster rings.

Another garnet only occasionally used for jewellery is a manganese aluminium silicate. This is the spessartite garnet; the name is derived from Spessart, the district in Bavaria where this stone is found. The few spessartite garnets of gem quality are usually orange-red in colour.

Natural glass

The natural glasses are not of great importance to the jeweller because they are only rarely cut as gemstones, but the exhibition of the work of the famous Swiss designer Gilbert Albert, in London in 1966, drew public attention to them, because Gilbert Albert used various types of natural glasses in some of his more lavish pieces.

There are two distinct types of natural glass: those which, like obsidian, are the result of the fusion of silica during the cooling of volcanic lava or of other earthly upheavals, and those which came to us from space.

Obsidian is usually black or grey. Like flint and jade it was important to early civilizations as a raw material for tools and weapons. The Aztecs and the American Indians sank deep shafts to recover obsidian. As well as using it for weapons and tools, the Aztecs also used obsidian for ornaments, and occasionally down through the ages it has been cut and used for jewellery.

The extra-terrestrial glasses arrived on Earth, it is believed, as a result of the disintegration of stars. The most famous of these glasses is moldavite, named after the Moldau river in Bohemia, where this mysterious glass was first found at the end of the eighteenth century. Similar glass has been found in many other places. It is often a rather interesting green colour, but is sometimes brown, sometimes black.

Jade

There are two quite different jades: nephrite which was at one time used for tools and weapons, and the more colourful jadeite.

Nephrite was used for tools and weapons because, like flint, a sharp edge is created when it is fractured. The Maoris of New Zealand used it, as did the stone-age people who inhabited the lake villages of Switzerland and the ancient peoples of Mexico, who also carved representations of their gods in this tough, fibrous material. The Chinese also used nephrite to make tools and weapons, and when metal superseded nephrite as a raw material for knives and axe-heads, they perpetuated their past by producing ritual weapons and a whole range of amulets from jade. In the early days, the Chinese carvers are believed to have cut their jade laboriously with bamboo drills and silica sand. Later, metal drills were used with powdered corundum, and later still carborundum was used as a cutting agent. At some time during the eighteenth century the Chinese carvers started to use jadeite instead of nephrite. They obtained their jadeite from the rivers of Turkestan.

Nephrite is a silicate of magnesium and calcium, and is found in a limited range of colours. It is usually either dark green, a mutton-fat white, a greenish-grey, or brown. Jadeite, a sodium aluminium silicate, is found in a much greater variety of colours, green, black, orange, mauve, yellow, blue, pink, and even tomato-red, brown and white.

Fine jade today fetches very high prices and so it is frequently simulated. There is a lot of bowenite serpentine sold as jade. This is a soapy green stone which is often quite wrongly described as 'new jade', a name given to it by the Chinese themselves, who now produce rather mannered and lifeless carvings from it. Another common jade simulant is the massive green grossular garnet. This is sometimes wrongly described as Transvaal jade. Sometimes jade doublets and triplets, employing a thin slice of natural jade bonded to a less precious material are found.

Lapis-lazuli

One of the earliest gemstones to be used in jewellery was lapis-lazuli. The ancient Egyptians used it in conjunction with turquoise and cornelian, *cloisonné*-set in their symbolic jewels at least as early as 3000 BC. Today lapis is more frequently used as a decorative mineral for boxwares and dishes. Lapis owes its beautiful blue colour to the presence of hadynite. Often also present in this complex mineral are sodalite, noselite and lazulite. Often lapis exhibits tiny gold-coloured specks, which are in fact tiny crystals of iron pyrite. Sources include Afghanistan, where the deposits have been worked on and off ever since the ancient Egyptians obtained their supplies from this area. It is also found in Chile, Burma, Pakistan, the USA and Canada. Synthetic lapis produced by Pierre Gilson that exhibits specks of iron pyrite, has recently made its appearance on the market. Glass imitations have been produced and natural lapis has been powdered and reconstituted.

Malachite

There has been a vogue in recent years for setting polished slabs of this beautiful green striated material in jewellery, though in the past its role has been rather as a mineral from which *objets d'art* were carved than as a jewellery stone. Malachite is a hydrated copper carbonate, and it has considerable commercial importance as copper ore. It is found in veins, in botryoidal masses and in the form of stalactites. It usually displays a series of concentric striations of various shades of green. Malachite is not infrequently found in conjunction with a blue mineral of similar habit, another copper carbonate, azurite. When the combined minerals are polished together the material is known as azurmalachite and is most attractive.

The main source of malachite in the past was the copper mines in the Urals in Russia. It has also been recovered from copper mines in Australia, Africa and North and South America.

Marcasite

The stone known in the jewellery trade as marcasite is really pyrite. Both marcasite and pyrite are sulphides of iron, but they have different crystal structures. In the eighteenth century, when marcasite first enjoyed a wide popularity, the difference between these two minerals was not recognized, so the small-faceted pyrites which the eighteenth century jewellers set in gold, silver and pinchbeck, were given the name marcasite.

Pyrite is found in Britain, but most of that used by the trade comes from the Jura mountains. There it is found in the form of cubic crystals, with the cubes interpenetrating one another at all angles. Often these crystals are of a golden colour, and sometimes actually contain traces of gold. It is pyrite which is known as 'fools' gold' – glinting among the pebbles in a pan it must momentarily have raised the hopes of many a prospector.

Pyrite is also found in massive form, and nodules of this massive pyrite are sometimes picked up in the south of England. Pyrite has been imitated both by cut steel and by faceted glass, and it was, in fact, the cheap imitations of marcasite jewellery which finally brought an end to the great post-war marcasite boom.

Opal

Opal, besides existing in its own right, is also the bonding material that cements together the microscopically fine quartz crystals of which chalcedony is composed. It is a hardened gel, consisting of microscopic silica spheres mingled with infinitely tiny droplets of water, and so is known as a hydrous compound.

Most of the opal found is of non-gem quality, a rather dull stone. The opal scattered with brilliant colour is much rarer, very rare indeed today. Australia has produced most of the gem opal in recent years, but many of the opal diggings in the Australian outback are worked out. From these diggings came the fine black opals, looking like the wings of tropical butterflies, as well as less spectacular but no less beautiful milky-white opals.

The iridescent colours in precious opal, which are like the colours seen in oily water, are considered to be due to interference effects, caused by the reflection of light from the minute spheres of amorphous silica of which opal is composed. These spheres are so small they can only be seen with the help of an electron microscope.

Besides the iridescent opal there is a beautiful translucent orange-coloured opal found in Mexico and other places known as fire opal. An apple-green translucent opal is also found in South America.

Much of the opal that is exported from Australia these days consists of opal doublets and opal triplets. The doublets are made by cementing a slice of opal to a base of common opal. The triplets also have a slice of rock crystal cemented over the top of the thin precious opal layer. The doublets when mounted can be difficult to detect, but the triplets have characteristic glassy appearance that once seen is easily recognized.

The French chemist and industrialist Pierre Gilson has also produced synthetic opal which can easily be mistaken for genuine opal (see the section on synthetic gemstones).

Peridot

Peridot is a magnesium sulphate. a green stone, the colour of which Oscar Wilde described as ‘pistachio green’. The island of Zeberget in the Red Sea, the traditional source of this stone, was once known as Topazios, which resulted in peridot being known as topaz in the early days.

Peridot, which is also found in Burma and the USA, is another of those stones which is getting rarer. This may to some extent be because the expeditions to the island of Zeberget, which used to bring back this mineral, nowadays visit the island less and less frequently.

The quartz gemstones

The most common of this vast family of gemstones is the colourless variety of quartz, known as rock crystal. Like all the other members of the family, rock crystal consists of crystalline silicon dioxide. The Greeks called rock crystal *krustallos*, meaning ice, and the stone certainly has an ice-like quality.

Crystals of colourless quartz that stand as high as four feet or more are found. Others are so small that they look like a sprinkling of frost on the mother rock. These crystals are found all over the world, including Britain. They are not very often cut as gemstones, but attractive beads are made from them, and the Chinese have a long tradition of rock crystal carving. They make snuff bottles and small dishes from this stone, and they have always had a particular affection for the variety known as Venus hair stone, which contains long needle-like rutile crystals. They also carve colourless quartz containing slender black tourmaline crystals, and quartz containing green actinolite needles.

To the jeweller, the most important varieties of quartz are the yellow quartz known as citrine, and amethyst, the purple variety. Citrine is the stone which is still sometimes wrongly described as quartz-topaz, or even just topaz, though the two minerals, quartz and topaz, belong to quite distinct species.

Citrine is a not unattractive yellow stone, varying in colour from a light golden-yellow to a red-yellow. The colour is determined by the amount of iron present. Natural citrine is not so common as its availability might suggest. Most of the so-called citrine in the trade is in fact amethyst, the purple variety of quartz, that has been heat-treated.

Amethyst suffered for a long time from the taint of Victorian popularity, but it is a very beautiful stone at its best, rich and royally purple, and it is now firmly back in favour. It is a stone which has had many legends woven round it, and has been used by more than one civilization as a symbol of dignity. Amethyst wine cups, or cups with amethysts placed in the bottom, were supposed by the ancient Greeks to protect those who drank from them from drunkenness. The name of this stone is, in fact, derived from a Greek word that means 'not drunken'. The Germans at one time used amethyst as a symbol of power, while the Roman Catholics and the Buddhists of Tibet have both used it as a symbol of sanctity.

Scottish regalia is often set with another form of quartz, the smoky brown quartz known as cairngorm. This is a fairly rare stone, but recently smoky quartz has been produced by irradiating amethyst; the colour change is said to be permanent. Another beautiful quartz is the cloudy-pink rose-quartz, used both for beads and carvings. This is usually found in massive form, but occasionally crystalline rose-quartz is found.

Some quartz has a multitude of fine yellow or blue asbestos fibres in it, which create a cat's-eye effect. This quartz cat's-eye is not to be confused with the more beautiful chrysoberyl cat's eye. Tiger's-eye or crocidolite is a massive quartz that also contains asbestos.

Recently, quartz has assumed a new role. It is used to control the quartz watches that most people now wear. One of its properties is that if an electric current is passed through a thin slab of it, it will pulsate at a regular rate. A physicist, W.A. Marrison, took advantage of this piezoelectric effect when he designed the first quartz-crystal clock in 1929, the clock which revolutionized our ideas of accurate timekeeping.

Besides the varieties of quartz already mentioned, there is a whole group of microcrystalline, or cryptocrystalline, quartz. These consist of microscopically fine quartz crystals bonded together by opal which is another silica gemstone. These types of microcrystalline quartz are usually known as chalcedony. Among them are the apple-green chrysoprase, the flesh-red cornelian, the black or black-and-white banded onyx (nowadays usually simulated by staining agate), and the dark green plasma,

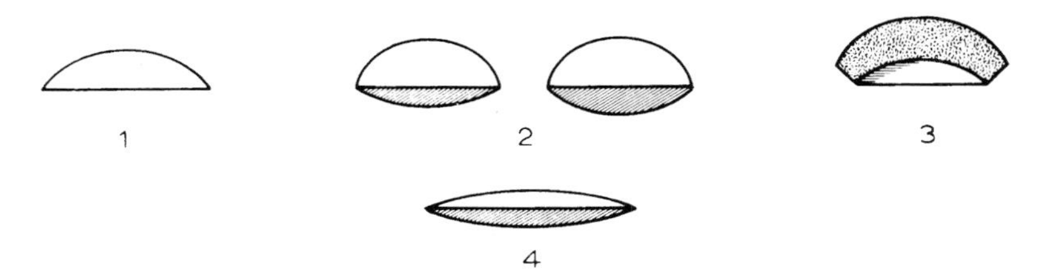

Figure 2.8 Different profiles of coloured stones cut cabochon: 1 – simple; 2 – double; 3 – hollow (section); 4 – tallow top

which when speckled with red flecks of jasper is called bloodstone. However, the most famous chalcedony is agate, used by the jewellery trade both in its banded and its dendritic form. The dendritic form is the so-called moss agate, the inclusions in which look like fronds and leaves. Agates are usually cut cabochon or as flat slabs (Figure 2.8). Staining is often employed to improve banded agates, and much of the agate used by the jewellery trade today is stained.

Rhodonite and rhodochrosite

These are the two most popular opaque red stones which are used for *objets d'art*, and, more recently, for jewellery. A number of designs nowadays feature beads or slabs of one or other of these two most attractive minerals.

Rhodonite is a silicate of manganese, exhibiting at its best a deep rose-red coloration. The red areas are usually divided by areas of black, which consist of oxidized manganese. The stone is moderately hard, about 6 on Mohs' scale, and is perhaps seen to best advantage when carved to form bowls. It is found in many areas in the USA, and in Russia, Mexico, Australia and Africa.

Rhodochrosite has a flesh-pink colour quite distinct from the colour of rhodonite, though it also owes its colour to manganese, being a manganese carbonate. Rhodochrosite is striated like agate or malachite. The striations result from its stalagmitic formation. Rhodochrosite is appreciably softer than rhodonite, only 4 on Mohs' scale. The main sources are Eastern Europe, India and the USA, but there was also an important find in Argentina in the 1930s.

Both rhodonite and rhodochrosite are very occasionally found in crystalline form. The beautiful translucent pink cubic crystals of rhodochrosite are particularly attractive. These crystals, although of no importance as gem material, are very desirable acquisitions by those who collect mineral specimens.

Sodalite

This attractive blue decorative stone is today widely used for *objets d'art* and occasionally in jewellery. Sodalite has been called 'poor man's lapis', and is, in fact, one of the constituent minerals of lapis-lazuli. However, it never attains the rich blue coloration characteristic of good-quality lapis.

Sodalite is a sodium aluminium silicate containing sodium chlorite and is moderately hard, 5½ to 6 on Mohs' scale. It is widely distributed, and is found in North and South America, India and Africa. Like other mainly massive minerals sodalite is found in crystalline form. The crystals are cubic and exhibit a range of colours including green and red as well as blue.

Sphene

Sphene, a silicate of titanium and calcium, is cut to produce green, yellow and brown gems. They are not very hard, only 5½ on Mohs' scale, but they show considerable fire when faceted. Sphene is found in Austria, Canada, Malagasy, the USA and Brazil, and is one of the very limited number of gem minerals found in the Swiss Alps.

Spinel

Spinel is a gem which occurs in several colours. The most usual are red, pink, orange-yellow, green and blue. It is a gem which crystallizes in the cubic system. It is fairly hard, 8 on the Mohs' scale, and therefore well suited for use in jewellery. It is found mainly in Sri Lanka, Burma, Thailand and the USA. The white or colourless spinels encountered in jewellery are almost invariably synthetic and they should not be described other than as 'synthetic spinels'. The term 'white spinel' sometimes seen in jewellers' shops is inadequate if the stones are, in fact, synthetics. The jeweller also has to be on the lookout for spinel doublets, including doublets with a natural spinel top in a variety of colours, sometimes bonded to a synthetic spinel base.

Topaz

It may at this stage seem unnecessary to stress the importance of making it clear to the public that topaz is not citrine, and that citrine is not topaz. However, that confusing description 'quartz-topaz' is sometimes applied to citrine, and some irresponsible jewellers flagrantly display citrine labelled 'topaz', or 'real topaz'. The only gem that can properly be called topaz is topaz.

Citrine is nothing more nor less than the yellow variety of quartz. It is crystalline silicon dioxide. Topaz is a fluosilicate of aluminium. It is much harder than quartz, much more brilliant and a much rarer stone.

Topaz, like most gemstones, is found in a variety of colours. The best known, the one that is confused with citrine, is the beautiful sherry-coloured topaz, most of which comes from Brazil.

There are also very attractive blue, green, yellow and pink topazes, and there is colourless topaz. Natural pink topaz is very rare, and most of the pink topaz that has been found has acquired its colour as a result of the heat-treatment of the brown stones. Much of the blue topaz on the market also owes its colour to heat treatment.

Besides being simulated by citrine, topaz is also simulated by synthetic corundum, tourmaline and aquamarine, and synthetic spinel doublets.

Tourmaline

The beautiful long crystals of tourmaline are found in many colours; blue, red, pink, green, yellow and brown. There is even an opaque black tourmaline. Some crystals of tourmaline are parti-coloured and are cut to produce parti-coloured stones. Some crystals also show different coloured centres like a piece of seaside rock when sectioned. In fact, no other gem is found in such a variety of colours as tourmaline.

The chemical composition of this stone is so complex that many gemmological textbooks do not even give a formula for it. It is basically another aluminium gemstone, a borosilicate of aluminium. However, many other ingredients are present, and it is the presence and the proportions of these which determine the colour.

Tourmaline is moderately hard (7 to 7½ on the Mohs' scale), and the stone may be transparent, translucent or even opaque. As with aquamarine the colour is often improved by heat-treatment. Russia, Brazil, Africa, Malagasy, Pakistan and the USA are all sources of tourmaline.

Transparent tourmalines are usually mixed-cut, and they can make very attractive gemstones, particularly the red and the acid-green tourmaline. Those tourmalines showing a cat's-eye effect are, of course, cut cabochon, and inferior tourmaline crystals are sometimes turned into beads.

Turquoise

There have been very few periods when this opaque blue stone has been out of favour. Often the demand has exceeded the supply and to fill the gap simulated turquoise has been made down the ages, going back to the days of ancient Egypt. It is thought, in fact, that turquoise was the first gemstone to be simulated. Jewellers must therefore be on guard against imitation turquoise made from glass, plastics and clay, and against natural stones whose colour has been improved by staining or waxing. Synthetic turquoise has also been produced by Pierre Gilson.

The name turquoise means Turkey stone, though most of this gemstone has always come not from Turkey but from one quite small mountain area in Persia. This sky-blue stone has a complicated chemical composition. It is, like opal, a hydrous compound, a hydrous copper aluminium phosphate containing iron. The colour, which has considerable bearing on the price of this stone, is due to the copper and the iron. Turquoise is usually cut cabochon and takes a good polish.

Zircon

Zircon has already been mentioned as the most convincing natural simulant of diamond. It is a silicate of zirconium found in a variety of colours. The most common natural colours are brown, green and blue. It is the brown stones that the Sinhalese heat in crude native furnaces to produce the colourless and blue zircons used by the trade.

Zoisite and tanzanite

The crystalline form of zoisite was not in the past of great interest to anyone except mineralogists. The usually, though not invariably, colourless orthorhombic crystals, though attractive in form, were seldom cut as gems. In 1967, however, a new find of zoisite crystals in a range of colours was made in the Meralani Hills in Tanzania. Some of the crystals were blue, and it was found that the others, the green, yellow and brown crystals, could be turned blue by heat-treatment. These stones, christened tanzanites by Henry Platt of Tiffany's, are of a most attractive mauve colour. They have become much in demand as ring stones despite the fact that they are not particularly hard (6 on Mohs' scale), inclined to be easily fractured, and particularly averse to ultrasonic cleaning. Since their first discovery the tanzanite mines have been nationalized, and supplies of good-quality stones have become sporadic.

Two other forms of zoisite are of importance, the massive green and pink varieties.

The massive green zoisite, which often contains hexagonal ruby crystals comes mainly from Tanzania, and is used mostly for bowls or as polished slabs intended as shelf specimens or table mats. The pink variety is called thulite, after Thule, the ancient name of Norway, from where it is obtained. It tends to be found most often as a variegated stone, the pink contrasted with white areas.

The organic gems

The so-called organic gems owe their existence to living organisms: pearls to oysters, mussels and so on, jet to primeval trees, and amber to prehistoric pines. By far the most important of the organic gems is, of course, the pearl.

Natural pearls

Natural pearls are those which grow in oysters and other molluscs, without assistance from man. Oysters are the best known source of pearls, which are, however, also found in mussels and in other shellfish such as the giant clam.

There are a number of different species of oyster. Many of the so-called oriental pearls, famous for their pink-white colour, their regular shape and fine lustre, are found in the *Pinctada vulgaris*, a small oyster which lives in the Persian Gulf and in the Gulf of Manaar. These oysters usually measure not more than 2½ to 3 in. across, and seldom produce pearls of more than 12 grains in weight. The grain used as a measure for pearls is 0.05gram (¼ metric carat). Bigger pearls are found in the *Pinctada margaritifera* which live in the waters off the northern coast of Australia, and some as big as 40 to 60 grains in the *Pinctada maxima*, also found in Australian waters and off the coast of Malaysia.

Freshwater pearls, found in *Unio* mussels found in our rivers, were treasured by Roman invaders of Britain, and have had many admirers during the centuries since the Romans left our shores. Chief among these were the Scots who used, and indeed still use, them in their jewels and their regalia. For although few of these freshwater pearls have the majesty or the lustre of the pearls found in oysters, they do have a charm of their own. They are sometimes very colourful. They are found in a variety of pastel shades, the grey and pink pearls being the prettiest of them.

The pearls found in oysters may also display subtle tinges of colour. Some have a pink, some a blue and some a green tinge. The *Pinctada martensi*, found round the shores of Japan and widely used for cultured pearl production, tends to produce a greenish pearl. The early cultured pearls had a distinct green tinge like that associated with the natural Japanese pearls. Later, however, cultured pearls of a better colour began to come out of Japan, once the Japanese farmers had evolved a technique for improving their pearls by dyeing and bleaching.

Oysters also produce beautiful, natural and cultured black pearls. 'Black' does not adequately describe these, as they are full of subtle colouring similar to that displayed by black opals. Black pearls, which fetch high prices, should be viewed with care, for some of those on the market were not black when they came out of the oyster. Their colour is due to artificial treatment, possibly with silver bromide, the light-sensitive chemical used in the coating of photographic films.

Some pearl fishing in the Persian Gulf and off Sri Lanka is still carried on sporadically in the traditional way, though there are virtually no full-time divers working the oyster beds nowadays. The divers used to go out to the oyster beds in boats. Most of those who fished in the Persian Gulf sailed there in the stately but often decrepit dhows of Bahrain. Before 1939, as many as 600 of these dhows tacked down from Bahrain to the reefs of the Gulf, some with perhaps 100 men on board. Today this fleet is decimated. The oil industry offers alternative employment and, understandably, there are few young men nowadays prepared to take up the arduous and dangerous work of pearl fishing.

Certainly, pearl fishing in the Indian Ocean was not an easy way to make a living. The divers sometimes went down to a depth of 90 ft, on the end of weighted ropes, to prise up the oysters from the rocks. An average of only one oyster in 40 contained a pearl. Though an experienced diver had some indication of the presence of pearls from the outward appearance of the oysters because those containing pearls tend to be misshapen, this was not an infallible guide. Each diver had to bring up as many as 500 oysters a day, the result of perhaps 30 dives of 1 to 1½ minutes duration, if the trip was to be profitable.

Off Australia the divers have more recently gone down in diving suits, and the future of pearl fishing probably depends on divers using today's sophisticated equipment, rather than upon the romantic and naked diver whose only equipment was a knife and a string bag hung around his neck.

The oysters brought up from the beds in the Persian Gulf and in the Gulf of Manaar were left to rot on the decks of fishing boats, and the pearls were eventually extracted. Often only one pearl was found in an oyster; in fact, the best are found singly. Sometimes there might be two, three or more found in the sun-rotted remains of one oyster, and, indeed, as many as 87 tiny seed pearls have been found in a single oyster.

The price of an oriental pearl is still, as it has been traditionally, calculated from a base price, which is arrived at by inspecting the pearl and assessing its quality. Its colour, its shape and the presence or absence of flaws are taken into account. If the standard base price is 5p, a good pearl might be assessed as having, say, a 50p base price, a particularly fine one a £1.50 base price. The weight of the pearl then enters into the calculations. This is squared to arrive at the actual price of the pearl. A 2-grain pearl with a 50p base price would he valued at £2 ($2 \times 2 = 4$, then $4 \times 50p = £2$). A 20-grain pearl of the same base price would be valued at £200 ($20 \times 20 = 400$, then $400 \times 50p = £200$).

Pearls are created by the secretory cells in the robe-like mantle enfolding the organism of the oyster. The primary function of these cells is to produce the two hinged shells in which the oyster lives. Some of these cells secrete a liquid which solidifies as a horny black or brown substance which becomes the outside of the shells. Other glands secrete the calcium carbonate, which solidifies in a crystalline form to build the iridescent smooth lining of the shell known to the layman as mother-of-pearl, and to the gemmologist as nacre. The small calcium carbonate crystals are incidentally embedded in, or attached to, each other by a substance known as conchiolin which is similar to the material of which our fingernails are composed.

Mother-of-pearl and pearls are both composed of the same substance, crystalline calcium carbonate, secreted by the self-same cells. No one is quite sure what causes these cells to perform their secondary role, that of producing pearls. Some people believe it is the result of a tumour, others that it is caused by a tiny worm sucked into the oyster during the process of opening and closing the shell to draw in sea-water containing the plankton on which the oyster feeds. Many people believe it is the result of a disease. Whatever the cause, once a nucleus of nacre is formed in the mantle of the oyster it continues to coat this nucleus with nacre, perhaps because it acts as an irritant. As a result a pearl grows inside the oyster layer by layer. A pearl cut in half and viewed through a lens shows distinct layers of nacre, so that it appears rather like a sliced onion or a tree trunk that has been sawn through. A strong lens reveals that the layers are made up of plate-like formations of calcium carbonate which resemble a terraced vineyard on a hillside. It is these plate-like formations that refract the light and give the pearl its lustre.

Cultured pearls

The ancient Chinese were the first people to discover that a bivalve will coat with layers of nacre any object inserted into it. They are known to have placed tiny carvings of the Buddha into mussels to produce little mother-of-pearl ornaments. It was the Japanese, however, who first used this technique to produce a simulant for natural pearl. They placed a mother-of-pearl bead in a little sac of mantle, and then placed this in the mantle of a live oyster (Figure 2.9). The man who first made pearl farming a commercial proposition, though he was not the inventor of the technique as many people believe, was that strange and dedicated man Kokitchi Mikimoto. The first spherical cultured pearls were produced by another Japanese, Tatsuhei Mise, in 1907. However, it was the government scientist, Tokiehi Misikawa, who developed the culturing technique which is still used today. Many of Mikimoto's fellow countrymen laughed at him in the early days, but his beautiful diving girls and his cultured pearls were soon to become famous. The first cultured pearls he produced were blisters, but before many years had passed he too was producing spherical cultured pearls.

The first round cultured pearls were seen in Britain during World War I, but it was not until the early 1920s that they began to arrive in any quantity, and to create a problem for the pearl trade. It was by no means easy in those days to distinguish between them and natural pearls, but soon gemmologists evolved techniques for identifying them. Drilled pearls intended for necklaces were tested with an endoscope, which made use of the fact that the spherical layers of nacre of which the natural pearl is composed reflect back light (Figure 2.10). The parallel layers of mother-of-pearl bead in the centre of a cultured pearl do not.

Today pearls are tested by X-ray examination. This reveals the mother-of-pearl bead in the centre of a cultured pearl (Figure 2.11). The development of the cultured pearl industry, which at first seemed to be a threat to the jewellery industry, turned out to be a boon. It came at a time when natural pearls were becoming ever rarer and more

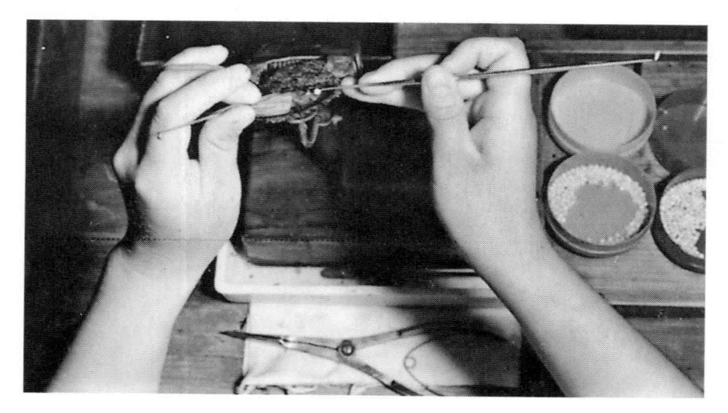

Figure 2.9 Mother-of-pearl beads being inserted into the mantle of an oyster to produce cultured pearls. This is a very delicate operation

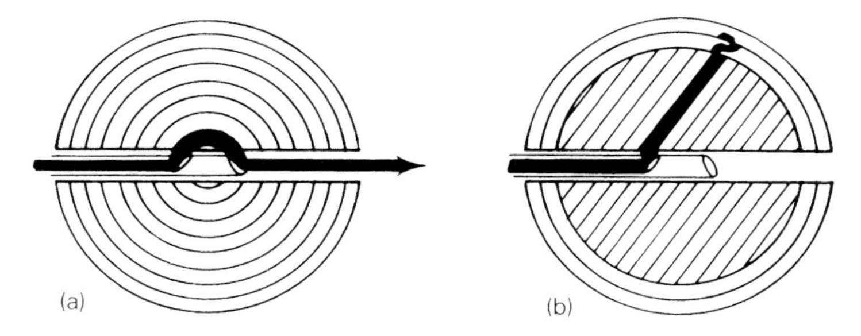

Figure 2.10 Testing pearls with an endoscope. In the natural pearl (a) the light passes straight through the drilled hole to be reflected back, while in the cultured pearl (b) the light is not reflected back but is dispersed

Figure 2.11 An X-ray picture of a cultured pearl necklet, revealing the presence of the nuclei (by courtesy of the late Robert Webster, F.G.A.)

expensive, and also when more and more people wanted to buy pearls. Soon millions of women in the West were wearing necklaces made up from Japanese cultured pearls.

Another type of cultured pearl has more recently appeared. This is the non-nucleated cultured pearl grown by the Japanese in a species of freshwater mussel living in Lake Biwa. The description Biwa pearls tends to be applied to all cultured freshwater pearls although most of them now come from China, rather than Japan. A tiny piece of mantle, which eventually disintegrates, is used as the nucleus which triggers the animal's nacre producing mechanism. Non-nucleated pearls have also been produced by Australia's rapidly developing cultured pearl industry, using the *Pinctada maxima*. The first non-nucleated cultured pearls to arrive in this country were characteristically barrel-shaped, but subsequently more spherical pearls have been produced. A number of different shaped pearls, known by names such as 'rice' and 'stick' also came on the market. These were available in a variety of colours, some pleasing, others rather garish when their colour, as is often the case, is the result of staining.

The Japanese pearl farmers used to destroy substandard or misshapen cultured pearls, but recently a number of jewellery designers have been incorporating baroque cultured pearls in their designs. The source of some of the finest cultured pearls is Hawaii, where they are being produced by expatriate Japanese. These beautiful 'South Sea' pearls are often strongly coloured. The peach-coloured and grey pearls are particularly attractive. This area is also the main source of black cultured pearls. The cause of the coloration is possibly traces of chemicals present in the waters around Hawaii.

Fortunately for the pearl farmers, oysters are very productive. In spring, when the temperature of the water in the Japanese bays reaches 24°C, the female oyster begins to lay eggs; one oyster produces as many as 5 to 6 million eggs a day. Those that are fertilized turn into larvae and then grow at an amazing speed from a tiny animal to one whose shell is 18 mm across in a matter of a month. At the end of that month they will weigh thirty times what they did at the beginning.

Experience has taught pearl farmers that oysters have to be in first-class physical condition before they will produce pearls of fine lustre. Lack-lustre oysters, it seems, produce dull pearls. While the young oysters are developing, the pearl farmers make the nuclei which will be placed in them. In the early days of culturing all kinds of things were used, but today the little spheres of about ⅛ inch in diameter that are inserted in the oysters are cut from the shells of specific mussels found in the Mississippi River in the USA. Pieces of this shell are polished until they are perfectly round, and then placed in little envelopes of mantle taken from live oysters. Employing what amounts to a delicate surgical operation the oyster is opened and the sphere and the piece of mantle inserted in the right part of the animal. The shock of the operation causes many oysters to reject the transplanted spheres, and after the operation they are X-rayed to make sure that the implantation has been successful. Most of the operators are women called *tamaire-san*, which is Japanese for pearl-pusher. The more skilled of these pearl-pushers will operate on a thousand oysters a day.

The oysters are then placed in nets again and returned to the sea. The bamboo rafts that once supported the nets have nowadays been replaced by garish but practical plastic floats. These may not be so picturesque, but they do make life easier for the pearl farmer, who, like any farmer, wages a continual war with nature. Marine worms bore into the shells, halt the secretion process that deposits the nacre on the spheres, and eventually kill the oysters. Eel, octopus and the larvae of other pests eat the oysters' flesh; the notorious red tide, a thick scum of dead plankton, can destroy whole pearl farms.

If the oysters survive the perils of the deep for two or three years they are then 'beached' during the winter when the animals cease to produce nacre. The pearls are removed and sorted. Far from all the pearls recovered are perfectly round and lustrous and of good colour. Perfect pearls are the exception rather than the rule. If this is borne in mind, it can be appreciated that a string of good quality cultured pearls represents very good value.

Though all cultured pearls, with the exception of the Biwas, consist of a nucleus over which an oyster has deposited layers of nacre, there are great differences in quality and colour. These affect the price that is asked by the farmer and ultimately by the jeweller. After the pearls have been beached they are first classified by size. The diameter of pearls is measured in millimetres and the sizes usually used for pearl necklaces range from 3 mm to 9 mm. Larger pearls than this are usually used for pendants, ear-rings, rings, and for setting in gold jewellery.

The effect of size on price is similar to that of all gems – the bigger the pearl the higher the price, and the rise in price from size to size is quite dramatic. A 16-inch string of 9 mm pearls would probably cost nine or ten times as much as a string of 3 mm pearls of the same length and quality. Pearl ear studs, incidentally, cost more than the price of two necklace pearls of the same size, because they have to be as identical as possible in size, shape and colour, and good matches are not easily found. Dealers normally ask at least 50 per cent more than the standard rate for that size of pearl for a matched pair of pearls.

Lustre, colour, shape and clarity are the considerations which determine the quality and value of a pearl. The most important of these is lustre. For a cultured pearl to have good lustre it must have an adequate covering of nacre over the nucleus, and pearls that have not been left in the oysters long enough lack lustre. Lack of lustre can also result from the failure of the cultivator to ensure that the oysters were healthy and kept under the right conditions. Under an electron microscope it is possible to see the difference in the structure of a pearl which is lustrous and one which is not. If the crystals of aragonite, which is the particular form of calcium carbonate from which the pearl is constructed, are regularly distributed and have strong hexagonal forms, then the pearl will be of good lustre. If the aragonite crystals are thinly scattered and lack clear outline or are irregularly deposited then the pearl will look lifeless. This is simply because it is those plates of aragonite crystals which break up the light into its primary colours and give a good pearl its lively quality. If these crystals are thinly scattered and do not present good sharp edges to the light, this play of light will not be present, or at best will be fugitive. An experienced pearl dealer, it is said, can tell a good pearl from a poor one by weighing it in his hand, and in fact lack-lustre pearls do weigh fractionally less than fine ones.

Colour, in the sense of the colour display resulting from the dispersion of light by the nacre, is an aspect of lustre. But cultured pearls also have different ground colours. Many of the shades are very subtle, but are usually identified as white, silver, pink, yellow, green, gold and bronze. Fancy colours also exist, of which blue and black are the most common. Some countries tend to show a preference for one shade rather than another. In the USA, for instance, they tend to prefer silver pearls, while in Britain pink ones are more popular. Some women feel that one shade suits their complexion rather than another.

Most of the cultured pearls produced in Japan tend to have a greenish hue when they come out of the oysters, but they are stained with a variety of chemicals before they come on the market. Pearls with slight blemishes can also be improved by treatment with hydrogen peroxide. These improvements are permanent and need not deter the purchaser.

The next thing to ensure when selecting cultured pearls is that the clarity of the pearls is good. 'Clarity' is used here in the sense of the absence or presence of flaws. The individual pearls should be inspected for surface marks and internal spots.

Artificial pearls

Before Japanese cultured pearls appeared on the market, people who could not afford natural pearls bought artificial pearls. These have been made for centuries. The Chinese produced artificial pearls at an early date. The French produced them as early as the seventeenth century, while Roger Griffiths in *A Description of the River Thames*, published in the middle of the eighteenth century, tells us that the fishermen who got their living from the river: 'take the fish merely for their scales which they sell to make beads'. The scales of fish are still the usual basic raw material of the artificial pearl industry, which today produces untold millions of artificial pearls every year. Suspended in lacquer, the tiny crystals which are found attached to these fish scales are coated on to glass beads, by dipping, to give them an iridescent appearance very similar indeed to the appearance of natural pearls. Artificial pearls produced by depositing fish scales on a mother of pearl bead are now widely produced and these are very convincing simulants.

The identification of artificial pearls is not difficult. The simplest test is to run pearls, suspected of being artificial, across the teeth. If the pearls feel smooth they are artificial. If the teeth detect the gritty texture of the edges of the plates of nacre, then the pearls are either natural or cultured. Artificial pearls used for necklaces also often show a flaking of the coating around the drill holes.

Amber

Amber is not a gemstone but an organic gem. Like the other organic gems, pearl and jet and coral, it owes its origin to something that was once alive – to an organism. The organism responsible for amber was a species of pine which once covered much of northern Europe, but no longer exists today. Amber is the fossilized resin which was exuded by those pines.

Amber is an attractive substance in itself, and is sometimes still more attractive because of the natural objects contained in it. When the resin ran down the trunks of those prehistoric pines it often entrapped insects, foliage, even small lizards, and these have been eternally preserved in a transparent golden tomb. At Kalingrad in Russia, where a great deal of amber has been found, there is a museum housing a collection of 50 000 specimens of prehistoric flora and fauna preserved in the gem.

Amber used to be plentiful along the east coast of the British Isles, and in the Bronze Age was exported. It can still be found among the pebbles along this coastline, but today only a few fragmentary pieces are likely to reward a beachcomber.

Amber is light in weight, and if rubbed it acquires an electric charge, and will then attract bits of paper as a magnet attracts iron filings. Both these characteristics are useful in differentiating between natural amber and some of its many simulants. Plastic imitations are produced in great quantities, and sometimes the manufacturers even enclose natural flies in this spurious amber. If a fragment of one of these plastic imitations is whittled off with a knife and burnt, the smell given off will be the acrid smell of burning plastic and not the aromatic smell of burning amber.

Glass imitations are also common, but these are markedly heavier than natural amber, and do not develop an electrical charge. More difficult to identify, unless one is used to handling amber, is pressed amber, small fragments of natural amber often being compressed together to form a lump of useable size. It is, however, possible, using a lens, to detect the joins between the compressed pieces.

Coral

Coral is a kind of tenement dwelling secreted by the coral polyps, for their own convenience. This communal dwelling place is made of calcium carbonate, and grows on the sea-bed like a forest of pink or white trees.

Coral has been dredged up from the bottom of the sea for centuries and was probably first used in jewellery in Hellenistic Greece. One of the main sources used to be the Mediterranean coast of Italy, and the craft of coral carving flourished in Naples. Coral carvers are to be found there to this day, but they now depend upon imports from Japan for most of the material which they turn into beads or carve into charms. The most desirable colours are the rich red variety and the delicate pink, known as 'angel-skin' coral.

Reconstituted coral, made by powdering and subsequently pressing natural material, has made its appearance on the market.

Jet

Jet, another organic gem, derives as does amber, from trees. It is fossilized wood, a close relative of coal. Jet is a native British gem, and its source is the area round Whitby in Yorkshire. Millions of years ago, trees there became covered by the sea, as a result of some shift of the Yorkshire coastline. The trees were enfolded in the mud of the sea-bed and gradually, as the mud turned into rock, they became fossilized.

Jet has been used for jewellery at least since the Bronze Age; some has been found in graves dating from that period. The Victorians rediscovered jet. It was the great demand for jet for mourning jewellery during the Victorian period that caused the Whitby area to be finally denuded. In recent years only the odd fragment has turned up to reward the patient collector. Jet has also been found in France and Spain.

This sombre substance is either black or brown in colour, and though it is quite soft it takes a good polish. Like amber it acquires an electrical charge when rubbed, and it burns like coal. It has been imitated by glass and by plastic.

Synthetic gemstones

Whether the existence of extremely plausible synthetic stones is a boon or a menace depends on one's point of view. Most retail jewellers understandably see synthetic gems other than in the form of watch jewels as a problem, and the more difficult they are to identify the greater the problem they represent. But today a wide range of man-made gems produced in the USA, Europe and the Far East exists, and the jeweller is left with no choice but to live with them, learn to cope with them and perhaps exploit them increasingly for profit.

Verneuil synthetic corundum

The commercial production of synthetic gemstones goes back to the end of the nineteenth century when, as the culmination of a number of small-scale and not particularly successful experiments, Auguste Verneuil, a French chemist, invented the flame-fusion process. This synthesis was carried out in the famous 'furnace' that still bears Verneuil's name. This furnace consists in simple terms of a hopper in which the chemicals to be fused are placed, with a trip hammer above it that taps the chemicals into an oxy-hydrogen blowpipe. Below this pipe is a 'candle', which can be lowered as the cylindrical boule of fused chemical grows on top of it.

The furnace was first used to produce synthetic corundum. The powder placed in the hopper was aluminium oxide with trace elements of other metallic oxides to produce differently coloured boules. The most important, and indeed the first, synthetic stones to be produced commercially as a result of this process were synthetic red corundums containing chromic oxide, that is to say synthetic rubies. Subsequently blue corundum was produced by the addition of titanium and iron oxides. Later, synthetic corundum came to be produced in a great variety of colours: white corundum, which sometimes masquerades as diamond, purple corundum simulating amethyst, yellow corundum simulating topaz, and even synthetic corundum simulating alexandrite, which shows a colour change in different types of light.

It is necessary to draw a distinction between a synthetic and a simulant. Synthetic red corundum has the same chemical composition as natural ruby, and similar physical characteristics. It is therefore correctly described as synthetic ruby. Synthetic purple corundum does not have either the same composition or the same characteristics as amethyst. All that it has in common with amethyst is that it is the same colour. So purple corundum is a synthetic simulant of amethyst. The same is true of the synthetic purplish-green corundum that has been sold as alexandrite. This is a synthetic simulant of alexandrite; natural alexandrite is not a corundum, but a chrysoberyl. Incidentally, a synthesis of chrysoberyl has been achieved, and a synthetic alexandrite does exist, but it has never been produced on a commercial scale. Perhaps the best-known synthetic simulant today is cubic zirconia which simulates diamond but has a quite different chemical composition and very different physical characteristics.

Of all the synthetic corundums, far and away the most useful has been synthetic ruby, because of its role as a watch jewel. Untold millions of watch jewels had been produced in the Verneuil furnace before alternative techniques came to be adopted in recent years.

Verneuil synthetic ruby has also been widely used in jewellery during the course of the present century. Fortunately Verneuil synthetic rubies, because of their method of growth in the form of circular boules, have fairly easily recognizable characteristics under magnification, in the form of curved lines and distinctive clouds of gas bubbles. Verneuil stones have, however, been reported which have been treated so that the characteristic curved structure is no longer visible and inclusions have been induced which resemble those that occur in natural ruby. Synthetic star rubies and star sapphires are also produced by the Verneuil process.

Synthetic spinel

One of Verneuil's students, L. Paris, was responsible for first producing synthetic spinel, but this material was not produced commercially until the 1930s. Synthetic spinel's role has been mainly as a simulant of other gemstones, because only red spinel has any importance as a gemstone.

Synthetic spinel is produced in boule form in the Verneuil furnace, and the colourless form of it was at one time a widely used synthetic simulant of diamond.

Spinel is a magnesium aluminium oxide. Though it is relatively hard (8 on Mohs' scale) it is of course much less hard than diamond. Synthetic spinels rarely display the characteristic curved bands associated with the Verneuil boules, but the stone has none of the fire of diamond, though it has considerable lustre.

Blue, green, pink and red synthetic spinels have been produced simulating such stones as blue zircon, green tourmaline and peridot. The red version is of course a true synthetic of natural red spinel. In the 1950s a synthetic spinel simulating lapis-lazuli was produced in Germany, some of this material even incorporated the characteristic gold-coloured specks of iron pyrite. A synthetic spinel imitating moonstone has also been produced.

Synthetic rutile

The last thirty years have seen rapid developments in the synthesis of gem minerals, and the production of synthetic simulants. One of the first simulants to make its appearance was synthetic rutile. This material was again produced in a blowpipe by the flame-fusion process, from titanium dioxide, but the technique of production is more complicated than that needed for the production of synthetic corundum. The boules that were produced were black, like the crystals of natural rutile seen as inclusions in rock crystal. Subsequent heat-treatment was necessary to turn these black boules blue, green or yellow. The only synthetic rutile likely to come the jeweller's way, however, is a very pale blue stone simulating diamond, which was produced in the past but is not currently being made. These stones had six times the colour dispersion of diamond and therefore had tremendous fire; they were in fact too good to be true. They were also comparatively soft, only 6 to 6½ on Mohs' scale.

Strontium titanate

A relatively new synthetic simulating diamond, first seen in 1955 and also produced by the flame-fusion process, is strontium titanate. This substance has no natural equivalent. Again, this stone is much too soft to be mistaken for diamond by anyone on their guard, being only 6 on Mohs' scale. It is, however, singly refractive like diamond, and has a great deal of fire, far more than that exhibited by diamond.

The original boules produced in the blowpipe furnace are, like synthetic rutile boules, black in colour and have to be heat-treated to turn them colourless. The resulting material when faceted is rather brittle, and the cut stones tend to become flawed if subjected to ultrasonic cleaning. Strontium titanate was often sold under the trade name 'Fabulite'.

YAG

Another simulant of diamond is YAG, standing for yttrium aluminium garnet. This is not in fact a synthetic garnet, but is so called because of its garnet-like structure. YAG is produced by what is called the 'pulling' technique, which consists of drawing a rotating seed crystal out of a melt of yttrium aluminium oxide.

These stones are nothing like so hard as diamond, only 8½ on Mohs' scale, but optically they are closer in character to diamond than either synthetic rutile or strontium titanate. They have usually been sold under a trade name, the best known of these in the UK was 'Diagem'. Coloured versions of YAG have been produced, including a green version simulating the rare demantoid garnet.

Cubic zirconia

Cubic zirconia, often called CZ, is one of the latest and the most persuasive of the synthetics which simulate diamond. Before the trade became aware of the existence of this synthetic a number of experienced dealers were taken in by it, so deceptive is it. Now that its physical properties are widely known and equipment has been devised to make identification relatively easy, CZ should present no great problem to any jeweller who exercises due caution.

A variety of electronic detection equipment has come on to the market to protect the jeweller from the threat posed by CZ. The easiest way to determine whether a colourless stone is a diamond or a cubic zirconia is to test the stone using one of these instruments. A diamond will readily scratch cubic zirconia because though this synthetic is relatively hard at 8½ on Mohs' scale, diamond is at least 300 times harder. Scratch testing is, however, inadvisable because diamond will also scratch diamond.

Another test, which can be applied to an unmounted stone is to test the specific gravity. Cubic zirconia is approximately one third heavier than diamond. It has a specific gravity of 5.65 compared with 3.52 for diamond. The simplest way to test a mounted stone is, however, to employ one of the relatively inexpensive heat-probe units which are now available and which, if the instructions are followed carefully, are very reliable.

Although the zirconium-based synthetic constitutes a problem for the trade, it has also proved to be very saleable in its own right and more and more jewellers are stocking jewellery set with it. It might be said to have provided the trade with a superior and very attractive form of costume jewellery.

Synthetic moissanite

Synthetic moissanite, a form of silicon carbide, is tougher than diamond, has a refractive index higher than diamond, higher dispersion, more fire and lustre than diamond. This produces a sparkle second only to diamond. Many of its properties match closely those of diamond.

The detection of synthetic moissanite has in recent years proved a problem for some retailers, partly due to the fact that these stones test as diamond on the normal thermal testers. The company producing synthetic moissanite suggests the stone is first examined on the thermal diamond tester and then on the moissanite u.v. light absorption machine. Synthetic mossanite is now available in white, green, blue and pink. It can be polished in the brilliant, princess, oval and the trilliant cuts. Diamond is single refractive whereas synthetic moissanite is strongly double refractive. The larger stones present no problem to the experienced jeweller but the stones under 25 points are not so easy to detect.

Hydrothermal synthesis

During the last war because of the difficulty of obtaining adequate supplies of natural quartz from Brazil, the hydrothermal synthesis of quartz became an important commercial process in the UK. The process involves the use of an autoclave, which has been described as 'a sophisticated pressure cooker'. The method dates back to experimental work carried out by G. Spezia in the first decade of this century. The technique consists of causing substances to crystallize out of mineral-rich solutions by superheating, that is by heating the liquid under pressure. Crushed quartz is placed in a chamber filled with water, and as a result of the superheating the water becomes mineral-rich. When the solution cools it becomes supersaturated and crystallizes out round seed crystals placed in the solution. The conditions closely resemble those under which natural quartz crystallized out in rock fissures and in spherical chambers to produce geodes.

Synthetic quartz produced by this method has become increasingly in demand since the quartz watch acquired its present popularity.

Hydrothermal synthesis has been used for the production of other synthetic stones, notably emerald, and more recently rubies. The American, Carroll Chatham, has stated that his 'Chatham-created emeralds' are produced by this process. Doubts have been expressed about this, however, and it is now generally believed that the Chatham stones result from a different process known as melt-diffusion, that is from chemicals heated in a crucible.

Synthetic emeralds

Whatever the method he used, there is no doubt that the Chatham Company produced fairly large quantities of very plausible synthetic emeralds with physical characteristics very close to those of natural emeralds. They are by no means easy to identify, though identification is possible by laboratory testing. Because good-quality natural emeralds are scarce, Chatham's synthetics, though far from cheap, have been used extensively in jewellery in the USA. As well as faceted stones, some jewellery designers have also used uncut Chatham synthetic crystals. Chatham has also produced other synthetic beryls, the blue aquamarine and the pink morganite, and synthetic rubies by either the hydrothermal or more probably a melt-diffusion process.

Sophisticated synthetic emeralds have also been produced in Europe, by Pierre Gilson in France, various people in Germany and more recently in Japan. No information about the methods of production has been released, but it is believed to be a flux-melt technique. An American firm, Linde, has also produced synthetic emeralds.

The chemists who have produced these plausible synthetic emeralds and rubies are continually experimenting with new synthetics, and there seems no reason to suppose that there is any limit to their ingenuity. Gilson has produced synthetic turquoise and lapis lazuli, and more recently synthetic opals that are remarkably similar to natural opals. These are produced by reproducing the microscopic spheres of which natural opal is composed and bonding them together. The size of the spheres incidentally, determines the colour of those characteristic flashes of red, blue and green in both natural and synthetic opal. So far the gemmologists, like those working in the Gem Testing Laboratory of Great Britain in London, have been able to find characteristics that enable them to distinguish from natural stones all the synthetics that have been produced. For the jeweller, however, without the know-how, the time or the equipment, the situation is very worrying, and likely to become more so in the future.

Synthetic diamond

Potentially the most important of all the syntheses is that of diamond, a synthesis probably first achieved in Russia in the late 1960s, and then in the Western world in

1970 by American GEC. Fortunately the trade seems unlikely to have to come to terms with these stones in the near future. Though GEC have produced stones of a carat and upwards in weight, they have stated that such diamonds were appreciably more expensive to make, by the use of very high pressure and temperature, than it costs to mine stones of similar size and quality. It seems unlikely that such diamonds of high clarity, which can be white or coloured as required, will be commercially synthesized in the near future. As the late Robert Webster rightly said, however: 'One cannot be complacent, for with the advances of science today there must surely come a time when synthetic diamond becomes a commercial reality.'

The fact that synthetic industrial diamonds are nowadays being produced in considerable quantities implies that there is a growing body of knowledge being accumulated about diamond synthesis. There seems every reason to suppose therefore that this will one day be applied to gem-diamond synthesis. For the present, however, the jeweller has only the simulants that may masquerade as diamond to worry about, and not synthetic diamond itself.

Modern cutting methods

The faceting of coloured gemstones has been traditionally carried out by hand on a lap. The crystalline material to be cut is inspected to see which way to orient it in order to maximize the colour present; the colour in coloured stones tending to be unevenly dispersed. Then the material is fixed into a stick called a peg. The lap, having been charged with a diamond paste that acts as the abrasive, is turned by means of a handle. The peg with the stone fixed in the end of it is inserted in one of a series of holes in an upright stake called a jam-peg. Then working entirely by eye the cutter grinds on the facets and subsequently polishes them.

One factory in Idar Oberstein in Germany some years ago mechanized the cutting of the less expensive stones such as amethyst. The stones are first orientated for colour, as in hand cutting, then attached to metal holders that fit into a continuous belt carrying the stones to an automatic grinding machine. This reduces all the material to a standard size. The standard-sized stones are then mounted in a holder, which looks like a large comb. The combs are placed in specially designed machines that are programmed by computer to cut all the facets on the stones. They then pass through similar machines that polish the facets. Any shape of stone can be produced, but the process is obviously wasteful of material. For this reason it is not likely to be applied to stones of high value. Besides being more economical, the technique allows additional facets to be cut at little more expense, resulting in stones of greater brilliance.

The same factory has also mechanized the polishing of agates, which is carried out against a revolving sandstone wheel by applying automatically fed stone-holders to the wheels. Previously this was also an operation carried out entirely by hand.

In a number of factories in Idar Oberstein nowadays the automatic polishing of gemstone beads can be seen. The drilling of these beads is then carried out ultrasonically. Hollow needles vibrate rather than cut their way through the gem material. Ultrasonics are also being employed to produce crests on seal-stones, once a job that could only be done laboriously and with infinite patience by a gem engraver using tiny dentist's drills.

Chapter 3

Antique silverware

The dating of a piece of silver by the hallmarks alone can be dangerous because faked hallmarks are not uncommon. Then pieces exist with feet or lids that do not belong to them, and many pieces of early silver were 'improved' during the nineteenth century. Fortunately, few fakers have had much historical sense and the improvers who repoussé-chased plain Georgian cream jugs were more concerned to appeal to the tastes of their Victorian customers than to deceive anyone, so that this added decoration is nearly always out of period. Those who replaced missing lids or damaged feet often made errors of style. It is obvious, therefore, that anyone dealing with antique silver will find it invaluable to develop an eye for period styles and not simply depend on the hallmark. Otherwise they are bound sooner or later to make mistakes.

There are many excellent books available which provide a good basic guide to period styles. There are fine collections in the principal museums which may be studied; probably the best is that in the Victoria and Albert Museum. There are also illustrated catalogues put out by the top-ranking auctioneers to attract buyers to their sales. Much can be learned about both style and value by trying to date the illustrated pieces and checking the result against the text, and by trying to value the pieces in advance of a sale.

The silver of the past was not created in isolation; it was made to appeal to the taste of the period in which it was produced. Silversmiths, like furniture makers and architects, were influenced by the manners and the fashions of their age, and some understanding of the influences and the social habits of the past is a great help in understanding antique silver.

Silver made before the reign of Elizabeth I is so rare that it needs only passing mention here. This silver made during the last years of the sixteenth and early years of the seventeenth centuries is Renaissance silver. The Renaissance was, artistically speaking, the rebirth of the classical traditions of Greece and Rome. It began in Italy in the fourteenth century and travelled slowly northwards through Europe. By the time it reached England it had lost much of its early conviction. Elizabethan craftsmen paid lip-service to it but for them its imagery had become a collection of clichés. The classical columns had lost their perfect proportions, the representations of human figures had become stiff and stylized, and the fruit and flowers and leaves were no longer vigorous growths, but had become flabby and lifeless parodies of nature (Figure 3. 1). Not that the overwrought steeple cups and standing salts that have survived from this period are devoid of beauty or charm. Some of them are very attractive, but they are the late flowerings of a great tradition and the products of an ostentatious and often shallow age (Figure 3.2).

Perhaps, however, we judge silver made during the reign of Elizabeth I and James

Figure 3.1 Standing salt, made in London in 1583, typifying the debased baroque style of much of the silver of the period

I harshly because we have so little by which to judge it. Much of it was sold and melted so that the noble families of England could buy powder and shot and men and horses to fight one another in the Civil War. Little of the sensible everyday silver with which earlier generations had decked their tables survived this holocaust. At the end of the Civil War there remained only the important pieces which their owners thought it worthwhile to hide against better days, and a few pieces that survived by accident.

The Civil War destroyed not only the silver of the past, but also virtually destroyed the craft of silversmithing. Silversmiths depended on the patronage of the great and noble families and for eighteen years the owners of the great houses spent their money with the armourers and not with the silversmiths. When the war ended the Lord Protector set his face against the ostentations of the past. Puritan England was a utilitarian age and as such did not prosper the silversmith (Figure 3.3).

When Charles II came into his own in 1660, he and his court were eager to live again like gentlemen. After years of privation abroad, they were anxious to surround themselves once more with the symbols of gracious living, and soon the sound of hammering was once again heard in the workshops of those silversmiths who had survived. But there were too few silversmiths and too many orders, and much Caroline silver, as the silver of the Restoration period is called, shows signs of being made in a hurry for a not-too-exacting clientele. Some of it is frankly shoddy.

The restored court of Charles II had acquired foreign tastes during its sojourn abroad, and it attracted Continental craftsmen to London to cater for these tastes. Also at this time English merchants were trading farther from home, bringing back tea and coffee from the East, and importing china and other wares decorated with the alien imagery of China and Persia. Chinoiserie became fashionable, and the silversmiths

Figure 3.2 Bell salt, made in 1599, illustrates the charming simplicity of some of the silver of the Elizabethan period (by courtesy of the Worshipful Company of Goldsmiths)

decorated their wares with peacocks, figurines and eastern looking trees, scribed with the swift strokes of the graver or chased laboriously on cups and bowls and dishes (Figure 3.4). This Caroline decoration looked like a child's conception of a distant land, and the silversmith's designs were but pallid copies of the lacquer decoration applied by Chinese craftsmen to the long-case clocks and cabinets of the period. Side by side with this chinoiserie style, Dutch craftsmen and their English contemporaries were appealing to the taste the court had acquired abroad by producing silver repoussé

Figure 3.3 A goblet of 1657 showing the plainness usually associated with the relatively few silverwares made during the Commonwealth period

Figure 3.4 A large Caroline tankard with typical Chinoiserie decoration

chased all over with flowers and birds. Big ragged tulips, so beloved by the Dutch, often dominated the designs (Figure 3.5).

In addition to producing new styles of decoration the silversmiths working in the second half of the seventeenth century produced new vessels for the new beverages. The first silver tea-pot appeared in the 1670s and the first coffee-pot in the 1680s. Porringers, too, were produced with handles in the form of female figures, often so stylized and so emaciated as to be scarcely recognizable for what they are. There was

Figure 3.5 A Charles II ginger jar (by courtesy of Sotheby & Co.)

Figure 3.6 Three silver-gilt, gadroon-decorated casters dated 1700

a great demand for tankards too, and the Caroline tankard was often a huge vessel holding sometimes as much as two or three pints, fitted with hinged lid, a bold scroll thumbpiece, an S-scroll handle and a big spreading foot.

The last years of the sixteenth century might be called the age of the gadroon and the flute. The gadroon was a lobed border, and this decoration was almost invariably applied to the silver made during the reign of William and Mary (Figure 3.6). The form of much of the silver of this period was derived from the balusters, the moulded columns making up the balustrades so much in evidence in the new baroque architecture of what might be called the second Renaissance, a second return to the classical past. The typical William and Mary baluster candlestick was gadrooned round the nozzle, knops and base. The tankards of the period had gadrooning round their domed lids and their bases. That curious punch-bowl of the period, the monteith, with its removable scalloped rim, usually had a fluted body and a gadrooned foot.

At this time there was a great influx of Huguenot craftsmen from France, whose influence on English silver was to be felt for almost a century. These were men who came to England to seek refuge from political persecution following the Revocation of the Edict of Nantes by Louis XIV in 1685. The Huguenots were highly skilled craftsmen and they flourished under the influence of the Age of Reason. The eighteenth century milord was a patron of the arts who believed that his age was a reincarnation of all that was good in ancient Rome. Huguenot silversmiths found him a generous patron, and his rigid classicism chastened their French ebullience. Adopting the baluster form, they expunged from it the fussy gadroons and produced the plain, stout, sensible silver of Queen Anne and the early Georges. Some of the silverwares made by Huguenot craftsmen for their English patrons did, however, exhibit a degree of continental decorative flair.

Many connoisseurs consider the silver produced by the Huguenot craftsmen and their English contemporaries during the first 35 years of the eighteenth century to be the finest of English silver, and it would certainly be hard to find anything more satisfying than those bold straight-sided coffee-pots (Figure 3.7) and chocolate-pots, those elegant casters and trencher salts, plain candlesticks and caddies. Undecorated, except for the moulded wires strengthening foot and lip, unenriched with the mercurial gilding so popular earlier, they depended for their appeal on shape alone, and upon the charm of the soft grey silver highlighted by the flickering flames of candles.

Figure 3.7 Straight-sided silver coffee-pot of the Queen Anne period, with handle at right angles to the spout

Shapes changed as the century advanced. Straight-sided cylinders and octagons gave place to the pear-shape, the pyriform vessels (Figure 3.8). A little more decoration crept in. Cast-chased lions' masks, palmer's shells and acanthus leaves appeared round the junctions of spouts and handles, while with the curved bodies needing the support of feet the silversmiths borrowed the baroque imagery of the furniture makers, employing their claws, spheres and scrolls. But the bodies of their vessels still remained unadorned, essays in pure form.

In the 1730s, however, the age of baroque gave away to the age of rococo. Plainness was replaced by a riot of decoration, symmetry gave way to asymmetry. This style came from France, and its chief exponent was a member of a refugee Huguenot family, Paul de Lamerie. There were great rococo silversmiths in Paris in the eighteenth century, but none of them was a greater exponent of the style than Lamerie, a fact reflected in the fantastic prices paid in the salerooms today for his work. He must be considered the greatest individual silversmith ever to have worked in this country. His control of rococo imagery was superb and, unlike some of the work of his contemporaries, his forms are never swamped by the riot of chasing. Despite the basic asymmetry of the style a Lamerie piece never looks lop-sided (Figure 3.10). His repoussé-chasing and cast-chasing, whether depicting chinoiseries, a style which at this time had a brief revival, or the masses of shells, fruits, flowers, foliage and cartouches of the true rococo, is unfailingly beautifully delineated and exquisite in detail.

The excavation at Pompeii and at Herculaneum in the middle years of the eighteenth century brought about yet another radical change of taste. The eighteenth century, as has been suggested, thought of itself as a reincarnation of the classical world. Up to the middle of the century it had seen this classical world through the eyes of the Italian Renaissance, which the eighteenth century gentleman had rediscovered

Figure 3.8 Pear-shaped silver tea-pot of the Queen Anne period with spoon handle decoration on the lid and a hinged cover on the spout

on the Grand Tour. He had seen the architecture of Palladio, the paintings of Titian and Tintoretto and the statues of Michelangelo at first hand, and had brought back some of the art and the bric-à-brac of the Renaissance as souvenirs of his visit. He had then set the craftsmen of England to cast his house, his statue-littered gardens and his household possessions in a Renaissance mould. The art and architecture unearthed at Pompeii and Herculaneum came as a shock and a revelation. It transpired that the Italian Renaissance had been as wrong about Roman art as the Gothic revivalists of Victorian England were to be misguided about the Middle Ages.

The Adam brothers seized their opportunity to found a new classical revival based

Figure 3.9 A simple bell made during the rococo period but showing no trace of rococo decoration. Made by Peter Archambo in 1739

Figure 3.10 A Paul de Lamerie silver-gilt cup and cover (1745), illustrating the asymmetrical decoration of the rococo style (by courtesy of Sotheby & Co.)

on real Roman models. They employed artists to make drawings among the ruins dug up by the eighteenth century archaeologists, and armed with these drawings they descended from their native Scotland on to an England eager for something new. And so the age of Adam began.

Robert Adam, who dominated his age as he dominated his all-but-forgotten brothers, decorated everything with the motifs from the friezes of Pompeii and Herculaneum, not only the façades and the ceilings of his houses, but even the furniture, the carpets and the silver which he designed to go into them. The Romans, it seemed, were great lovers of the honeysuckle, the anthemiom as they called it, and in every nook and cranny of such houses as Kenwood in Hampstead and Syon Park in Middlesex are to be found the stylized flowers of this climbing plant.

Robert Adam did not design a great deal of silver, but he focused the attention of the silversmiths of his day on the Roman style, and they produced vessels inspired by the slender footed Roman urns which had been unearthed. They decorated their wares with the honeysuckle, the husks, the swags and ribbon bows that had once added beauty to Roman walls (Figure 3.12).

The age of Adam, too, was the great age of piercing. The silversmiths had rediscovered the piercing-saw earlier in the century. Prior to that, piercing had been done with punches. Now they ran riot with this new tool. Pierced dishes were produced in their hundreds, and the blue-glass liners of sugar basins and cruet holders peeped through the pierced metal.

Figure 3.11 Four drawings illustrating the change of style in the coffee-pot during the eighteenth century: 1, Queen Anne; 2, George II, the body now curved in at the base, and the lid much flatter; 3, the body became pear-shaped (1760); 4, a pot influenced by the classical revival urn from the 1780s

Nineteenth century silver

In the nineteenth century everything was romanticised – the past, distant lands, rural life, nature; probably as a reaction to the materialism of the industrial revolution. Another result of industrial growth during the century was the emergence of an ever-increasing middle class with aspirations and the spending power to fulfil them. This created an unprecedented demand for silverwares which in turn resulted in large factories being built in both Birmingham and Sheffield, and the development of mass production techniques. This in turn resulted in the greater affordability of both silverwares and silver-plated wares from which the 1830s rapidly replaced Old Sheffield plate as a substitute for sterling.

Styles

Past centuries had witnessed relatively few changes in style. In the eighteenth century for instance the honest simplicity of the Queen Anne period had been replaced by the rococo, which in turn had been superseded by the classicism of the Adam period. The nineteenth century on the other hand saw a plethora of styles succeeding one another and often overlapping. The neo-classical was closely followed by the neo-renaissance. The neo-gothic and the neo-rococo gave way briefly to japanalia, chinoiserie, neo-Egyptian arts and crafts and finally the art nouveau. Not only did these fashions overlap but the imagery from a number of different styles not infrequently appeared side by side on the same piece.

The silverwares produced between 1800 and 1900 can be considered under three

Figure 3.12 An inkstand of 1794 influenced by the Adam style

headings. There are the magniloquent wares produced by famous makers for royal and noble patrons which today fetch astronomic sums in the auction rooms. The best known of these makers is Paul Storr. His decorative pieces on the neo-baroque style were satisfyingly heavy and superbly chased and incorporating castings immaculately modelled (Figure 3.13). They were indeed better than almost any silverware from an earlier period, produced as many of them were to cater to the lavish taste of the spend-

Figure 3.13 One of a pair of soup tureens made by Paul Storr at the beginning of the 1820s. It comes from the Devonshire collection and is produced by permission of the trustees of the Chatsworth Settlement. It illustrates the neo-baroque style, for which Storr is famous, and the superb quality of his work

Figure 3.14 A teapot made in Sheffield in 1828 well illustrates the mundane silverwares produced for a middle class clientele throughout the century

thrift Prince Regent. Storr was not alone, however, in producing silver of outstanding quality. Supplying the leading retailer of the period, Rundell Bridge and Rundell and their competitors, Wakelin and Garrard among them, were a number of workshops employing skilled craftsmen. Among these were the workshops run by Digby Scott, Benjamin and James Smith, William Bateman and Edward Farrell.

Then there were the factory made pieces aimed at the middle class clientele, well enough made for the most part but often paying no more than lip service to the prevailing fashion by incorporating a decorative wire border or a decorative finial. A number of these factories, which in their heyday employed many hundreds of men and women, were still producing fifty years ago. Even in the 1950s they continued to produce the same designs that were in their catalogues three quarters of a century earlier, struck from the same tools in the same thundering presses.

Then there were small wares, useful little items made by specialist makers such as Nathaniel Mills, which today appeal to collectors who often pay high prices for the rare or the unusual. With little demand for tea-sets and large tablewares these days it is often smallwares which the retail jeweller is most likely to be concerned with and which in his role as valuer he is most likely to be called upon to appriase.

It is virtually impossible to list all the types of smallwares produced over the period from 1800 onwards. Among the more familiar are the various boxwares, wine labels, caddy spoons, mustard pots and casters, posy holders, taper sticks, milk and cream jugs, vesta cases and a whole range of flatwares such as sugar tongs, grape scissors, fish slices and crumb scoops, and this list is nowhere nearly a comprehensive one.

Boxwares

Boxwares were produced for a variety of uses as well as in a diversity of styles. In the early part of the century the demand for snuff boxes increased dramatically with a number of specialist makers prospering as a result, Nathaniel Mills, Joseph Taylor, the Linwoods and Samuel Pemberton among them. As with so many antiques the attribution to a well known maker adds to the price and where smallwares are concerned the unusual and the rare command a premium. Whereas a standard machine engraved snuff box might fetch as little as £150 at an auction, a repoussé castle top from Mills' workshop with a rare subject could fetch from £800 to £1000 (Figure 3.15). Other

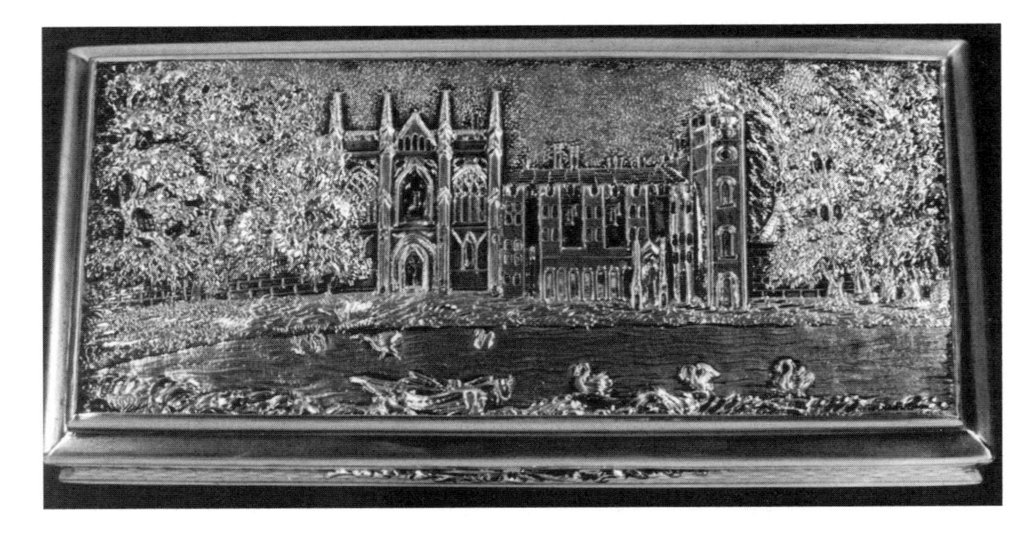

Figure 3.15 A typical example of the castle top snuff boxes made by Nathaniel Mills from the 1820s onwards

boxwares in demand during this period included those little vinaigrettes designed to hold aromatic vinegars to cure the vapours to which nineteenth century women seem to have been particularly subject. Then there are card cases designed to hold the calling cards left when calling on a neighbour. Also to be found are sovereign cases, counter cases for gamblers, toothpick boxes, cigar cases and from the end of the century cigarette cases. Then there are vesta cases, which seem to have a particular fascination for collectors, probably because they are so varied in style and decoration. Made to hold those early matches with wax stems and phosphorus heads, these little silver boxes were usually oblong in shape with rounded corners and were designed either to be carried in a pocket or dangled from a watch chain. Examples decorated with enamel painting are particularly sought after by collectors, but this is yet another area fraught with problems for a valuer as cases can vary in value from as little as £30 to as much as £500 or more for an interesting enamelled example made by a well-known maker such as Sampson Mordan of London.

Wine labels and caddy spoons

Two other much sought-after collectables are wine labels and caddy spoons. Wine labels have a long history, the earliest examples dating from the 1730s, but in common with so many smallwares, demand greatly increased from late in the eighteenth century when Matthew Boulton was a major producer and still more so in the nineteenth century. This is yet another minefield for the retailer undertaking valuations. A fairly standard label might be valued at £50 and three shell form examples made by Benjamin and James Smith in 1811 could well realize at auction in excess of £1500. Other sets by a known maker could even realise more than this. Names are again important because prestigious makers were invariably very good makers with a flair for design. Another important factor where a label is concerned is the design and the name inscribed on it. Out of a hundred or so names which have

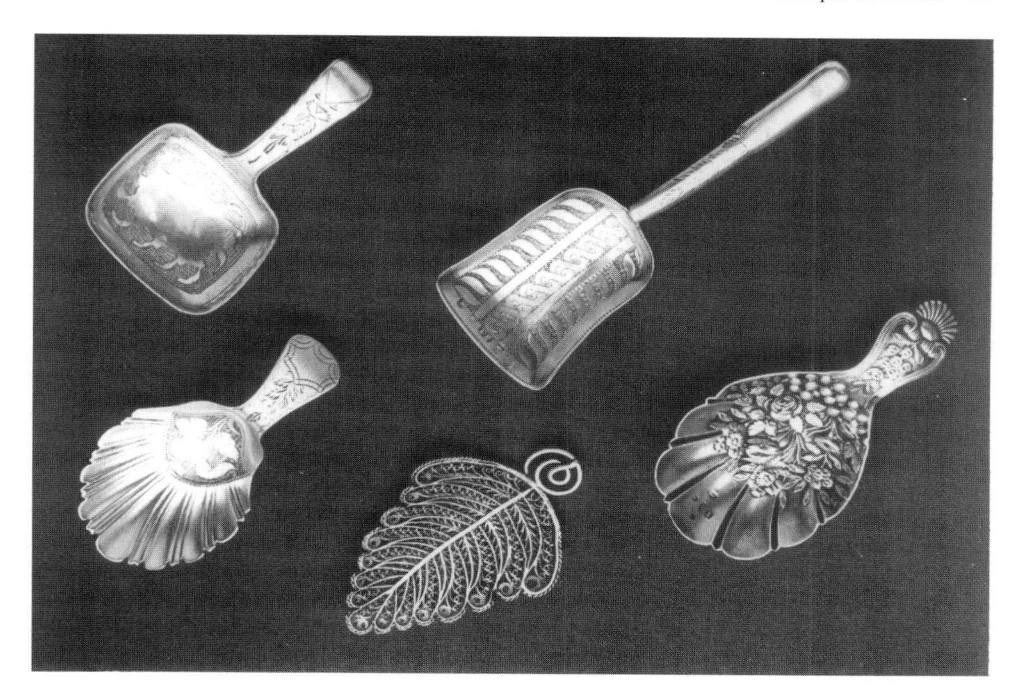

Figure 3.16 This collection of caddy spoons serves to illustrate the infinite variety of patterns which were produced

been recorded some are very rare and therefore that much more desirable to a collector.

Originally in the eighteenth century, the lids of tea caddies were used to measure out the tea, but by the 1790s caddy design had changed and there arose a demand for little spoons to ladle the tea into the teapot. In the years that followed thousands of these little spoons were produced in an infinite number of designs (Figure 3.16). I nterestingly they were specifically mentioned in the Hallmarking Act of 1790 and this has resulted in their having the added attraction for collectors of unusually well preserved hallmarks. Again, as with so many of these smallwares, values vary from around £50 to as much as £500 or £600 depending on rarity, desirability and condition.

Decoration

The nineteenth century could be described as the age of decoration. The surfaces of nineteenth century silver are often a riot of repousé or flat chasing, engraving, engine turning or piercing. This is especially true of the smallwares, little pieces such as posy holders being invariably covered in a riot of decoration. Sometimes this decoration is appropriate. Berry spoons may have bowls decorated with repoussé berries. Grape scissors and wine labels are overgrown with vine motifs, fish slices have blades engraved with representations of fish. However, this is not by any means always the case. The Victorians had a penchant for figurative objects, peppers take the form of people in fancy dress, salts are in the form of ducks, clowns support tapersticks or are transmogrified into sugar snips. Then the surfaces of many smallwares are peopled with representations of soldiers and sailors, huntsmen and dancing girls, repousé chased, engraved or enamelled. It is indeed the endless variety of these smallwares which is very much part of their attraction for the collector and their saleability as gifts.

This has necessarily been a brief summary of the styles of the past and the reasons why these styles came into existence. The dividing lines between one period and another were not sharply defined. While the pioneers of a new style were exploring its possibilities, the more conservative craftsmen went along happily in the old way. Also the provinces tended to lag behind London. But the provinces also produced their own local variations on the fashions of the Metropolis. They also catered for local tastes, making such pieces as the Scottish quaich and the Irish dish ring that were not made in London. Plain pieces have been made in every age. There were, for instance, little tumbler cups made in the seventeenth century. There are also decorative pieces dating from plain periods; the two-handled cups and covers of the first quarter of the eighteenth century often had a wealth of cut-card, strap-work or spoon-handle decoration on body and lid, as well as very decorative handles. Nevertheless, most pieces of silver are true to their period, and to anyone familiar with the silver of a particular period there is something unmistakable about the proportions, the shape, and the detail, so that they have only to look at the hallmarks for confirmation of what they have already divined.

Chapter 4

Valuation of jewellery

The dictionary definition of assessment is 'value, evaluate or estimate'. This makes it the correct word to describe the process undertaken in the assessment of jewellery. The jewellery trade is composed of numerous areas – beside manufacturers and retailers there are pawnbrokers, while the retail sector comprises both privately owned and multiple companies. Each of these sectors will require different practical skills and knowledge by those who are employed in them when called upon to evaluate jewellery. For instance, the pawnbroker will need up-to-date knowledge of current second-hand values and an ability to calculate their possible resale price, whereas someone employed in a firm manufacturing jewellery would need to consider the prices of the raw materials including the gemstones, project production costs and normal manufacturers' profit margins when assessing a piece of jewellery.

Pawnbroking

Pawnbroking in Britain is a growing business but it is an industry which has been operating for over 2000 years. It comes under the Consumer Credit Act and each outlet requires a separate Consumer Credit Licence (Category A) before being allowed to operate. These are obtained through the Office of Fair Trading. The industry has its own professional association – The Pawnbrokers Association – which has been in operation for over 100 years. It is estimated that there are around 1000 outlets in the UK.

People employed in this branch of the trade need to be able to read and understand hallmarks; test any unknown precious metal with acids; and estimate qualities, weigh the metals, grade and measure diamonds, and assess coloured gemstones. Computers are used by some companies to network with other branches to provide the actual loan values for a given piece of jewellery under assessment. Where these exist the employee needs only to key in the relevant assessment details of the item to be given the loan amount and the projected retail value.

Retailers

Retailing has over many years carried out the assessment or valuation of jewellery for their customers which has been accepted by various other professional organizations such as insurance companies and solicitors. Indeed, with the increase of trade and the public's awareness of the grading of diamonds, coloured gemstones and precious metals, the demand for valuations of jewellery has increased. The assessment of jewellery must, however, have a specific purpose for it to be of use to the recipient.

The most common form of assessment undertaken by the retail jeweller is that of valuations for insurance purposes. This can be one of the most important services provided by the jeweller, which he can use to his benefit. It is a service that the mail order or discount store does not provide, thus allowing the jeweller the opportunity to demonstrate his skill and knowledge. Also occasionally it allows a jeweller to handle beautiful and unusual pieces, rare gemstones and fine craftmanship which otherwise might not pass through his hands. It is profitable too in that it involves only a limited outlay of capital. Providing such a service may further lead to the customer subsequently making a purchase from the company.

In order to carry out this service the jeweller will need to possess or have access to reference textbooks and gem-testing instruments, have a good knowledge of designers and design periods, understand hallmarks and have a thorough knowledge of diamonds and coloured stones and to regularly update this knowledge. He will also need to have access to catalogues giving trade values for precious metals, diamonds and other coloured gemstones. Obviously, suitable premises and adequate security are essential and the safe keeping of confidential information is critical.

A customer bringing in their jewellery for valuation needs to be asked the specific purpose of the valuation, whether it is for insurance, probate (Confirmation of Will in Scotland) or divorce. There are different guidelines laid down for carrying out each of these types of valuation. Some customers assume that a value for an item has been enshrined in stone, one that all jewellers should agree on. The truth is somewhat different.

The commonest type of valuation is for insurance replacement and this valuation is based on the replacement cost to the client, that is the retail price, including value added tax, with a reasonably equivalent item.

The condition of an item will materially affect the value placed on it. If it is damaged in any way when it is taken in by the jeweller it must be noted. It may be necessary to provide a valuation which is based on the costs to recreate the item to its original design. If this is the case it should be clearly stated on the valuation because jewellery made by a well-known designer or a leading jewellery house would command a considerably higher price than the stones and metal present would suggest.

Almost all professional valuations carry explanatory notes related to the terms used and the conditions that apply to the valuation. This list also includes disclaimers. These notes and disclaimers provide the jeweller with a way of explaining to the customer the finer differences between the types of valuations for insurance. The letters NRV, SHRV, ARV and FRV may be placed alongside a valuation description. Definitions of these and other terms are now given.

NRV (new replacement value) This value reflects the average current new replacement price of the item. It does not cover a one-off re-creation facsimile of the original. This term is usually applied to items which have been made during the last fifty years.

SHRV (secondhand replacement value) This value reflects the average current secondhand replacement price of a similar item. This value is usually applied to items, which have been manufactured during the periods 1910–1951.

ARV (antique replacement value) This value is calculated with the intention of allowing the customer to replace the item with another antique. This value is usually given to articles manufactured prior to 1910.

FRV (facsimile replacement value) This value reflects the valuer's estimate of the likely current replacement cost of re-creating a hand made facsimile of the original with the same quality materials as used by the earlier craftsman.

Valuation for probate The prices allocated to a valuation for probate are based on what the pieces would fetch if sold at auction after the death of the owner. The prices put on such pieces are relatively low compared with the insurance replacement values. Also the condition of the item will affect the value. The figures stated would be what the pieces would raise in the open market. It needs to be clearly stated whether the jeweller has any interest in purchasing the items should they come on the open market or not as the case may be.

Valuation for loan This type of valuation is usually requested as proof of security in respect of a loan either from a bank or other lending agency. It is critical both for the lender and borrower that any figures quoted are correct and these must be at a level no higher than a probate value. This can sometimes be difficult to judge because of fluctuating gold and gemstone prices.

Private sale This type of assessment is used when a customer is wanting to sell an item of jewellery to a private person. An example of how this can operate would be if the value the item would raise from the trade was £30 to £40, whereas the jeweller might consider the replacement value for insurance to be £100. The sale-between-parties valuation would then be around £65 to £70, so both the seller and buyer benefits by the transaction. Should the jeweller, carrying out this type of valuation, wish to make an offer to buy the item this must be clearly stated on the valuation.

A valuation for any purpose will comprise three or more elements. These are: the purpose of the valuation, a specific description of the item or items, a clear critical assessment, and a price. The purpose is confirmed by the client whereas the description depends on the physical characteristics of the jewellery. This must take into account the hallmark, and the nature and qualities of any diamonds or coloured gemstones present. The assessment of their qualities such as the colour and quality grades of the diamonds would be part of this process of appraisal. This should be detailed enough for the item to be readily distinguished from similar items of jewellery. The details of the statement should be capable of assisting the police in the identification of the piece should it be stolen. This implies that the valuer needs to have access to a range of gemstone testing equipment as well as the various diamond grading aids that are available to the trade. It is common practice to use a 10× lens for the evaluation of diamonds, coloured stones and organic gems. Some valuers, however, employ a microscope in order to find the inclusions first and then use the 10× lens. It is becoming common practice to have independent diamond grading certificates supplied with higher quality stones that weigh a carat and over and a number of jewellery manufacturers are nowadays supplying their clients with Gemological Institute of America grading certificates. This reduces the subjectivity of the assessment of the diamond by the valuer and serves to confirm his original findings. Because most of the items brought in for valuation for insurance or probate valuations are finished pieces of jewellery it is very difficult to accurately assess the colour and actual weight of the diamonds or other gemstones without unsetting the stones – an option which is not practical. At best this must result in an estimation of these factors, unless certification of the colour, weight and quality of the stones accompanies the item it is wise to state these details as estimations.

Putting a price on jewellery is usually done by calculating all the details of the diamonds or the coloured gemstones present including their estimated weights, the carat quality and the gram weight of the mounts, and setting this out in a linear format. Alongside each one the relevant prices from various trade manufacturing price lists are added. The manufacturers are really the people at the sharp end of the valuation spectrum because they are in direct contact with the diamond and coloured stone markets. Once these figures have been calculated it is then necessary to add a manufacturer's profit figure, a retailer's mark-up and the current VAT. This formula would obviously not be suitable for all the different types of valuation, but is used expressly for the valuation of jewellery for insurance purposes. It is prudent for a valuer to keep records of his working notes as they can prove invaluable when carrying out a valuation of the same item a few years later. They can also be very helpful should a problem arise with the description or weights specified.

Examples of valuation statements

Figures 4.1 and 4.2 show valuation statements from John Bodenham, the manufacturer of the pieces of jewellery. Specific elements, such as customers details, descripton of the item, value and the purpose of the valuation, signature of the valuer, and date of the assessment, all need to be given. If trade certification has been obtained for the item then reference to this and the organization giving the qualification should be noted on the insurance valuation form.

The quality of these pieces of jewellery and the presentation of the valuation certificate demonstrate the professionalism of the manufacturing jeweller, who is a Fellow of the Gemmolgical Association of Great Britain and holder of the Diamond Grading Award.

It is not recommended that one looks to enter the field of valuing jewellery for any of the purposes given in this chapter unless the applicant has at least five years' experience in the trade. Some gemmological knowledge, preferably qualified to FGA standard, access to a range of gemmological instruments and received instruction in the skills of carrying out the process of valuing jewellery are essential requisites.

In 1987 the National Association of Goldsmiths launched the Registered Valuers Scheme. This was designed to provide the jewellery trade with a professionally recognized body of people who undertake to carry out valuations under a code of conduct laid down by the Association.

Mr. J. Bloggs
Avenue Road
Anytown
Anywhere
AE19 5AN

Date

Valuation for Insurance Purposes

I have examined the following item of jewellery:

Brooch: consisting of two precision cut, thistle shaped amethysts and approximately 2.5 carats of pave set round brilliant cut diamonds, mounted in hallmarked 18ct. white and yellow gold. Coloured photograph supplied.

I would consider the current replacement value of the item, for insurance purposes only, inclusive of VAT to be:

£4500.00 Four Thousand and Five Hundred Pounds.

Appraiser
John Bodenham F.G.A. D.G.A.
Bodenham & Shorthouse

Figure 4.1

Mr. J. Bloggs
Avenue Road
Anytown
Anywhere
AE19 5AN

Date

Valuation for Insurance Purposes

I have examined the following item of jewellery:

Ring: A fine quality oval shaped brilliant cut diamond, weighing 1.5 carats, described on G.I.A certificate number X12478 as colour F, clarity SI, claw set and mounted as a ring in Hallmarked 18 carat white gold. The diamond is complimented by two tapered baguette cut diamonds, one flush set to each shoulder of the ring. (Weight of baguette cut stones – 0.36 carats.) Total diamond weight 1.36 carats.

I would consider the current replacement value of the item, for insurance purposes only, inclusive of VAT to be:

£12500.00 Twelve Thousand and Five Hundred pounds.

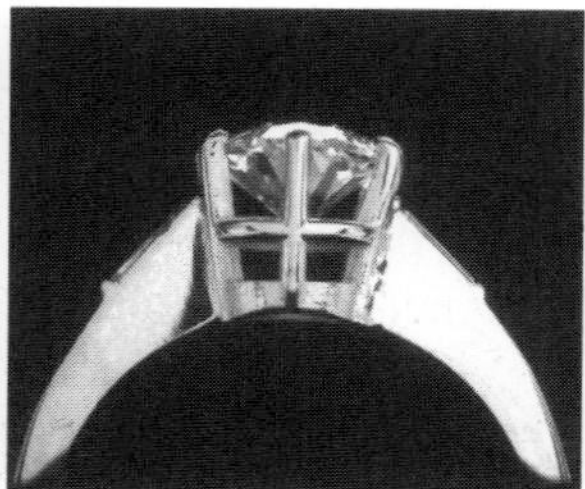

Appraiser

John Bodenham F.G.A. D.G.A.

Bodenham & Shorthouse

Chapter 5

Manufacture of silverware

Hollowares

The bodies of silver vessels can be made in four ways. They can be hand-raised, folded, spun, or formed between dies in a press. Hand-raised silver is stronger, and most people contend that it is also more beautiful than silver produced either by spinning or by stamping. The hand-raising of silver is, however, an expensive process, for not only does it demand the skill of a highly trained craftsman, but it also takes a long time.

Hand raising

The ancient skills of the silversmith have been handed down from generation to generation of craftsmen, and the tools as well as the technique of hand-raising go back as far as recorded history. Modern smiths, like their predecessors in centuries past, use a section of a tree trunk with shallow depressions sunk into it to start shaping the body of a vessel. They take a disc of sheet silver, which they have cut to size, place it over the depression in the trunk and begin to hammer it with a ball-faced hammer. They hammer round and round the silver disc in courses, working gradually towards the centre. By the time the centre is reached the flat disc has been raised to a saucer shape.

Annealing

The hammering changes the atomic structure of the metal, the silver becoming hard and brittle. When this happens the smith anneals the silver by heating it with a torch until it glows a dull red, then quenching it in water. Doing this restores the malleability of the metal. The operation has to be repeated many times as the work progresses.

At the stake

When smiths start raising the saucer shape to form the shape of the body of a coffee-pot or a jug, they work over small anvils known as raising stakes. These are either firmly fixed in a tree-trunk, or more usually in a vice. With the saucer-shaped piece of silver supported on the stake they begin to hammer again, working this time from the centre outwards in the direction of the edge. They hammer course by course, stopping to anneal when necessary, and the shape rises up under the hammer blows. It looks simple, just as the throwing of pottery looks far simpler than it is. The silversmith's problems are indeed not greatly dissimilar from those of the potter. The silversmith's craft is slower and more painstaking, but both craftsmen are concerned to push their raw material in the way in which they want it to go, to get the walls of their vessel evenly thick, the shape true, and the strength where it is needed.

Planishing

At the end of a day of hard work the body of the vessel which the silversmith set out to make will be formed. It will, though, bear the marks of the hammer all over it, regular and even marks if the smith has done his work well, but too obtrusive to be acceptable. The next job, therefore, is to smooth out these hammer marks, which is done by more hammering. The smith goes over the whole surface with a flat-faced planishing hammer. This hammering also leaves its marks, but the regular pattern of small indentations left by the planishing hammer is one of the things which gives hand-raised silver its character.

Wires

Wires are often used to strengthen the lip of a vessel or provide the foot. Today standard pattern wires can be bought from the bullion dealers, but many silversmiths still make their own. Judith Banister, in her book *Old English Silver*, described the process of wire-drawing as having: 'a ring of the medieval torture chamber about it'. Certainly the machine, the draw-bench, used to draw wire looks not unlike one of those racks on which prisoners were stretched to loosen their tongues. Wire-drawing is, in fact, a stretching process. A length of silver rod, which has been reduced in diameter by passing it back and forth between grooved rolls, is drawn through the progressively smaller holes that have been drilled in a draw-plate. The rod gets thinner and longer as it is drawn through ever smaller holes, and so becomes wire of the required diameter. Fancy wires are made either by casting or by rolling.

Casting

Not only wires, but also decorative details such as those placed around spout and handle junctions, and also the feet and finials of vessels, were traditionally made by sand casting. A wooden or metal model of the part to be cast is first made. Then the two halves of a metal box, with a pouring hole at one end, are filled with oiled sand. The model is placed in one half of the box, both surfaces of the sand are dusted with charcoal so that they will part easily, and then the top and bottom of the box are brought together and the sand is dried. The two halves of the box are now parted, and the model removed. A channel is formed in the sand to connect the impression left by the model with the mouth of the metal box. The two halves of the box are next brought together again and clamped. Molten silver is now poured into the mouth of the box and flows into the hollow area in the sand left by the model.

The metal is allowed to cool and the casting box is opened again and the silver impression of the model taken out. Any flash, as the metal straying round the edges of the impression is called, is removed and so is the sprue, the tang of the metal that sets in the pouring channel. The casting is then brightened up by chasing. Increasingly nowadays, however, such silver components are lost-wax cast (see Chapter 7).

Chasing

Hand-raised silver is decorated either by flat-chasing, cast-chasing, repoussé-chasing, engraving, texturing, piercing or enamelling.

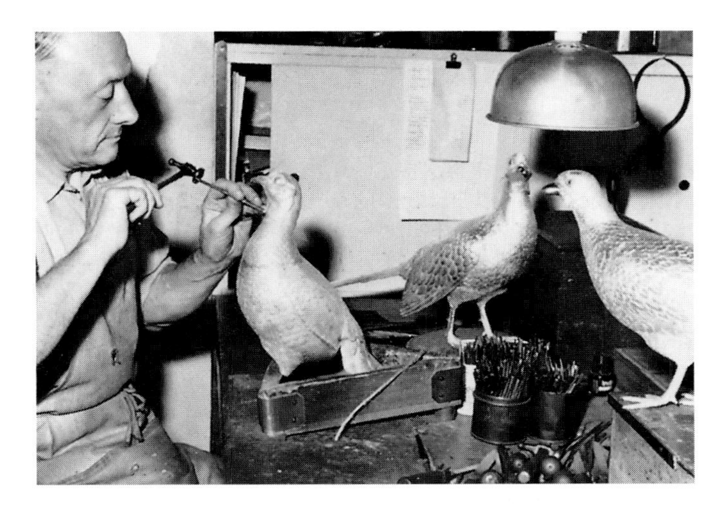

Figure 5.1 The chaser at work. The chasing of one life-sized model of a pheasant takes 175 hours

Chasing is the name given to the process of decorating silver with a hammer and punches. To flat-chase the walls of a vessel, it has to be supported either by resting it on a bed of pitch or more often by actually filling the vessel with pitch, and placing it on a sand-filled leather bag. The pattern is marked out on the silver and then it is outlined with a fine punch. Then the chaser begins to fill in the detail. The right punch is chosen from the hundreds or even thousands standing in round tins on the bench. The punch is placed on the silver and given a deft blow with the hammer. The punch pushes the metal aside and forms a small indentation in the surface. The chaser moves the punch, deals it another blow and so creates a channel. Patiently, blow by blow, a pattern of lines is built-up. One of the best known chasers in the trade had a collection of 6000 punches, some of them dating back to the reign of William and Mary. His hammer was 170 years old, and some idea of the work entailed in chasing may be gained from the fact that it took him 175 hours to chase a life-sized model of a pheasant (Figure 5.1).

Not only can the chaser create channels in the silver but he can also, by the use of special matting punches, texture the metal between the lines.

Cast-chased decoration is the application of chased-up castings to a piece of silver. Repoussé-chasing is relief decoration that has subsequently been chased up (Figure 5.3). The relief pattern is first hammered up from inside. To do this on an enclosed vessel such as a tea-pot, a special tool called a snarling-iron has to be used. This is a Z-shaped piece of steel one end of which is domed. The non-domed end is secured in a vice, and the vessel is put over the domed end. By hitting the arm of the iron the craftsman drives the dome against the inside of the silver and bows it outwards. It can be imagined what skill was required to create one of those deeply embossed patterns of fruit and flowers on a Caroline jug, working by remote control with a snarling-iron. Once the repoussé pattern has been raised on the silver, the chaser then puts in the detail, using the same techniques as for flat-chasing.

Engraving

Although superficially it resembles chasing, engraving is carried out in quite a differ-

Figure 5.2 The chaser painstakingly creates, a fraction of an inch at a time, the pattern on to the body of the vessel with hammer and punches

ent way. It is done with a graver, a pointed piece of steel rod fitted into a handle. The engraver cuts away metal with this tool, whereas the chaser pushes metal aside with a punch. The engraver cuts lines with a single rapid stroke, as a pencil line is drawn. The chaser has to create lines laboriously, blow by blow, advancing perhaps a tenth of an inch each time the punch is moved.

The character of these two kinds of decoration is implicit in the way in which they are carried out. Engraving is sharp-edged and free-flowing at its best (Figure 5.4). Chasing is a softer and stiffer form of decoration. Chasing is more varied in texture because the chaser can easily change the weight of the lines by changing punches.

Texturing

Modern hand-raised silverwares are sometimes decorated with texturing. The texturing is applied with specially made hammers with indented faces.

Figure 5.3 Cup with repoussé-chasing and cast-chased handles. This piece dates from 1658

Figure 5.4 An engraving on a piece dating from the reign of Charles II shows the free-flowing nature of this decoration (by courtesy of Sotheby & Co.)

Saw-piercing

Saw-piercing is a very ancient craft, but was not used by the English silversmiths until the eighteenth century, and not extensively used as a method of decoration until the second half of that century. Earlier the piercing featured on the top of casters was punched out and not sawn. Saw-piercing is used to decorate fruit dishes, the covers of rose bowls, the tops of casters, and so on. The piercing-saw is like a fret-saw, and the technique is the same as fretwork. After the design has been marked out – usually by powdering a pricked paper pattern with chalk – the piercer drills a hole in the silver, inserts the saw-blade and begins to follow the intricate curves of the design (Figure 5.5).

Enamel

Enamel is glass containing quantities of metallic salts to give the different colours. The presence of iron oxide gives the glass a green colour, gold chloride gives a red,

Figure 5.5 Saw-piercing. The handles and border on this piece were cast and then chased up

and so on. These metallic oxides give transparent colours, the addition of others, such as tin oxide, give opaque colours. To create the enamel the glass is powdered and mixed with the powdered oxide. The mixture is suspended in oil or water and then painted on to the silver. It is then fused in an oven and finally polished with pumice. A number of coats of enamel are usually given and each one is separately fired.

Enamelling is a very old craft, and different forms of enamel decoration have been used at different periods. Those Anglo-Saxon jewels seen in museums are often decorated with cloisonné enamel. In this process a raised design was created by the soldering of metal strips on to the surface of the article to create little cells called cloisons, and these cloisons were then filled with enamel of different colours. Many sixteenth century vessels and watch dials were decorated by a somewhat different type of enamel work, known as champlevé enamelling. Here, metal was gouged away to create a recessed design, and the recessed areas were then filled with enamel.

Sometimes reference is made to other types of enamelling referred to in books and in saleroom catalogues, such as basse-taille, plique-à-jour and grisaille. Basse-taille is a form of champlevé enamelling, but the metal is cut away to different depths to produce a shaded effect. Plique-à-jour is like cloisonné, but the cloisons have no backs, so that a stained-glass-window effect is produced. Grisaille enamelling is painted enamelling, but only shades of grey are used to create the design.

The seventeenth century saw the great flowering of the art of painting in enamels, a craft still occasionally used today to decorate watch cases, brush-sets and compacts. A matt white enamel was laid over the surface and then allegorical pictures and flower sprays were painted over this. One colour at a time was laid, each colour being fired separately, so that a painted enamel may require ten or more firings.

Folding

The bodies of the straight-sided coffee and chocolate pots were made by a process known as folding. Sheet silver was bent up over a former and the joint was then soldered. The technique is still used to produce reproductions of early eighteenth century pots.

In the factory

Reference has already been made to the development of mass production in the silver trade in the nineteenth century, and today most of the silverwares, and all the silver-plated wares sold by the retail jeweller, are made in the factory rather than in the craft workshop. They are spun or formed between dies in the press and not hand-raised under the hammer. The bodies of vessels are produced in a matter of minutes in a lathe or by three or four blows of the force of the press, instead of by thousands of hammer blows, and so this process also takes a matter of minutes. Few press shops survive in Europe today, however.

If, though, the making of the vessel itself is today a rapid and a not particularly skilled operation, the making of the chucks or dies involved is a highly skilled, lengthy and very expensive business. Diesinkers work in toughened steel, tough enough to stand up to heavy and repeated blows. They work from a copper model of the vessel that the dies will eventually reproduce. First of all they square up a pair of steel blocks, mill the surfaces to very exacting tolerances, and polish the blocks to a high finish. Taking their measurements from the copper model, they cut out, or 'sink', an impression of the outside form of the vessel into one of the blocks, carving away the steel and then polishing the inner surfaces. This recessed tool is called the female half of the tool and is concave. The male half of the tool is convex and reproduces the inside form of the vessel. The female half is fixed into the bed of the press, and the male tool is fixed into the force of the press, the weighted hammer that descends when the press is operated. If a circle of silver sheet covers the female tool it will be forced by the male tool to take the shape of the original copper model from which the tools were made. The force of the press will 'draw' the body of the vessel. If the vessel is a deep shape, and has to be deep-drawn, it will not be possible to form it in one operation between one pair of tools. It may have to be drawn by a succession of blows in the press between a succession of progressively deeper tools. If the vessel is of a complicated shape it may be necessary to stamp the body in two halves, and then to solder the two halves together.

Spinning

This is the third method of making the body of a silver vessel and stands halfway between hand-raising and stamping. Spinning consists of forcing the silver over a chuck in a lathe. It is a skilled craft, but is much quicker than hand-raising. Spinning does not strengthen the metal as does hammering, but it is possible by this technique to produce flowing curves that have a beauty of their own. The spinner first makes a chuck in wood. This chuck represents the inside shape of the vessel to be spun. It is mounted in a spinning lathe, and a disc of silver sheet is placed against it. The spinner now pushes the silver sheet over the revolving chuck using a tool that consists of a chisel-like burnisher fixed in a long wooden handle. The long handle fits under the spinner's arm and gives him the necessary leverage to force the metal over the chuck. Split chucks are used for vessels of an enclosed form, so that the various components can be withdrawn through a narrow neck.

Decoration from the tools

Silverwares made by spinning, and those which are formed in the press, may be decorated in exactly the same ways as hand-raised silverwares, or decoration can alternatively be put on in the course of the pressing operations. It is possible to simulate chasing by using tools that have a pattern raised on the female half of the tool and the same pattern recessed on the other half. Similarly, saw-piercing can be simulated by using dies. Imitation chased decoration or imitation saw-piercing is, however, rather lifeless when compared with the real thing. It is too regular and lacking human variation, but, of course, it is very much cheaper to produce. Because decorative silver has been out of fashion in recent years, and because the silver trade is shrinking, only a handful of craftsmen remain today who can carry out the decorative processes by hand. In the foreseeable future, therefore, more and more machine applied decoration may have to be accepted.

The decorative wires used on such patterns as Celtic are produced by rolling silver between hardened steel rollers on which the pattern of the wire has been cut out as bas-relief just as a female tool is cut out.

Electroforming

The technique of electroforming has been used by some silversmiths in recent years to produce decorative silverwares.

Electroforming is not a new technique. It has been used by museums over the years to reproduce models of antiques, but its use by silversmiths as a production technique is a more recent development. The technique consists of first making a nonmetallic mould, nowadays usually of an epoxy resin. This is then placed in an electroplating bath where silver is deposited on the mould, reproducing every surface detail. The process therefore avoids the production of expensive dies, or such labour-intensive operations as chasing or engraving. A famous example of the use of this process was the production by Louis Osman of the 24 carat gold crown used for the investiture of Prince Charles as Prince of Wales in 1969. Figure 5.6 shows an epoxy mould, together with two complicated dishes produced from it by electroforming. It clearly demonstrates the saving of labour, compared with producing such dishes by more conventional techniques.

Small wares

Silver cigarette boxes, compacts, and silver-back brushes, can also be made either by hand or in the press. Hand-made wares have the advantage of greater strength and beauty, and the disadvantage of a higher price. The strength of these wares is the result of hand-hammering. A cigarette box that has been flat-hammered by hand is much less likely to bow or dent than one stamped out in the press, but this hand-hammering is a long, skilful and therefore expensive, process. Usually much more care is given to the details when making hand-made wares. Often, for instance, hand-made boxes have beautifully made flush hinges that are all but imperceptible when the box is closed. It is by pointing out such details as this, as well as the greater strength of the article, that the salesman can justify the considerably higher price of the hand-made over the machine-made product.

Figure 5.6 Pattern and two dishes produced by electroforming (by courtesy of The Worshipful Company of Goldsmiths)

Engine-turning

This is a type of decoration usually reserved for boxes and brush-sets. To carry out this operation the part to be decorated is fixed with pitch to a block which is then clamped on the carriage of the engine-turning machine. The cutting tool of the machine is not unlike a graver, and indeed this is an engraving operation, for the tool cuts away the silver. The tool is brought to the work by the engine-turner, and the work is moved against the tool. For straight-line engine-turning the piece simply moves up and down against the tool, and when each cut has been completed the work is moved sideways, so that a row of furrows is cut right across, say, the back of a cigarette case. To create the wavy lines of the well-known barley pattern the work is vibrated as it moves against the tool. The use of eccentric cams makes it possible to create these wavy patterns.

Sometimes engine-turned wares have layers of transparent enamel applied over the pattern.

Finishing

Once the component parts of a silverware have been produced they have to be soldered together. Silver solders, which are up to the sterling standard, are bought from the refiners, and solders that melt at different temperatures are available so that a second joint can be produced close to one that has been soldered previously without the risk of melting the first joint again. Now the silverware is ready to be sent to the assay office.

Back from the assay office a silverware still has to have one very important operation carried out upon it – polishing. At this stage it will be milky-white in colour and non-reflective. If a piece is badly scratched it may have to be given a preliminary finish against a grinding wheel, but usually the first polishing operation consists of mopping or buffing. A mixture of oil and sand is thrown on to the piece as a polishing medium and the piece is then held against a rotating leather mop. Even this, however, is a rather too abrasive form of polishing for hand-raised silver. It would most certainly remove the desirable patina produced by the planishing marks. Usually, therefore, the first polishing operation on hand-raised silver is done on a calico mop, using Tripoli powder as a medium. Matt finishes are produced by using special mops.

A felt mop and jeweller's rouge, which is powdered hematite, is used to give the silver its final brilliant polish. A very high finish is sometimes imparted to small articles by a process known as burnishing, which consists of rubbing the silver with a small steel chisel. Brown ale is commonly used as a polishing medium in this process.

Quality

Machine-made silverwares vary considerably in quality. At their best they are little inferior to hand-raised silverwares. At their worst they can be frightful, made from the thinnest gauge silver, decorated with shoddy imitation chasing or crude repetitive pierced scroll patterns thumped out in the press. Solder joints are sometimes rough, and the bright finish which is imparted by the polishing fails to hide the unevenness of the surfaces. The decorative wires and crude cast feet produced from bad models may be poorly finished.

Flatware and cutlery

Flatware, the tools used for eating that do not cut, can be made by handcraft methods, but today only a tiny fraction of the total annual production of spoons and forks are hand-forged.

Hand-forging

This is the traditional method of making silver flatware. The craftsman who sets out to hand-forge a silver spoon or fork takes a piece of silver bar and beats it over an anvil with a hammer. It is his hammering which gives the hand-forged spoon or fork its strength, in the same way that the hammer blows used to raise hollowares, or the flat-hammering of a silver tray, give these strength. You can drive the tines of a hand-forged fork into a plank without bending them, a test of strength that few, if any, mass-produced forks would withstand.

When the smith has flattened out the silver bar, he roughly shapes the piece by hammering it over a recessed die known as a swaging block. The bowl of a spoon is formed in much the same way that a smith begins to raise the body of a tea-pot over a recess in a tree-trunk, but using in this case a recessed metal block. This process gives the spoon-bowl a distinctive thick edge, which is a sure indication that a spoon has been hand-forged (Figure 5.7).

The next process is to impart the pattern to the handle of the piece, and to give it

Figure 5.7(a) Hand-forged spoon – stages in the forming

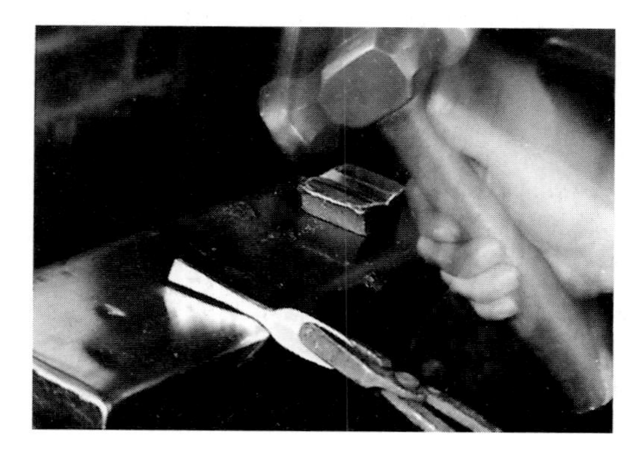

Figure 5.7(b) The spoon beginning to take shape under the hammer

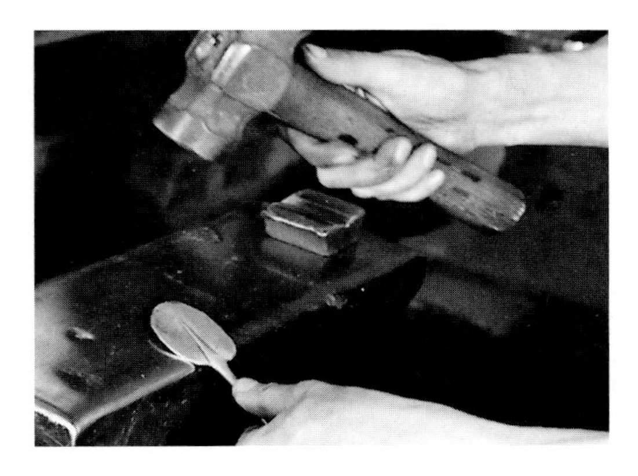

Figure 5.7(c) The craftsman spreading out the silver with his hammer to form the bowl of the spoon

its final shape by die-stamping it in a drop-forge (Figure 5.8). This, too, is really a hammering process. The hammer is the heavy 'monkey', the descending weight of the drop-forge.

The hand-forging of flatware survives only to cater for special orders from a handful of customers to whom price is no object. But it makes it possible to perpetuate some of the beautiful old patterns, such as chased-vine, for which the demand is so small that it would be impossible to recover the cost of producing a set of press tools.

Figure 5.8 Pattern being struck on to the handle end of a blank in a drop-forge

Flatware from the factory

In a flatware factory, silver flatware is produced from silver sheet, e.p.n.s. flatware from nickel silver sheet, and stainless-steel flatware from the appropriate grade of stainless-steel sheet, usually 18/8. From the sheet the blanks are cut out in the first of a series of stamping operations. The shape of the blank already suggests the form of the final product, but the metal is the same thickness all over. The next operation is to cross-roll the part of the spoon which will become the bowl and that part of the fork which will become the tines (Figures 5.9(a)-(c)). Heavy rollers spread out the metal and reduce it to the right thickness. After that the spoon is bowled and the prongs, or tines, of the fork are cut out in a press (Figure 5.9(c)). Further press operations impart the patterns to the handles, and the curves to the profiles of the pieces. Between these operations the blanks have, of course, to be annealed regularly to restore ductility to the work-hardened metal.

After these various forming operations there follows a complicated series of finishing operations. The edges of the pieces are smoothed off against grinding wheels, and the spaces between the tines are cleaned out (Figure 5.10). Increasingly these days these operations are carried out mechanically. Next, the pieces are polished with a mixture of sand and oil on revolving leather mops.

The famous women, the 'buffers', who have done this work in the factories of Sheffield, are a dying race, because today few young girls care to take up this work

Figure 5.9(a) Production of forks and spoons by mass-production techniques – stages in the process

Figure 5.9(b) Male and female dies used to form spoons and forks in the power presses of a modern flatware factory

(Figure 5.11(a)). For this reason, also for speed and economy, an increasing amount of flatware is today being polished on automatic machines. The pieces are clamped in long racks and positioned against a battery of revolving mops. After one end of the flatware has been polished the pieces are turned round in the racks and presented to the mops again. The pieces are finally mopped with felt mops on other machines. Few people would deny that the buffers do a better job than the machine, but there seems to be no alternative to accepting a slightly lower standard of polishing in the future (Figure 5.11(b)).

After polishing, nickel-silver spoons and forks go to the plating shop, where their coating of silver is electrolytically deposited on them. In modern factories this process too has been mechanised (Figure 5.12). The spoons and forks are loaded on jigs, and on these they travel automatically from bath to bath, their journey dictated by a computer program. First they go through a series of degreasing and washing processes to

Figure 5.9(c) Bowling a spoon in the press

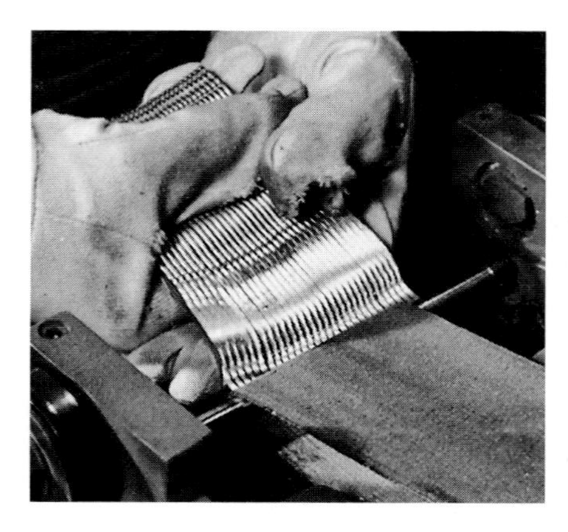

Figure 5.10 One of a series of grinding processes to smooth the edges of the pieces and to clean between the tines of the forks

ensure that they are absolutely clean. Dirt or grease on the article would prevent the silver from adhering properly to the base metal core. When the articles have passed through the cleaning baths they arrive at the plating bath. Control of the current density and of the time-cycle of the automatic plating plant makes certain that a controlled thickness of silver is plated on each piece, and regular tests are made to ensure that the plating is of the desired thickness and that the adhesion is good.

There is still, however, no generally adhered-to standard, though now a number of factories produce plated flatware to standards laid down by the British Standards

Figure 5.11(a) A 'buffer' hand-polishing flatware in a Sheffield factory

Figure 5.11(b) Mechanised polishing of flatware on Claypole machines

Institution. These cover not only the deposit of silver, but stipulate strength, resistance to corrosion etc., which must be adhered to if the flatware is to be sold with a kite mark and the appropriate BSI Standard number. There are two standards laid down for silver deposits, 'standard thickness' and 'special thickness'. There are different minimum deposits specified for place setting and serving pieces. There are also standards laid down for cutlery. These standards do not have the force of law behind them, but because the standards serve to reassure the public that they are buying reputable products which should give them good service for many years it is in the best interests of a reputable manufacturer to adopt them and for a retailer to sell products bearing the kite mark. Anyone wanting to acquaint themselves with the details of these standards should obtain a copy of BS5577:1984 from the British Standards Institution.

After plating, the racks with their burden of flatware continue their journey on the electrically-powered roundabout, which dips them into cleaning baths and into rinsing

Figure 5.12 Mechanised plating – the blanks are loaded on jigs

baths, until finally they reach the end of their journey. They are then dried in sawdust, given a final brilliant polish with jewellers' rouge and packed.

Knives

The domestic knife was revolutionized in 1914 by the discovery of stainless steel by Harry Brearley. This new metal presented some initial problems of working, but before long it had entirely replaced carbon steel as the raw material of the domestic blade. The knife blade is fashioned from a length of stainless-steel rod, which is die-stamped in a drop-forge, after being brought to red heat (Figures 5.13(a) and (b)). In this operation the tang and the bolster are formed. The tang is the rod which will eventually fit into a hole in the knife handle and secure the two together; the bolster is the collar separating the blade from the handle. After these have been formed the metal is reheated and the blade rolled out. Then it is given its final shape between blanking tools in a press.

The blade now has to be turned into a cutting tool. It has to be forged, or hardened, so that it will take and hold an edge. Formerly this was done by hand, but today mechanical hammers are increasingly replacing hand-hammering for this work. Hammering makes metal brittle, and a brittle blade would be of no use. So the brittleness must now be taken out of the metal by a process known as tempering. The blade is heated in a furnace to a high temperature and cooled, usually in a blast of air. It will now bend without fear of cracking. It is, however, possible to buy cheap knives the blades of which have not been properly tempered, or not tempered at all, that can be snapped in two just by pressing them firmly on a table.

A variety of handles are used for domestic knives today. The old celluloid handles, called Xylonite, that imitated ivory, are still produced, but today nylon is more often used. Laminated wooden handles bonded with water-resistant synthetic resin glues

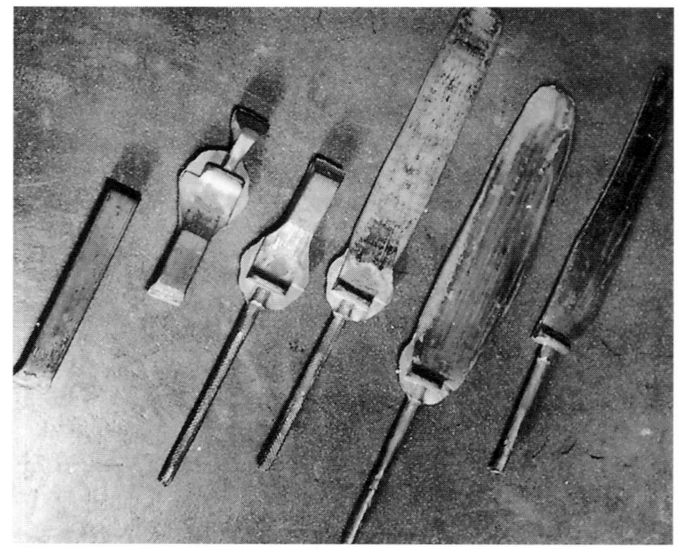

Figure 5.13(a) Blade-making – stages in the making from stainless-steel rod

Figure 5.13(b) Blade-making – die-stamping in a drop-forge after heating the metal red-hot

are used, and so occasionally are ceramic handles. All these handles, however, have one fault in common. The adhesion of handle and blade depends on the 'cement' used to bond them together.

Although modern adhesives and modern techniques have made it much more likely than in earlier days that these knives will stand up to domestic abuse, it is expecting much of any bonding agent to withstand regular immersion in boiling water, while the washing-up machine represents a new threat to this bond. It can be appreciated therefore why the one-piece knife and the hollow-handled knife, which is hard-soldered to the blade, have become more and more popular.

The one-piece knife is made all of a piece. The handle is hot forged from the bar before the blade is rolled out. Hollow handles are made from two stampings struck with the pattern of the flatware they are made to match. These are then soldered together and hard-soldered on to the blade.

Working in stainless steel

Most of what has been written about silver and e.p.n.s. flatware applies equally to stainless steel flatware. The only difference is that stainless steel, being a much harder metal than the other two, is much harder on the tools. It is much more difficult though not impossible with very heavy presses, to draw very deep shapes in stainless steel. Thus there is a tendency for the bowls of stainless-steel spoons to be much shallower than those of silver or e.p.n.s. patterns. They were made in this way to begin with because it was easier and cheaper, but then shallow bowls became fashionable for a time, an interesting example of economic considerations starting a fashion.

When it comes to forming deep-bodied hollowares in stainless steel the problem is, of course, much greater. Experiments have been made to form them by other means than between dies in a huge press. One of these new methods is explosive forming. The explosion of a small charge has been used to drive liquid inside a cup of stainless steel with great force, persuading it to take the shape of a female die. It seems likely that we may see forming of this type becoming more common in factories should stainless steel hollowares again become popular.

Working in pewter

The best pewterwares are cast. The pewter is melted and poured into gun-metal moulds, and many of the moulds used today date back to the times, a hundred years ago and more, when pewterware was used in every home in the land. Pewter can also he spun or stamped in the press. Pewter stamping, is no different from stamping silver or nickel silver, but pewter spinning sometimes differs from silver or nickel-silver spinning because some craftsmen do not use chucks. They 'spin on the wind' as they say. The reason why cast pewter is better than either spun or stamped pewter is that the rolling out of the metal into sheet form, as the necessary preparation for either of these processes, breaks up the atomic structure of the metal, which results in a softer weaker vessel than one made by casting.

Chapter 6

Jewellery past and present

In the normal course of business the retail jeweller and his staff are not likely to come across much, if any, jewellery made before the eighteenth century. However, because during the nineteenth century the designers of jewellery so frequently harked back to the past, it will be useful to the jeweller to have some knowledge of the styles of earlier periods, if only to understand better this later revivalism. Also the tremendous public interest in the jewellery of the past fostered by a number of major London exhibitions, notably that of the treasures of Tutankhamun, makes it desirable for those engaged in the jewellery trade to have some knowledge of the jewellery of earlier times.

The jewellery of the past has been subject to the same influences as architecture, furniture and silverwares. In broad terms, it was gothic during the Middle Ages, baroque during the Renaissance, rococo during the middle years of the eighteenth century, and a little of everything during the nineteenth century. It was Eastern in derivation during those periods when the mysterious East was in fashion. At times when people were looking back to the Middle Ages with a nostalgic longing for the good old days, the jewellers produced gothic revival jewellery to cater for customers who lived in houses vaguely resembling castles. But the jeweller has always been very much of an individualist, often making a piece with a particular customer in mind, and trying to surprise that customer with his ingenuity. A classic example of such a jeweller was Carl Fabergé. Every year he produced his famous Easter eggs for Alexander III and Nicholas II of Russia. Apart from the different surprise contained in each egg, the style of each new egg was different from the one made the previous year.

It has always been the task of leading jewellers in every age to please princes and wealthy patrons by a display of their skills and their imagination, and the jeweller has never been hampered by considerations of utility. An architect has to create a house that can be lived in. A coffee-pot is a utensil and must function as such. A chair must be comfortable. To some extent the famous Bauhaus dictum that 'form must follow function' has always applied. But jewellery has no function in this sense. It is pure adornment, and in the past it was always accepted as such. People just didn't expect it to be practical. No king would ever have worn a hat that burdened him like a crown. Few women, even for fashion's sake, would have worn those weighty parures had they not been jewels (Figure 6.1). So the jeweller had more freedom than other craftsmen, and made full use of it. The jewellery of the past is often a fantasy of complexity, and it is bewildering in its variety. This is, of course, why so many of the histories of jewellery that have been compiled are a little bewildering. It is difficult to bring order to such chaos.

Figure 6.1 Early seventeenth century jewellery with complex floral motifs

Ancient jewellery

From a very early date Nomadic men and women almost certainly adorned themselves with stones or seeds threaded on to strings, or attached decorative objects to such clothes as they wore. The earliest systematically manufactured jewellery, however, was produced in those areas where man first settled into communities and grew crops, notably in the Nile valley in Egypt and in Sumeria in the land between the two rivers, the Tigris and the Euphrates.

The earliest such jewellery to survive probably dates from about 4000 BC and came from royal tombs in Egypt. By the third millenium BC a distinctive style of jewellery, wrought from gold and set with gems, was being made in Egypt. This style was to persist for thousands of years. Most of this jewellery had an obvious mystical significance. It is clear, too, from the evidence of surviving tomb paintings that large goldsmiths' workshops existed in ancient Egypt by this time, and those who worked in them had acquired considerable skills. These early goldsmiths were able to solder, anneal and refine gold. They had evolved the chaser's art of decorating with the chisel and employed the technique of lost-wax casting. The lapidaries of ancient Egypt were also able to polish and shape relatively hard gem material with great precision, no small accomplishment when the only grinding compound available to them was probably silica sand, a medium which the Chinese also used in the early days to fashion nephrite.

It is believed that from its first discovery in Egypt, in the alluvia on the banks of the Nile, gold was thought to symbolize the sun. Therefore gold became the prerogative of the kings of Egypt, who were supposed to be descendants of the sun god Ra. Gold is thought to have been used in the first instance to produce symbolic discs, which ensured that the wearer enjoyed the protection of the sun god. This sun symbol was incorporated in many of the talismanic jewels made in Egypt over a period of some three thousand years. The gems that the Egyptians used (cornelian, lapis-lazuli and turquoise) may also have been thought to have magical properties, though we have no evidence of this. The ancient Egyptians cut these gem materials into variously shaped slabs and set them cloisonné into gold settings to produce a whole range of amulets depicting the complex hierarchy of Egyptian deities. The amulets worn by the royal

Figure 6.2 Ancient Egyptian pectoral depicting the vulture goddess, her head replaced by a portrait of Tutankhamun; gold, cloisonné are set with turquoise, lapis and cornelian (by courtesy of Cairo Museum)

family and their court during their lifetime were buried with them in their tombs. The kings, queens, princes and princesses who wore them were sometimes themselves represented on their jewels in the company of such deities as the Ureas serpent, the dung beetle or scarab, and the falcon god, who they believed protected them from their enemies and the malign fates. A typical example of such a jewel was a pectoral found in the tomb of Tutankhamun (Figure 6.2).

This is in the form of a vulture with outstretched wings, but the head of the bird is replaced by the head of the young king. Also in an early pectoral from the tomb of King Senwasnet III, the king is represented by his name in a cartouche, protected by the vulture goddess Nekhbet above it.

The magical properties of jewellery and the belief that gold set with gemstones can protect their wearer, have a very long history. The association of jewels with royalty and religion, which is equally deeply rooted in the past, still persists. We still have our royal regalia and our papal rings 6000 years or more after the priest-kings of Egypt first appeared before their people decked in golden jewels.

In the early eras of civilization jewellery became a symbol of privilege throughout the ancient world. This tradition spread from Egypt and Sumeria, to the Troad and to Maecenae and Minoan Crete, to the Persian Empire, to ancient Greece and to Etruscan Italy. In pre-Columbian Mexico and Peru the tradition existed too. The Spanish conquistadors found shiploads of golden treasures in the palaces and temples of the Aztecs and the Incas, and sent them back to their king in Europe. Here too, in South and Central America, gold was considered symbolic of the sun. It was known to the Incas as 'the sweat of the sun'.

Aside from the jewellery of ancient Egypt, the early jewellery that has had the greatest influence on the work of subsequent generations of goldsmiths is that of the Etruscans and the Hellenistic Greeks. Both Etruscan and Hellenistic jewels were copied, or used as a basis of their designs, by the Castellani family who had their workships in Rome in the 1800s. The influence of the Castellanis, father and sons, is to be seen in jewellery produced in the latter half of the nineteenth century.

The jewellery that the Etruscans produced in Italy between the seventh and first centuries BC is characterized by granulated decoration. A great deal of nonsense has been written about the mystery of granular decoration having been discovered by the Etruscans and lost when Etruria was overrun by the Romans. Granular decoration was used by the ancient Egyptians, the Sumerians and the Minoans long before the Etruscans adopted it.

The technique would present no great problems to a modern craftsman. The granules can be readily produced by pouring gold on to powdered charcoal, and the problem of making them adhere to a gold background can be solved in a number of ways. A mixture of copper oxide and fish glue provides a method of sticking the granules in position. Heat is then applied, which drives off the fish glue as a gas and causes the copper to bond the granules to the gold surface by brazing. Another method is to dissolve copper in sulphuric acid and dip the granules in the resulting solution. The acid is driven off by heating, leaving the granules covered with copper. Flux is then applied to the gold background; the little copper-covered golden balls are arranged on it, and heat is again applied (in this case from underneath), producing a neat brazing.

The great achievement of the Etruscans was the fineness of their granular work. They applied as many as a hundred granules to each linear inch of one of their fibulae, the complex safety-pins of the ancient world. Just as we are inclined to underestimate the technical ingenuity of the craftsmen of the past, so too we tend to underestimate the patient application of which they were capable and which this jewellery illustrates so dramatically.

Greek filigree

The same metallurgical skill, artistic flair and patience that produced Etruscan jewellery also made it possible for the Hellenistic goldsmiths of the fourth century BC to produce those intricate confections of delicate golden wire that we call filigree. They applied filigree to great collars featuring Heraclean knots, to penannular bracelets and to earrings with terminals in the form of sheep, goats and lions derived from the art of the nomadic peoples living on the northern borders of the Persian Empire.

The Greek goldsmiths, incidentally, obtained more than motifs from Persia. The gold they wrought came from there as well, either as a result of melting down pillaged treasure or from the exploitation of Persian gold mines after Alexander conquered that great empire.

The Alexandrian conquests also provided the Greek goldsmiths with the right intellectual climate in which to work. Many of Alexander's officers married high-ranking Persian women of taste and discernment and brought them back home to Greece. They set the fashions and encouraged the crafts, goldsmithing among them, that flourished in the Hellenistic period.

Roman gems

The Roman goldsmith never achieved the same dexterity with gold that the Etruscans and the Greeks had done. Rome's contribution to the history of jewellery was the re-awakening of interest in gemstones. The Romans obtained their gems from every corner of their huge empire, and by trade, even from beyond its boundaries. They revived the marriage between gold and gemstones that had characterized the jewellery of ancient Egypt. The Romans were also probably the first collectors of gems for their own sake, many rich Romans assembling considerable collections of stones.

Jewels of the dark ages

Throughout the so-called dark ages, after the Roman Empire was swept away by

Barbarian invaders, the jeweller's art surprisingly continued to flourish. The jewellers of Europe from AD 500 onwards made jewels for the kings and tribal chiefs, jewels usually set with garnets. In these jewels the stones were polished flat and set cloisonné as the ancient Egyptians had set their stones. The earliest examples of such jewellery to survive come from Romania. The garnets probably came from the same deposits in Czechoslovakia that are still producing today. The style soon spread throughout Europe and into Scandinavia. The finest surviving examples of this style of jewellery are the pieces from the royal ship-burial on the banks of the Deben in Suffolk. Those pieces, from what is known as the Sutton Hoo ship-burial, were probably made in the seventh century AD (Figure 6.3) by one of the most accomplished English goldsmiths who ever lived.

Before this tradition had been developed on the continent, the Celtic goldsmiths in bronze-age Britain had evolved a distinctive goldsmithing culture of their own, producing such technically accomplished pieces as the Ipswich torques and the golden clothing-clasps and golden breastplate that can be seen in the collection of the British Museum.

Medieval

The jewels made in the 'Gothic' period, which were to inspire so many nineteenth century designs, were often functional. The goldsmiths of the period made brooches to hold cloaks at the shoulder and also belt buckles. Most of the brooches were more or less abstract in design, made from gold and roughly cabochoned gems. Some,

Figure 6.3 Shoulder clasps from the Sutton Hoo ship-burial, gold set cloisonné with garnets and glass. This type of decoration was widely used in Europe from the fifth century AD onwards; these clasps were made in the seventh century (by courtesy of the British Museum)

however, were in the form of miniature pictures in a circular frame, like vignettes created to illustrate some romantic medieval poem. Others were in the form of the heraldic household badges worn by those in the service of the great lords of the period. Some had religious motifs, like the famous Founder's Jewel of about 1400, now at New College, Oxford, which depicts saints standing in niches under Gothic arches.

Medieval jewellers also made a great many rings, usually set with roughly cabochoned or primitively faceted gems, which were often cruder versions of earlier Roman styles. Some of the shanks of these rings bore inscriptions in black letters. Some were made from gold only and carved with motifs familiar to us from Gothic architecture. Also from this period came those marriage rings carved to depict two clasped hands.

Figure 6.4 Portrait of Catherine of Aragon painted early in the sixteenth century shows her wearing typical jewellery of the period (by courtesy of the National Portrait Gallery)

Renaissance pendants

The Renaissance was an age in which the jeweller prospered and the jewels of this period were again to provide inspiration for many nineteenth century designs. The type of Renaissance jewel that is most typical, and that which attracted much attention from the nineteenth century revivalists, is the pendant. The sixteenth century pendant usually took the form of an allegorical openwork design in gold, set with baroque pearls or other gems and enriched with enamels. The Canning Jewel in the Jewel Room at the Victoria and Albert Museum is a typical though lavish example of the style. It consists of a merman brandishing a scimitar and holding a scrollwork shield. The merman's body consists of a large baroque pearl. His green enamelled tail is set with diamonds and a large cabochon ruby. More large baroque pearls hang pendant from the piece. Other popular motifs for pendants were lizards, ships, birds, dragons and mermaids. This period also produced elaborate necklaces and collars of gold, pearls and gemstones, and equally elaborate hair ornaments. Indeed contemporary paintings suggest that the pendants were often worn on the side of the head. Strings of pearls or chains interspersed with gems were attached to the pendants and bound up the elaborate coiffures.

The rings of this period were often elaborate too. Many continental rings particularly featured enamel-work. This too was the great age of the signet ring. The Romans probably invented the signet ring, and it is known to have been in use in England since Anglo-Saxon times. Examples earlier than the sixteenth century are rare, but large numbers of beautifully carved heraldic signets were produced in the Renaissance period, to judge from the number that have survived.

Flowers and arabesques

The seventeenth century was the age of delicacy, of bows and arabesques and floral themes. People at that time were fascinated by flowers. Many private and public gardens like the Jardin des Plantes in Paris date from this period, and many books were published filled with detailed botanical drawings.

Many pattern-books containing designs based on botanical themes were also produced at this time. The influence of this concern with botany can be seen on the backs of the pendant miniature-cases containing portraits of their loved ones worn by so many of the men and women of fashion of the period. These cases were usually decorated with delicate arabesques of flowers and foliage, carried out in champlevé enamels. White enamel was an integral element in most of the designs. The same delicacy of delineation and the same effective use of white enamel can be seen in a French necklace from the middle of the century in the Victoria and Albert Museum. This pretty necklace consists of a series of small ribbon bows enamelled white and pastel blue on either side of a large central bow. In the middle of the central bow two flowers, also enamelled in white and blue, are placed, and below it hang a pearl and an irregular pear-shaped amethyst.

Inventions and influences

To understand fully the jewellery of the early eighteenth century it is necessary to understand why jewellery is important to people who have acquired new wealth and a new social importance. In many periods, and in some countries even as late as the eighteenth century, the wearing of jewellery was restricted to a privileged few. Not only were the few the only people who could afford to pay for skills of jewellers and the precious materials in which they worked, but also people who were not members of a royal household, or members of a royal court at least, were often specifically forbidden by sumptuary laws from wearing jewellery. So jewellery came to be a symbol of social success, a mark of esteem. It was something which people who had 'arrived' socially wore as a matter of right, and which people who were just arriving in a class-conscious society like that of eighteenth century England, naturally wanted to possess. Jewellery still has this aura. It has a significance beyond its intrinsic value.

The early years of the eighteenth century saw the breaking down of the old order in which the landed gentry were the sole privileged class. As the century progressed the circle of the jeweller's customers gradually widened and the jewellery trade grew and prospered as a result of the eager demand from the new merchant middle class. A seventeenth century invention also stimulated the growth of the trade in the eighteenth century. This was the invention, or perhaps more accurately the re-invention, of paste. The ancient Egyptians had made simulants of gemstones in glass, but in 1676 George Ravenscroft discovered that the addition of lead oxide to flint glass made it possible to produce cut stones which were plausible imitations of the real thing. Ravenscroft's discovery proved to be well timed. The late seventeenth and early eighteenth centuries saw the birth of a new middle class, which made its money from trade. Some of the members of this new class could afford diamonds, rubies and sapphires as easily as the landed gentry, but there were many whose fortunes were not as large as their pretensions. They nevertheless wanted to put on a show at Ranelagh Gardens, in the Assembly Rooms at Bath, and at the candlelit evenings in the salons of their friends. Ravenscroft's discovery, which was exploited by the French jeweller George Frédérie Strass made it possible for them to do this without impoverishing themselves.

The eighteenth century saw a great change of emphasis in jewellery. In the past the mount had been all-important and the stones subsidiary. In the early years of the eighteenth century the stones become all-important, and the older style of mount, often enriched with enamel, gave place to something altogether lighter. It was yet another seventeenth century invention that was responsible for this – the invention of the brilliant-cut. This new cut revealed the full beauty of the diamond by capitalizing on its unique powers of reflection and refraction (Figure 6.5). So the diamonds, or paste imitations of diamonds, became the feature of the piece, instead of rather dull highlights in the general design as they had been previously.

The eighteenth century

New styles always owe at least some debt to the past. This was certainly true of the innovations of the eighteenth century, a century which revered the classical past, believing it to be the well-spring of reason and order. The eighteenth century gentleman sought to live his life in classical surroundings. All decoration was drawn from what he believed to be the fount of classical culture, the Renaissance art of Italy, which was a declared re-birth of the cultures of the Roman and Greek civilizations.

This then was the cultural background against which eighteenth century jewellery was produced. A popular piece in the early years of the eighteenth century was the stomacher, a bodice ornament covering most of the front of a dress. Joan Evans illustrates a particularly interesting example in *A History of Jewellery*. This was made about 1710, probably in England, and consists of a bow brooch from which hang three pendants. The bottom pendant is in the form of a cross, a popular motif at this period. The whole piece is set with large pastes, and the light mount is decorated with scroll-

Figure 6.5 An eighteenth century diamond rivière once belonging to Marie Antoinette

ing acanthus leaves. The design is perfectly symmetrical. Very similar designs are to be found in early eighteenth century plaster work and on the furniture of the period, though the cross motif seems to have been reserved for jewellery. The acanthus, which comes from the Corinthian capital, was the most popular motif of the first half of the eighteenth century. It adorned everything.

This piece is interesting in a number of ways. It might be thought unusual to find such a lavish and beautifully-made piece set with paste, but the maker of costume jewellery often, in those days, made the real jewellery as well, and made imitations to look as nearly as possible like the real thing. Another interesting feature is that though this piece well illustrates the subservience of the mount to the stones, which resulted from the invention of brilliant-cutting, it is in fact set with rose-cut pastes. Despite the optical superiority of the brilliant-cut, the pretty rose-cut with its flat base, and pointed crown, cut with triangular facets, remained popular right up to the end of the nineteenth century.

Finally, this piece is interesting because, being set with paste, it has survived to give us a good idea of what the lavish jewellery of the period looked like. Had it been set with diamonds it is highly unlikely that it would have stayed out of the 'melting pot' for two and half centuries. Someone, sooner or later, would have felt that the diamonds were worthy of a more fashionable setting. Few diamond pieces survived the century in which they were made, and we have for the most part to make our judgements on the jewellery of the past on the basis of the less important pieces set with small coloured stones and on the basis of imitation jewellery of this kind.

About 1735 the classical baroque symmetry, of which this stomacher is an example, gave place to a new, and asymmetric, fashion from France. The rococo (it derives from the French word for pebble work) fashion pervaded jewellery as it did most other applied arts at this period. Rococo design was much freer than the baroque which it superseded. It derived its imagery direct from nature, not from Renaissance ornament. A typical rococo piece is a riot of imagery, shells and scrolls, flowers and foliage, the whole design often held together by a central cartouche. The fashion was not long-lived in England, and not a great deal of jewellery in this style has survived.

About the middle of the eighteenth century coloured stones, which had been out of fashion for 50 years, became very popular again. The rococo phase, whatever else it had achieved, had freed design from the classical conventions. It had made naturalism acceptable, and in its wake came a period when jewellers sought to reproduce nature in gold and coloured stones. The 1750s and 1760s saw the production of flower brooches, roses with petals of topaz, leaves of emerald and buds set with rubies and many of the stones had foil placed behind them to improve their colour. Another popular subject at this time was the feather (Figure 6.6). The feather brooches of the eighteenth century set with small diamonds are among the most charming pieces of jewellery created between 1700 and 1800.

By this time the making of imitation jewellery was becoming a separate craft, and new materials had been discovered and were being exploited. The most interesting of these was pinchbeck. This copper-zinc alloy provided a plausible and inexpensive substitute for gold, and it was used until the nineteenth century when rolled-gold and electroplated gold replaced it.

Iron pyrite, inaccurately called marcasite, was first used in the eighteenth century as a diamond simulant, but before long it became accepted as a gemstone in its own right. It was sometimes set in gold plaques, but more usually it was pavé-set in silver.

Marcasite was simulated, later in the century, by cut-steel. Cut steel nails were faceted to resemble the rose-cut marcasites, and these nails were then riveted to the mounts. This was an expensive way of doing things, and before long cut-steel jewellery was being die-stamped all of a piece in a press. Matthew Boulton, who founded his famous factory at Soho Hill in Birmingham in 1762, was one of those who made cut-steel buckles to decorate the shoes of the fashionable men and women of the day.

A minor innovation belonging to the middle years of the eighteenth century was a paste imitation of opal. Quite a lot of jewellery was set with it, and although not a very

Figure 6.6 Diamond-set feathers which were made in the second half of the eighteenth century

convincing imitation, its appearance in obviously genuine period mounts may puzzle anyone who does not know about it.

All the standard forms of jewellery, necklaces, earrings and brooches were worn during the eighteenth century, but this age also saw the production of pieces peculiar to the times. One of the most fascinating of these peculiarly eighteenth century pieces was the chatelaine. The chatelaine was originally the lady of the castle who, during the Middle Ages, carried all her keys on a chain suspended from her girdle. Subsequently the chain she wore became known as a chatelaine. During the eighteenth century new uses were found for the chatelaine, and it became so ornate that it ceased to resemble a key chain, and developed into a complicated pendant. The main new use for chatelaines was as watch chains, but they were also used for étui, the decorative sewing cases of the period. The watch, or étui, was suspended from a clip attached to the lower end of the chatelaine, and at the top a hook was soldered on so that the chatelaine could be hung from a girdle. Sometimes a brooch pin was soldered on instead. In addition to the clip to hold the watch or étui, other clips were usually provided so that perhaps a watch key, or a seal, could also be attached.

The chatelaine that was used as a watch chain followed the fashion in watch cases. In the middle years of the century when repoussé-chased cases were in fashion the chatelaines consisted of a series of repoussé-chased plaques joined by short lengths of chain. Towards the end of the century, when those pretty enamelled watch cases with a circlet of half pearls round the outside were high fashion, the plaques on the chatelaines were enamelled to match the enamel decoration on the watch case. Chatelaines were made in gold and silver and in base metal, usually pinchbeck.

An enormous number of buckles must have been produced between 1700 and 1800. Set with diamond, paste, marcasite or cut-steel, they were used not only for shoes, but also for belts, and to buckle velvet bracelets and collars.

In the last quarter of the eighteenth century came a great classical revival movement in all the applied arts, sparked off by the discoveries of Herculaneum and Pompeii. One of the results of this movement was a demand for classical cameos, and it was at this period that Josiah Wedgwood began to produce his 'jasper' cameos at his famous pottery in Staffordshire. They were used for necklaces, bracelets and brooches and are once again being produced today.

The nineteenth century

The turn of the century does not always bring with it a radical change of style, and the styles which we associate with the nineteenth century were already emerging in the eighteenth. A reaction against the Age of Reason had begun well before 1800. The romantic revolution and the industrial revolution both had their beginnings in the eighteenth century, and it was these two revolutions, one artistic the other materialistic, that were to affect everything made in the nineteenth century. The industrial revolution was to result in the mass production of cheap wares, jewellery among them, for a new buying public.

A new stratum of society came into existence in the nineteenth century, a stratum below the rich merchant middle class that had grown increasingly powerful during the eighteenth century. From this new class, which the sociologists call the lower middle class, were to spring the *nouveau riche* of Victorian England, while its less successful members were to live comfortable lives in those staid Victorian villas still to be seen on the outskirts of our towns.

This new class, like the merchant middle class before it, was eager to ape its so-called betters. It was to supply this new class with inexpensive copies of the worldly possessions of the rich that men, women and children laboured for long hours in the stygian gloom of the new factories in Birmingham and Sheffield. Cause and effect were interwoven. The industrial revolution created this new society, and the demands of the new society kept the wheels turning and the wealth flowing.

The nineteenth century was proud enough of its material achievements, but the Victorians wanted something else from life beyond success. They wanted to escape sometimes from their own mundane surroundings, so they fostered the romantic revolution, and the romantic revolution brought to those monotonous new streets a breath of distant places and earlier times. The artists and craftsmen presented their customers with a romanticized view of the world. They did not show the Victorians an India of teeming poverty, but an endless parade of panoplied elephants on which maharajahs swayed in golden howdahs. Medieval England was presented as a fairy tale time, peopled with courtly knights and gracefully wimpled ladies.

As the century progressed the romantic images became mingled and confused. Minarets grew out of gothic castles which were entered through Byzantine gateways. Sir Matthew Digby Wyatt, writing in 1851, observed that: 'We design and execute in every conceivable style. We are equally at home in the reproduction of Classical and Byzantine . . . Etruscan ware and Majolica; we can execute Chinese or Athenian with the same facility. . . .' The late Ernie Bradford who quotes this passage from Wyatt in his book *English Victorian Jewellery*, comments: 'It is not that the collector or the student is likely to be deceived into thinking that a piece of Victorian "Gothic" enamelling or "Renaissance" pendant is other than nineteenth century in manufacture.' He then goes on to define the tell-tale differences as 'regularity of finish' and 'a certain mechanical coldness'. This is true, but more importantly these pieces look wrong, as all revivalist art looks wrong.

A man working in a period when a style is alive will have a fundamentally different attitude from that of a craftsman reproducing the superficialities of that style centuries after it has died out. This difference of approach can be seen at a glance. The lack of conviction, the lack of understanding in the composition, sometimes even in a single line of chasing can be perceived. This is not to say that a latterday copy of a past style cannot have a charm of its own, in the way that eighteenth century chinoiseries, which no one would ever imagine were made in China, have their own charm. Nowadays, we are beginning to discover more and more to charm us in nineteenth century art and craftsmanship.

The nineteenth century jeweller did not just raid the lumber rooms of the past, and look to lands afar for inspiration. The romantic revolution was more than anything else a return to nature as the prime source of artistic inspiration. Its exponents conceived of it as a great turning away from the artificialities of the last age. Much of the jewellery of the nineteenth century was taken from nature, but the Victorians had their own way of looking at nature. To most Victorian artists it was no harsh cycle of life and death, but an ever-continuing sunny summer afternoon. So the flower jewellery of the period depicts full-blown symmetrical flowers, stiff and orderly. There is no pretence of reality about these gem-set sprays, even when every detail, every vein and stamen are reproduced.

They were consummate craftsmen, these Victorian jewellers, and some of their flower jewellery has never been rivalled for sheer technical virtuosity. And though this jewellery is stiff, there is something endearing about the simple, almost childlike, approach. Simplicity is not a quality which is usually associated with Victorian art, but there is an underlying emotional simplicity about much of their work, and some-

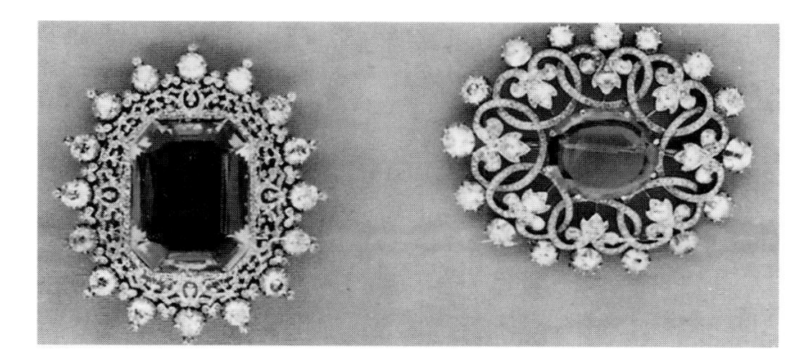

Figure 6.7 An example of the charming romantic jewellery of the early nineteenth century

times too they produced designs which had a great simplicity of line (Figure 6.7). In this connection those brooches consisting of a large stone, usually an amethyst, surrounded by a circlet of tiny pearls come to mind. Then there were the rivières, those necklaces that consisted of a single row of graduated stones in the most self-effacing of settings.

It is extremely difficult to place the styles of the nineteenth century in any chronological order, or to date a particular piece within 20 years. It is sometimes possible to pinpoint the inception of a new style, but styles tended to linger on, and to be revived from time to time. That curious jewellery made from human hair for instance was first made in the eighteenth century, when it was used as mourning jewellery. This hair jewellery was still being advertised in 1850. Then there was the great classical revival. The discovery of the ruins of great classical cities in Italy had been the inspiration of the Adam style back in the eighteenth century. It was not, though, until the first decade of the nineteenth century that the clearing of the ash and the basaltic debris from Pompeii was anything like completed, and the ruins became the showplace they have been ever since. This brought about a second classical revival at the beginning of the new century, and a great vogue for classical jewellery. At no period during the nineteenth century were classical motifs entirely out of fashion.

The cameo was very popular with the Victorians. The earlier cameos had, of course, been carved from gemstones, usually agates, but the typical Victorian cameo was carved from shell. These shell cameos, which were cut from mollusc-shell, were imported in very large numbers from Italy to be set by English jewellers. Some of them were exquisitely carved, classical figures clothed in flowing drapes were depicted in fine detail. But the majority of shell cameos were of poor quality, mass produced for a mass market.

Another result of the popularity of Pompeii was the neo-Etruscan style. Gold filigree work and granulation in the Etruscan manner were probably first seen in England in the 1820s. Castellani, the greatest exponent of the style, had set up a workshop in Rome in 1814, and Etruscan work retained its popularity for over half a century. It was still very much in evidence at the Paris Exhibition of 1867. Granulated work, the production of a pattern of minute grains of gold on a gold surface, is a technique which has intrigued goldsmiths down the centuries since the Etruscans perfected it. Castellani and a few of his fellow Italians managed to master the technique, and their work created a considerable stir both in Paris and in London, but the process of making this jewellery was too difficult and too painstaking to be readily copied at that

Figure 6.8 Granulated gold ear-rings in the Etruscan style, made by Carlo Giuliano, one of the followers of the Castellanis (by courtesy of Christies)

period, and therefore only a limited amount of granulated jewellery was produced in other workshops (Figure 6.8).

The beginning of yet another nineteenth century fashion can be dated from the time when Queen Victoria discovered the Highlands, and began to make her regular pilgrimages to Balmoral. Thousands of mezzotints of Highland cattle and stags at bay were hung in drawing rooms and parlours all over England, and the jewellers began producing the Scottish jewellery that has been popular ever since: those mounted grouse claws and the big silver brooches chased with Celtic motifs and set with a single cairngorm or amethyst. It was this interest in all things Scottish, fostered by Victoria's love of the Highlands, which also focused attention on the pretty freshwater pearls found in the mussels of the Tay and the Spey and in rivers in the north of England and Wales.

For much of her reign Victoria was a Queen in mourning. In this, too, her subjects followed her example and sombre mourning jewellery became a feature of the Victorian scene after the death of Albert. So great was the demand for the sleek black jet, that the deposits around Whitby, which had first been worked in Neolithic times, were virtually exhausted in the second half of the nineteenth century. Albert, in his turn, created a jewellery fashion, by wearing a heavy gold watch chain, and this heavy linked chain has borne his name ever since.

The most magnificent work of the period consisted of suites, or parures as they were called, set with richly-coloured amethysts or glossy garnets. Diamond jewellery was produced as well as lavish suites set with rubies and emeralds. But just as they were prodigal in their use of styles, so too they used the whole palette of gem materials. Mineral collecting was a popular Victorian hobby, and clearly the jeweller found no resistance to the lesser known gems among his customers. Every conceivable type of stone is found in Victorian jewellery. Now that many stones have become hard to find in nature this jewellery has become a major source of fine examples of such stones as demantoid garnet and peridot.

Certain stones at certain times had a particular vogue. Turquoise was popular during the second half of the century, and was often used as an alternative to pearls for those circlets that framed a large stone, an enamel or a cameo. Coral was very popular in the first half of the century. Coral from the Mediterranean was beautifully carved in the workshops of Naples, and the Neopolitan carvers enjoyed a flourishing export trade with Britain. The big opal finds in Australia in the 1820s provided the Victorian jewellers with beautiful black opals to mount, and the discovery of the great pipes in Africa at the same period created a new interest in diamonds.

The variety of stones used by the Victorian jewellers was rivalled by the variety of cuts produced by the Victorian lapidaries. The diamond saw was a nineteenth century invention, and the whole cutting industry became more mechanized at this time. The cutters made full use of their new equipment and produced beautiful variations on the brilliant-cut, like the marquise. They sometimes complicated the brilliant-cut by multiplying the facets, and the pendeloque, which already had had a vogue in the eighteenth century, now enjoyed a revival. But perhaps the cut which was particularly Victorian was the cabochon. Victorian cabochons range from tiny symmetrical turquoises to huge tear drop cabochon rubies. Strangely, in the last two decades of the century, when the mathematics of diamond cutting were first beginning to be appreciated (though they were not finally set out until Marcel Tolkowsky published his *Diamond Design* in 1919) there was a great vogue for the old rose-cut.

The settings and the mounts used by the Victorian jewellers were as varied as the stones set in them. Claw, millegrain, thread, pavé and box were all used for nineteenth century jewellery, while every technique was introduced to elaborate the mounts. Jewellers engraved and chased and carved gold. They used the coloured golds now readily available, often as inlays. In addition to painted enamel work they revived the old enamelling techniques of the Middle Ages: cloisonné, champlevé and encrusted enamelling, the more sophisticated basse-taille, which was the placing of translucent enamel over a relief design, and plique-à-jour, which consisted of removing the backing from behind the enamel to let the light shine through it as through a stained glass window.

The nineteenth century saw a great increase in the mass production of jewellery. The fly-presses in the Birmingham factories bumped out jewellery components in their thousands. Gold jewellery was stamped out in this way, but mass-production techniques really came into their own in the production of imitation jewellery, in particular rolled-gold jewellery. The end of the century saw the foundation of the great rolled-gold factories of Pforzheim in Germany. Most of these factories began as specialist producers of chain, but soon turned over to manufacturing rolled-gold jewellery that sold throughout the world. The meteoric growth of these factories during the 1880s and 1890s was an indication of the popularity of their products.

A great deal of paste jewellery was made in the nineteenth century, marcasite also was popular, imitation pearls were made in increasing numbers, mosaics were imported from France and Italy, while the imitation cameos made by Wedgwood, and the glass cameos invented by the Scotsman James Tassie, rivalled the mollusc cameos from Italy. However, the imitation jewellery of the Victorians differs from much of the costume jewellery of our own day in that it was intended to be a replica of the real thing, a substitute for what was beyond the pocket of the purchaser.

Art nouveau

In the closing years of the nineteenth century another artistic revolution began. It grew out of the enthusiasms and beliefs of the Arts and Crafts Society, whose figurehead was the talented William Morris. Their avowed aim was to reassert the standards of craftsmanship which they felt were being eroded by the growing industrialization of methods of production. They achieved only a limited success, for in the materialistic society of the later Victorian period their idealism found little response. The only really important outcome of the movement was to introduce the *art nouveau* style into this country.

Among the leading exponents of the *art nouveau* style were Lalique and Fonquet in Paris and Tiffany in New York. *Art nouveau* has been described by Graham Hughes as: 'the most exotic jewellery style ever', and in the hands of an artist of the calibre of Lalique it certainly was. The *art nouveau* movement, like most new art movements, was described by its followers as a return to naturalism. Its motifs, the most common among them being tulips, daffodils, butterflies, wasps, serpents and lizards, were not essentially different from those used by the earlier Victorians. However, an *art nouveau* flower is something very different from a flower delineated by a mid-Victorian craftsman. The *art nouveau* flower is not a static idealization but a lively and sensual creation not infrequently with evil overtones.

The movement petered out, as so many art movements do, when its motifs became clichés in the hands of uninspired copyists. The tulips and daffodils lingered on to decorate dado friezes and twine up and down cast-iron fireplaces. It ended up being somewhat of a joke, and only in the past few years has the best work of the *art nouveau* jewellers really received its due appreciation as some of the finest jewellery that has ever been produced. Their reputation was finally re-established by a magnificent exhibition at Goldsmiths Hall in 1987.

Twentieth century

The Edwardian period saw the production of a profusion of formal jewellery set with magnificent stones, immaculately cut. It was a period of great wealth when entertainment among the rich was still almost as formal as it had been at Louis XIV's Versailles. Great necklaces and tiaras of matching brilliant-cut diamonds, emeralds, rubies and sapphires were produced to a standard of workmanship unrivalled before or since. It was at this time that houses such as Cartier, Boucheron and Asprey came into prominence. When the jewellery made by such famous houses in the first forty years of this century is seen in the salerooms it seems to belong not to the threshold of our own times, but to some long-forgotten age. So much of our conception of jewellery has been changed by fashion and economic considerations in the past few decades.

Precious jewellery in the period 1900–1939 was in some ways like the jewellery made in the eighteenth century. All the emphasis was on the stones, and the less setting there was on view the better. This jewellery was geometric in design and beautifully composed, though it lacked any great artistic merit. It is hardly surprising, therefore, that some latter-day designers have described it derisively as 'portable wealth'.

The exception was the *art deco* jewellery produced in the 1920s and 1930s, which has recently become fashionable again. Over the past few years a large number of *art deco* style pieces have been produced. Otherwise innovations were few. The most important was the clip, followed by the double clip. In the period immediately after World War II this lavish style of jewellery lingered on for a decade, different only in that the later designs tended to be asymmetrical rather than symmetrical.

The more modest jewellery of the period continued to be made to late nineteenth century designs. These lingered on up to the beginning of the 1914–1918 war. Bar brooches, for instance, which were so popular in the 1890s, continued to be made in

Figure 6.9 Typical of jewellery produced between 1900 and 1939; fine coloured stones like the emeralds featured in these pieces are surrounded by equally fine diamonds

their thousands. Indeed, in the form of riding crops impaling fox masks, they were still very much in vogue in the 1930s and enjoyed a brief revival in the 1970s. Then in the twenties cultured-pearl necklets and ear-rings became jewels for all people and all occasions.

The design revolution

In the 1950s a revolution in jewellery design took place. The existence in Britain of 100 per cent purchase tax inhibited the production of new designs in the old lavish style, but a group of young designers, Andrew Grima prominent among them, conceived jewellery that would appeal on the basis of its artistic content, rather than its intrinsic worth.

This new jewellery was sculptural in concept, and was indeed sometimes described as sculpture in miniature. It certainly owed more to the work of Henry Moore and Reg Butler than to the jewellery traditions of the past. All it really had in common with earlier jewellery was that it was fashioned from gold and gemstones. It was indeed very much *gold* jewellery, such gemstones as were employed were highlights in the design and no more than that. The new designers disliked the sleek shininess of polished gold, and they textured it or finished it so that its surface resembled that of a nugget recovered from alluvia. More and more, too, the new designers turned their backs on faceted stones, and set their pieces with uncut stones, tourmaline and amethyst crystals, uncut emeralds and dioptase and slices of agate and chalcedony (Figure 6.10).

Figure 6.10 Pendant consisting of a watermelon tourmaline crystal and decorated with sticks of pink and green tourmaline scattered with bricks of 18-carat gold and square-cut diamonds on a gold necklet; typical of Andrew Grima's designs

The initial reception of this new jewellery by the trade was lukewarm, because it seemed likely to appeal only to a limited public. Gradually, however, people began to respond to these new designs, and more and more established manufacturers adopted texturing and set their pieces with unfaceted minerals.

At the same time another form of decoration, diamond milling, which broke up the surfaces of the gold into brightly polished facets made its appearance. Eventually virtually all jewellery offered to the jeweller, from watch bracelets to gem rings, was either textured or diamond-milled.

In 1974 a new fashion was born, to cater for a new interest in jewellery among the increasing number of young women who began to buy jewellery for themselves instead of waiting to be given it in the time-honoured way. This new fashion was born in Italy; it was based on chain, which Italian goldsmiths produce in a great variety of patterns. Fancy gold and silver chains were made up into necklets and bracelets, sometimes interspersed with beads cut from a variety of gem materials – chiefly coral, and agate dyed black to simulate onyx, but also sodalite, lapis, rhodonite, rhodochrosite and jade. These chain necklets often featured pendants carved from the same materials, sometimes framed with small diamonds (Figure 6.11). Rapidly, as it became clear that this new fashion was enjoying a phenomenal success, a whole range of inexpensive jewellery featuring these decorative minerals was produced in France, Germany and in the UK. The Idar Oberstein cutters were soon hard put to cater for the demand for beads and polished pieces of these materials. Even those factories equipped with ultrasonic bead-drilling machines were quoting long delivery dates, while the factories making chain were finding similar problems in meeting the unprecedented demand. Throughout the world, it seemed, more and more people were buying jewellery, and the jewellery they wanted was real jewellery of an informal kind that could be worn with anything on any occasion.

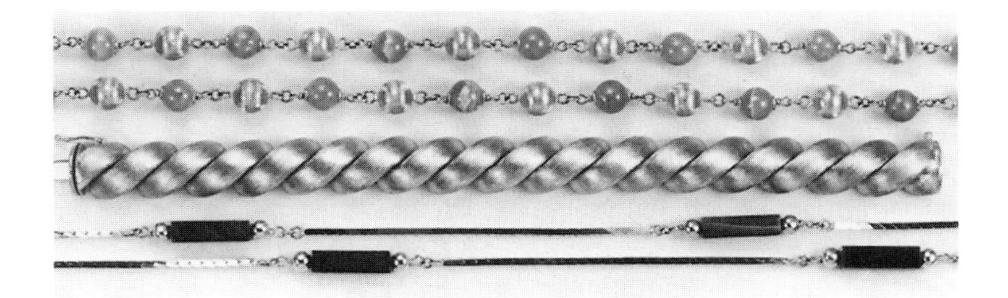

Figure 6.11 Modern Italian jewellery, showing use of fancy chain and gemstone beads; necklace beads (top) are rhodochrosite; bracelet illustrates texturing of gold

Subsequently, recent years, the rise in the price of gold and a world recession led to the creation of very delicate jewellery set with tiny stones, a case of economics dictating fashion. More recently there has been a demand for somewhat heavier pieces because the public had had bad experiences with the flimsy jewellery of the early 1980s. Another change which has recently taken place has been that today many more firms employ in-house designers than was the case a few years ago. De Beers and the former World Gold Council had been, to some degree, responsible for this because they influenced public taste by their promotional activities. The UK, however, remains a very cost conscious market, with many shops prospering by selling at discount prices.

Chapter 7

Manufacture of jewellery

Man and machine

Not so many years ago a great gulf existed between the fine jewellery made by handcraft methods and the cheap and usually mundane jewellery made by machine techniques. Today the gulf has narrowed for three reasons. The jewellery trade is making more imaginative use of the machine techniques available to it. A shortage of skilled labour has forced the firms in the trade as a whole to use machinery to achieve an adequate turnover. The spread of wealth into many more pockets has created an increased demand for interestingly designed middle-class jewellery. As a result of these three factors well-designed, well-made jewellery is being mass produced today, and more and more manufacturers are exploring every possibility of replacing hand work by machine work (Figure 7.1). One manufacturer, once well known for his fine handmade pieces, carried this so far that he described his firm as 'engineers in precious metals'. He is not untypical.

The design

Every piece of jewellery made begins as a design. It is no coincidence that most of the great craft workshops of the past were set up and run by designers of more than average ability. Design is after all very important, because if the original conception is poor, all the skills of all the craftsmen in the world can never make a good thing of it. This was a fact often lost sight of in the first half of the twentieth century. At that period jewellery making was changing from a craft to a business, run not by a designer-craftsman but by a businessman who probably had never sat at a bench or a drawing board. Sometimes such firms bought in good designs from Paris, but more often than not they employed an underpaid and underprivileged draughtsman to produce an endless succession of variations on time-worn themes.

The great revival of public interest in jewellery which has taken place in the last twenty years has been in great measure due to the re-emergence of the designer as a key figure in the industry. There is no doubt that the importation of Italian jewellery into the UK, jewellery made in the workshops of designer–craftsmen in the town of Vicenza and at Valenza on the banks of the Po, was at least partly responsible for this revival of interest. This jewellery was not only made to a price that many people could afford, but it was well designed. It was fresh and it was interesting. The success with which this Italian jewellery met made it possible for a new generation of designer–craftsmen in the UK to set up their own workshops with the prospect of selling what they produced. It also emphasized to the established firms in the UK the importance of the designer.

Figure 7.1 Light engineering techniques are being increasingly applied to the production of jewellery. This picture shows automated assembly in a German watch bracelet factory

There are exceptions to any rule, but to be a really successful designer of jewellery it is necessary to be trained as a jeweller. Only someone who has worked in precious metals can fully understand their limitations and appreciate their possibilities. And it seems, too, that only a jeweller can have the right feeling for jewellery. Painters and sculptors have designed jewellery in the past, but almost never successfully. Jewellery is not just a work of art, a piece of miniature sculpture – it is a part of dress, a fashion accessory. It is also many other things to many people. It is a symbol of love, a mystic symbol, a symbol of wealth and a symbol of security. All this jewellery designers have to understand. They must understand, too, the functional problems: necklaces must hang right; earrings must be comfortable to wear; rings must not ladder nylons; and brooches must not tear flesh. In addition to all this they must have inspiration. They must be capable of creating something both different and beautiful, and they must develop a personal style. Their work must also bear the signature of their personalities. It should be recognizable, as a piece of Lamerie silver can be recognized, even before looking at the hallmark. A good jewellery designer is, in fact, a remarkable person who at long last is again being given his due.

Designers work in different ways. Some of them, particularly those who work for houses making pieces set with important stones, make a wax model and stick the actual stones that will be used into it. Some designers make the roughest of sketches and work out the interpretation with the person at the bench. Others get an idea, sketch and sketch away until they are satisfied with the design, and then produce a drawing so detailed that any craftsman who had to make the piece would be able to see at a glance how it should be made. Nearly always, however, there is consultation between designer and craftsman during the making. Any drawing, no matter how detailed, is subject to interpretation, and a craftsman could alter the whole feeling of a piece in the course of making it.

The designer who designs for a mass market needs a different approach from the designer who creates one-off pieces. The one-off piece bears only the cost of the raw materials and the craftsman's time, and it has to find only one customer. The mass-produced piece may well have to bear the cost of £1000 worth of tools, which will employ the valuable time of the toolroom for as much as a month, though nowadays lost wax casting obviates the necessity for making expensive tools. More importantly this jewellery must find many, many customers because the manufacturer will produce dozens, if not hundreds of one design.

A failure to produce a suitable design means that the factory has laboured in vain. Designers for the mass market must therefore be students of public taste. They must be prophets, too. It is not enough to be able to see a trend and follow it. Nowadays jewellery has become a fashion commodity and one fashion ousts another with alarming speed. The time lag from the conception of a design to the appearance of the first piece in the firm's sample tray could be as long as 18 months or more. Designers of mass-produced pieces must therefore have the ability to anticipate the taste of tomorrow. But equally they must not leap too far ahead, otherwise the public will not buy and the design will be a commercial failure.

The responsibility for producing a commercially successful design often makes the designer over-cautious. The complex impulses that persuade people to buy jewellery nowadays, often result in the designer of mass-produced pieces lagging behind public taste, while the adventurous designer of one-off pieces has the opposite problem. Those who can afford to buy important jewellery are usually the older, more conservative people, and the people who are looking for value for money. It is the young who are the makers of manners these days, the leaders of fashion, the eager buyers of the new and the different. The teenagers are not outraged by novelty, and the wife of the up-and-coming and newly wealthy young executive is much more adventurous than were her parents; more concerned with what is effective now than what will be acceptable in ten years' time and less bothered about potential resale value.

The designer, having solved these complex sociological, aesthetic and commercial problems, and having produced a design, leaves the scene except as a consultant. The craftsman at the bench or the production manager of the factory now takes over.

The craftsman at the bench

The craftsman who is to make the one-off piece, called the mounter, is issued with the necessary raw materials, such as gold sheet, gold wire, gold tube, ready-made pins perhaps, and, of course, any stones that will eventually be set into the mount.

The raw material may be bought in ready sized form from a bullion dealer, and so may the joints and catches, and even the settings for the stones – all those things known as 'findings'. Some manufacturers, however, alloy their own gold, rolling it out between mirror-polished flattening rolls to make sheet. They roll out rod between grooved rolls to make wire, and then draw it down to size. Most firms, although they buy in sheet, rod, wire and tube, roll and draw it to the required size and shape. They require such a variety of dimensioned material that it would be neither economic nor feasible to buy material specifically dimensioned for each job they undertake.

Rationalization in the jewellery trade always finds itself at loggerheads with diversity. A craft workshop thinks in terms of producing dozens of a line at the most, and the large mechanized factories, who do go in for longer runs, nevertheless usually have hundreds of patterns in current production. Take for instance a firm producing wedding rings from drawn wire, and consider the many different sections a wedding ring can have, square rectangular, oval, D-shaped and so on, as well as the different weights and sizes. So in a wedding ring factory there are grooved rolls, and the walls are hung with draw-plates, with holes of gradually diminishing size, the holes shaped to correspond with the different sectioned wires that the firm uses. The ring maker buys in rod of the different carat qualities and from this produces the great variety of wires required.

Having been issued with the raw materials mounters set about fashioning them. Already, from looking at the design, they have decided how this can best be done and what materials will be needed. The tools used are as old as the craft, the same tools can be seen in drawings and paintings of jewellers' workshops hundreds or even thousands of years ago. Mounters use hammers, the basic tool of any smith, whether they work in iron or in gold. They use drills, and many craftsmen still favour the sensitive bow-drill, invented no one knows how many centuries ago. They use piercing-saws with slender blades, gravers and files, all tools as old as history. They use a gas-torch, it is true, which is a relatively modern invention, but this is only a different form of fire from that used by the smith through the ages.

The mounter will first perhaps cut out the outline of the piece from sheet gold with a saw (Figure 7.2). He may then well want to dome the sheet. This is done over a metal block, a doming block which looks rather like a large dice with hollow depressions of different sizes on the different faces. Using a punch and a hammer, the mounter places the head of the punch on the metal which has been positioned over one of these depressions, and then taps the punch with the hammer. Successive blows steadily drive down the metal until it follows the shape of the depression. In order to dome or depress smaller areas the gold sheet is placed on a bed of lead and a hammer and punch is used in the same way. The lead will give under the blows of the hammer, but will still support the gold and prevent distortion.

Very rarely is a piece of jewellery made in one piece. Small pieces will be cut out and shaped up, and soldered on to the main structure. Wires will be bent and soldered on.

The design may call for part of the gold sheet to be cut away, and this will be done with the piercing-saw. Where stones are to be set, collets will be soldered on, or the sheet will have holes drilled in it to receive the stones.

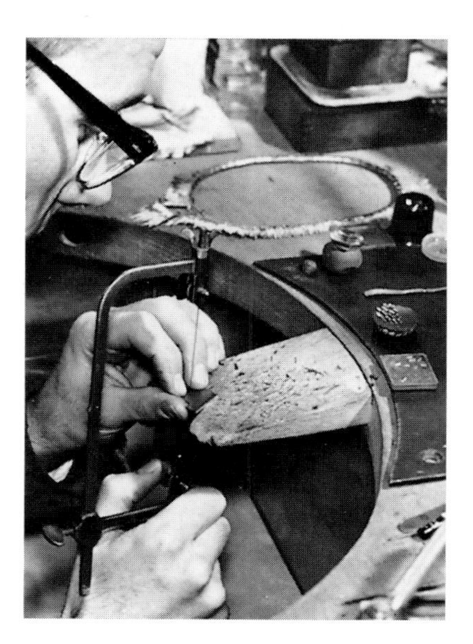

Figure 7.2 Making jewellery by hand. The mounter cutting the outline of a brooch with a saw

Figure 7.3 Assembling a hand-made bracelet

Every piece calls for different skills, those that the mounter acquired during his five-year apprenticeship and has been improving on ever since. One, which he will have to employ on every piece he makes, is his skill in soldering. The goldsmith's skill with the blow-torch is just as important as his skill with the hammer and at every bench in the workshop a blow-torch rests on its stand, kept alight always ready for use.

The soldering of jewellery consists of joining two pieces of gold or platinum together by heating a metal of the same quality, but with a lower melting point, and then letting it cool so that it forms a bond (Figure 7.3). This solder has to be of the same quality as the metal it joins, because with a few exceptions this is demanded by the hallmarking laws. The joint is fluxed with borax to make the solder flow; a little piece of solder is placed in position and heated with the torch. The skill is to get the temperature just right, and to use just enough solder to give a neat but strong joint.

When the mounter has finished work, having decorated perhaps some parts of the piece with texturing or engraving which is imparted to the surface with strokes of a steel graver, it goes forward to the finishing shop. Here perhaps some parts of the design are given a mirror brightness by polishing them with a diamond on a milling machine. Then the piece is polished. First it is either roughed against a revolving wire brush, or it is mopped, that is to say it is held against a revolving felt mop. A coarse rouge compound is used as a polishing agent. The polishing is finished against smaller mops revolving on their spindles. and for this process a finer rouge is used. Interior surfaces that are hard to get at are polished by drawing over them a thread loaded with rouge. The pieces are then cleaned in a cleaning bath. An ultrasonic bath is often used these days.

If a particular colour of gold is required, or if as sometimes happens there are slight differences in the shades of the gold wire and the sheet, it is a common practice to gold plate the piece. White gold jewellery is nearly always rhodium-plated to improve its colour. This practice does cause problems if the surface plating becomes worn away, however, because the white gold underneath has a decidedly yellow tinge compared with any remaining rhodium.

The setter's skill

The mount is now ready to have the stones set in it. A setter's work looks deceptively simple, as do so many jobs in the hands of a skilled craftsman. When they are seen opening up the claws of a claw setting, which looks like a miniature crown, cutting little seatings inside the claws to hold the girdle of the stone with a file or a dentist's drill, picking up the stone with a wax-stick, placing it between the claws and bending them over to hold it, it all looks very easy (Figure 7.4(a), (b)). But the stone must be perfectly flat, the claws must be evenly bent, just tight enough to grasp the stone, but not tight enough to damage it. Many stones, though they are hard are also brittle. Emerald particularly could easily be chipped or fractured by a heavy-handed setter,

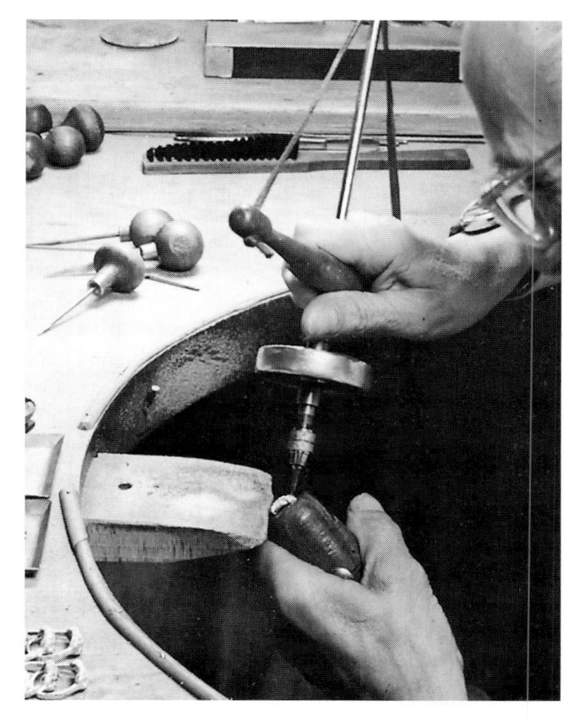

Figure 7.4(a) Stone setting using an ancient bow-drill to drill a mount to receive the stones

Figure 7.4(b) Stone setting. Using a scorper the setter pushes the metal up and over the girdle of the stone

while opal is soft enough to be bruised by the pressure applied by the scorper to the claw.

There are many ways of securing a stone in a metal mount. In the cheapest of imitation jewellery the paste stones are just glued into the holes prepared to receive them. In most of the jewellery sold by the jeweller, however, the stones are set in some form of collet. Most engagement ring stones are set in claw settings. These are usually lost-wax cast nowadays. Those required for important pieces, however, are still made in the workshop by bending up a strip of metal so that it forms a section of a cone. This is then pierced with a piercing-saw to create the claws. A little bezel is then soldered to the foot of the claws to allow the setting to be firmly soldered on to the shank. To create illusion settings a flat sheet of polished metal with slotted edges is soldered between the claws, and a tiny stone set in a hole in the middle of the plate. This gives the illusion that the stone is bigger than it in fact is. More sophisticated illusion settings nowadays employ a diamond milled plate which has highly reflective facets.

A common variation on the claw setting is peg setting. This, as its name implies, has little pegs to hold the stone, and it has the advantage of hiding less of the stone than the conventional claw setting. The girdle of the stone is held by the sides of the pegs, instead of the claws being bent over the stone. Pavé setting is used where the designer has ordained that a lot of small stones are to be sprinkled over the surface of the metal. The stones are dropped into holes drilled into the mount, and the setter then raises little grains of metal round the perimeter of each stone to hold it in place.

The rub-over setting is that used for signet ring stones. The stone is dropped into a cup of metal and the rim of the cup is rubbed over it to produce a narrow gold frame round the edge of the stone.

Millegrain setting is achieved by drawing a little serrated hardened steel wheel round the metal close to the stone to produce a little circlet of grains. Thread-setting is done in the same way but a plain wheel is used to create a thread of gold to hold the stone. Pearls are usually set by being drilled part way through and then attached to a metal peg with an adhesive.

Mechanical methods

There are many ways nowadays in which hand-work can be replaced by factory processes in order to reduce production costs. Apart from the fact that labour is in short supply, it must be appreciated that labour costs these days represent a large part of the production bill. Unless a piece contains important stones, the cost of raw materials will represent only a relatively small proportion of the total cost.

Stamping

Two techniques increasingly used today to economize on labour and labour costs are stamping and lost-wax casting. Stamping is not new to the jewellery trade; it has been used for thousands of years to produce inexpensive jewellery. In the past, however, jewellery produced by stamping mostly looked a poor machine-made imitation of hand craftsmanship. Today, the use of much more intricate tools and more imaginative designing have led to the production of stamped jewellery hardly distinguishable from the hand-made article. The textured gold watch bracelets made by leading firms illustrate this. Originally, the hand-made appearance achieved by these firms was the result of a lot of hand finishing, but now hand-work is being gradually reduced to a minimum by improved tooling. A bracelet can be made from components stamped out in the press and then given a hand-made look by hand texturing. However, it has now been proved possible to produce much the same effect entirely from the tools. When the bracelet has been polished it can be extremely difficult for anyone, except an expert, to see that it was not hand made. The major drawback to the machine-made product is that hundreds of identical bracelets will have to be produced to pay for the tools. If such production methods are used exclusiveness has to be sacrificed.

Different skills

The production of jewellery by mass-production techniques does not take away the skill from jewellery making; it just means that the skill is vested in different hands. It is the toolmaker and the model maker who are the skilled craftsmen in the modern jewellery factory, and the works manager who needs the skill and training to see at a glance how best a piece can be made and how best quality and economy can be reconciled. When the works manager has done this, the die-sinker begins to sink the design in blocks of hardened steel, working to the very fine limits essential for efficient production. Die-sinkers work with a mixture of traditional hand tools and machine tools, files and gravers, modern milling machines and spark erosion machines.

Mass-production techniques are more exacting than are hand methods. A thousandth of an inch or two error here and there are of no great importance in a hand-made piece. The goldsmith can offer up the components and make minor adjustments in a few minutes. In mass production thousandths are important. Production does not end in the press shop. The piece has to be assembled. If this assembly has to be entrusted to highly skilled goldsmiths, and if they have to waste much time in making minor adjustments, then all the economies resulting from stamping out the components will be cancelled out. The assembly of mass-produced components calls for mass-production methods, the employment of semi-skilled labour aided by ingenious jigs, which de-skill and speed up the work. The components from the presses must fit these jigs exactly or they must be scrapped or expensively corrected.

If many jewellery manufacturers have moved out of the world of handcraft into the world of precision engineering, many more firms are making use of an old process, revived not that long ago, which needs less expenditure on capital equipment than a modern press-shop and its attendant toolroom. This process, which is also more flexible than stamping, is known as the lost-wax process first used by the ancient Egyptians thousands of years ago.

Lost-wax process

The capital equipment required to carry out this process in a modern workshop is very modest. All that is required is a small vulcanizing press, a wax injector, the equipment for preparing and pouring the 'investment', a centrifuge and an oven.

The production cycle begins with the making of a model of the piece to be produced. This model is made in exactly the same way as a one-off piece of jewellery, with the ancient tools and techniques of the goldsmith's craft. The model may cost as much as £100 to make, but this is far less than the cost of a complicated set of dies needed to make the same piece of stamping. More recently an even less expensive method of making models has been developed. The model maker works in wax, carving it with tiny blades. The wax models are then reproduced in metal by the lost-wax technique.

Once made the model is placed in the middle of a small square frame (Figure 7.5(a)). Rubber is then packed round it, after which the frame is put into a vulcanizing press. There the rubber is heated so that it flows round the model and is vulcanized into a solid block. This rubber mould is then sliced in two and the model removed. The two halves of the rubber mould are put together and the channel created to link up the outside of the rubber block with the recess in the middle of it. The channel is now placed over the nozzle of a wax injector. The wax injector is a can containing molten wax. By pressing the nozzle at the base of it molten wax can be spurted under pressure into the rubber mould (Figure 7.5(b)). When the rubber mould is opened again a wax replica of the original model is found inside it. When a number of these wax replicas have been created, they are mounted on a base to form a 'tree' (Figure 7.5(c)). This tree is then placed in a metal flask. An investment, consisting mainly of crystobalite which is mixed with water to form a white liquid looking rather

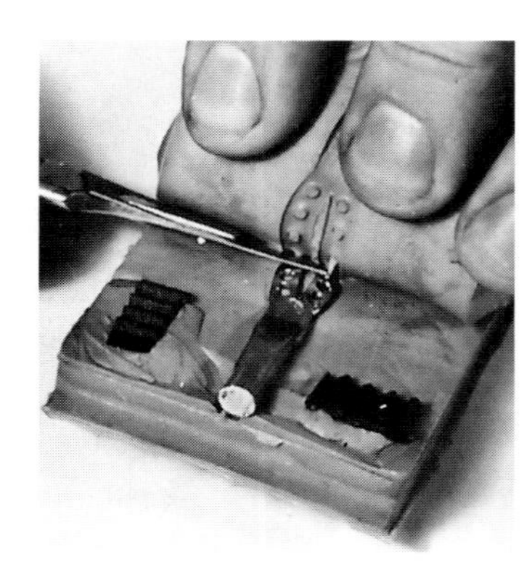

Figure 7.5(a) Lost-wax casting – making a hollow rubber mould

Figure 7.5(b) The rubber mould is placed over the nozzle of a wax injector to produce a wax replica of the original model

Figure 7.5(c) Wax replicas built up to form a 'tree' (by courtesy of the Mond Nickel Co. Ltd.)

Figure 7.5(d) An investment is poured round a wax tree in a metal cylinder. After heating in an oven the wax melts and is 'lost'

like and setting like liquid plaster of Paris, is poured into the flask. The wax tree is now totally invested in a solid cylinder (Figure 7.5(d)).

The next step is to lose the wax, the operation which gives the process its name. This is done by placing the investment in an oven, where the wax becomes liquid again and pours out of a hole in the base. The result is a hollow plaster mould. This is mounted on a centrifuge, which consists of a rotary arm mounted in the centre of a metal shield, and rotated by an electric or a clockwork motor. On one end of the arm is a small crucible made of refractory material, similar to that used to make firebricks, that will stand great heat without cracking. The crucible has a hole at its outside end. The investment is fixed alongside the crucible so that the hole in the crucible corresponds with the pour hole in the investment.

The quantity of gold needed for filling the investment is calculated, placed in the crucible, and melted with a blow-torch (Figure 7.5(e)). When the gold is molten, the centrifuge is rotated and the gold is flung by centrifugal force into the mould.

After cooling, the investment is broken away from the golden tree in its centre, and the final traces of investment are removed with hydrofluoric acid.

The individual gold replicas of the original model are now broken away from the tree and cleaned up. From this one investment alone a dozen or more perfect copies may result. Lost-wax casting is, however, a skilled operation. An inexperienced caster will probably produce many porous castings. Indeed even a skilled and experienced caster will turn out a proportion of castings that have to be scrapped because the metal contains bubbles which would either marr or weaken the finished piece.

Sometimes complicated designs have to be soldered up from a number of components made by lost-wax casting, but it is much easier to produce complicated

Figure 7.5(e) The cylinder containing the investment is placed in the centrifuge against a crucible into which gold is placed, and melted with a blow-torch

three-dimensional designs in this way than in the press. It is also quite simple to produce ring mounts for gem-set rings, complete with claw settings. A typical firm now makes 90 per cent of its gem rings by this process, and though they are slightly inferior in finish to the firm's hand-made mounts but they cost only a quarter of the price to make. The old techniques of sand casting and casting in cuttlefish still survive in the jewellery trade, but they are not much used nowadays.

It has been implied perhaps that manufacturing jewellers are faced with a simple choice of methods. They can, perhaps, use hand work, or press-tools, or employ lost-wax casting. In fact they can, and often do use all three of these methods to make one piece of jewellery. For instance, it is a common practice to use a lost-wax head and solder a drawn wire shank on to it to produce a gem ring. The drawn wire shank has more elasticity than has a cast one, and will better stand up to the stretching entailed in sizing it. If the ring is of a fancy design, a stamping could be incorporated in the head. Finally the skill of a traditional craftsman will have to be called upon to set it with a stone or stones.

Texturing

Texturing revolutionized jewellery a few years ago. Though it is currently less popular than it was, it will almost certainly come back in to fashion again. It is always hard to differentiate between cause and effect in these matters, but texture was either the result of a revolt against the bright and shiny, or it was the cause of this revolt. In the jewellery trade, texturing began on the Continent. The leading modern designers borrowed texture from the painters and sculptors, for whom it has always been as important as colour and form. After a few years the designers of Swiss watch cases borrowed the idea in their turn from the *avant garde* jewellery designers. Texturing made its first commercial appearance in the UK as a method of watch case decoration. Soon, texturing was found on almost every piece of jewellery coming on to the market. The texturing revolution came at a most convenient moment. Mass production was threatening to result in monotony. Texturing made it possible to make one pattern look like a dozen different patterns. It brought variety to the machine age.

Figure 7.6 Two bracelet watches illustrating various textured finishes on cases and bracelets

Texturing can be achieved in many ways. Wrinkled textures can be produced by heating gold up to a critical temperature and just letting the surface of the metal flow, or by letting molten gold fall on to a smooth gold surface and bond itself to it. The textile texture of shantung engraving can be produced by hand or by machine engraving. Bark finish can be hand cut or produced on a diamond milling machine or ground into the gold with dentist's drills.

Texturing can also be applied from the tools in a press, and of course if a hand-made model used for lost-wax casting is hand textured, this texturing will be reproduced on the surface of the jewellery resulting from the process.

The manufacturers of textured articles are inclined to be secretive about methods used to produce their finishes. Most textures can, however, be produced by fitting different tools into a dentist's drill. Perhaps, anyway, the precise method by which the scores of different finishes seen on jewellery are achieved is not important to the person behind the counter for whom this book has been written. If the technical intricacies were not important, however, the results produced were very important indeed. One result was to turn what once was rather dull merchandise into something interesting and therefore saleable. Gold watch bracelets once made a relatively unimportant contribution to a retail jeweller's turnover. Texturing turned this line into an exciting quick-turnover item before economic factors regretably reduced the demand for gold watches.

Diamond milling

What texturing did for the watch bracelet and the cuff-link, diamond milling did for the wedding ring. Diamond milling, as it is used for decoration on jewellery, is an alternative to engine-turning (Figures 7.7(a) and (b)). A diamond tool is brought to the work, and as a result of the way the milling machine is set up it produces a pattern of sharp bright cuts in the metal. The shape of these cuts depends on the size and shape of the diamond used, the angle at which it is presented to the work and the ratio of the gearing used.

Before diamond milling, the ranges of wedding ring manufacturers were very limited, and a retail jeweller might stock as few as half a dozen patterns. Today, the designs available from the manufacturers are counted in hundreds, and retailers carry much bigger ranges. More recently, textured wedding rings have also become popular. Originally the texture was cut into these, but they are now sometimes produced by casting the whole ring from a textured model. Currently plain wedding rings seem to be taking an increasing share of the market again, although patterned rings continue to sell in considerable numbers.

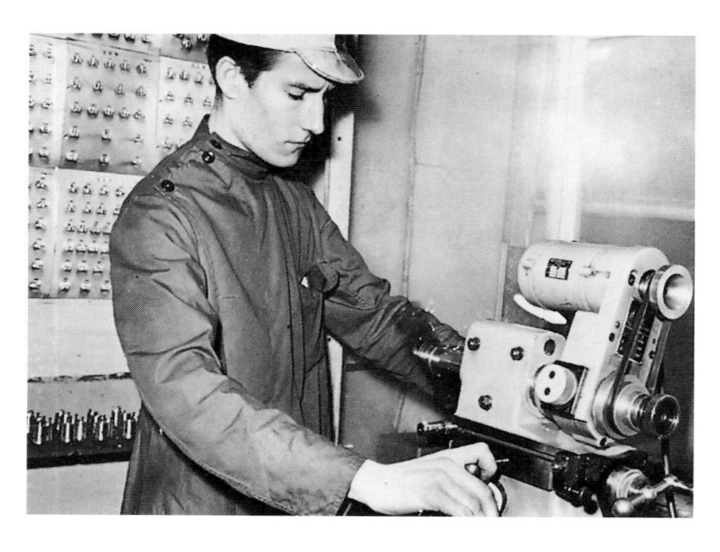

Figure 7.7(a) A diamond milling operator setting up the machine into which a diamond tool is fixed

Figure 7.7(b) Typical examples of diamond milling applied to wedding rings

Varying the finish is not the only way in which variety can be given to the ranges of manufacturers using mass-production techniques. By varying the stones in a standard mount, or building up stamped or lost-wax cast components in different ways, a range of jewellery can be achieved from a relatively few standard units.

Traditional decoration

Side by side with the new processes of decoration, traditional methods of decorating jewellery are used. Enamelling of various kinds is still used as a jewellery decoration. Niello wares, imported from the Far East, have been widely sold in recent years. These are produced by cutting a pattern into the metal and filling it with a black niello compound consisting of silver, copper, sulphur and borax. Hand engraving still provides an effective method of decorating jewellery, and inlaying patterns of coloured gold is another form of decoration. Tiny pieces of different coloured gold are fitted together like the pieces of a jig-saw and soldered on to a gold surface. Then attempts have been made from time to time to revive the ancient art of granular work.

As fashions go in cycles, all the techniques employed in the past will almost certainly be revived, even if they are currently out of fashion.

Chain

Chain has always been an important component of jewellery. The largest of the Pforzheim's jewellery firms all began as chain makers in the nineteenth century. Like most of the things sold by the jeweller chain can be made by man or machine.

To produce hand-made chain, wire is bent up on the bench to form the links, and the chain is laboriously built up link by link, each joint being subsequently soldered. It takes a chain maker about a day to link 85 gold links by hand to produce about 7 ft of chain.

By machine some 50 ft of similar chain can be produced in an hour. It is fascinating to watch the chain-making machines working away unattended. The wire is fed to the tools, which bend it to form a link, hold the link in position and form the next one through it. It all happens so quickly that the machine has to be slowed down for the process to be demonstrated.

There are many types of chain, too many to do more than mention the most common here. There is for example Milanese, an Italian chain as its name suggests, which consists of interwoven rows of links, like a miniature version of the chain-mail worn by the warriors who fought at the Battle of Hastings.

Brazilian chain is a round chain which consists of a series of linked cups that fit into one another to form a sinuous golden snake. For this reason it is sometimes known as snake chain. Brick bracelets consist of little squares or oblongs of gold cut from seamless tube of square section, which are wired together internally to produce a broad flat chain. There are also many variations on trace. These all have different names. Belcher has wide links;' curb has twisted links; alma has broad-ribbed links and fetter has long links. There are also a vast variety of chains produced consisting of mixed links.

Chapter 8

Hallmarks on gold, silver and platinum

Hallmarking

In the Middle Ages crafts were controlled by craft guilds who fixed prices and wages, insisted on reasonable working conditions and looked after the sick and needy members of the craft. They also ensured that journeymen were skilled in their craft before they would allow them to practise it, and they demanded that everything made by any member of the guild was up to standard. They maintained that bad workmanship or the use of poor materials by the individual could undermine the reputation of the whole craft.

In London

The Goldsmith's Guild of London was one of the earliest of these guilds to be formed. This was already in existence in Norman times, and by the thirteenth century it was 'a numerous and powerful body'. But not powerful enough, it would seem, or not sufficiently jealous of its reputation, to exercise sufficient control over its members. Nothing, of course, could have been easier in those days than to defraud the public by adulterating gold and silver with too much alloy, and the temptation proved too great for some of the members of the guild to resist. By 1238, things were getting serious, and 'as a result of numerous frauds perpetrated by some of the members of the craft', Henry III made an Order in Council, in which he commanded the Mayor and Aldermen of the City to choose six 'discreet goldsmiths' to superintend the craft. It was the successors of these six discreet members of the guild who were, in 1300, recognized by Edward I as the guardians of the craft, and entrusted with the job of assaying 'every manner of vessel of silver' before it was offered for sale. To show that they had carried out this assay, a word which derived from the French word *assai*, meaning to essay or to try, the guardians marked the pieces that had been through their hands with a punch showing a leopard's head. They were also required to assay wares of gold to see that these too were of the required standard.

The standard demanded for silver under the Act of Edward I, which gave these powers to the guardians of the craft, was 'of the sterling alloy or better, at the pleasure of him to whom the work belongeth'. Those who worked in gold had to 'work no worse gold than of the Touch of Paris', which was 19⅕ carat.

This then was the beginning of the practice of assaying and hallmarking, the oldest form of consumer protection in existence today. The old guild became the Worshipful Company of Goldsmiths and received its first Royal Charter from Edward III in 1327 and this body remains responsible for hallmarking in London.

Under the Act of 1363 every master goldsmith was required to stamp with their

Figure 8.1 Set of marks on a valuable piece of antique silver, sent to the London Assay Office in 1741 by Paul de Lamerie

own mark each piece of gold and silver that they made. From that date London-made silverwares began to bear two of the four marks found on modern pieces. One of these was the Warden's mark, the leopard's head, sometimes called the King's mark, the other the maker's mark. To begin with, makers' marks were usually in the form of heraldic devices, probably taken from the creaking signs that hung above the goldsmiths' shops. Initials became common during the latter half of the sixteenth century. Only later, at the end of the seventeenth century, did it become customary to use letters. Under a statute of William III of 1696, the first two letters of the maker's surname were used. Then from 1720 the goldsmith's initials were used. This is the type of maker's mark used to this day (Figure 8.1).

The practice of punching a silver or goldware with a separate mark in the form of a letter of the alphabet, to show in which year it was submitted for assay, began in 1478. Since the research carried out into the date letter system by Octavius Morgan in the 1850s, the existence of these date letters has been an invaluable aid to collectors and dealers in antique plate because they make it possible to ascertain at a glance exactly when a piece of plate was made.

The last of the four marks found on London silverwares made its first appearance in 1544. This is the quality, or standard, mark. It took the form of a lion standing sideways in a shield with its head turned to the left, the lion passant guardant of heraldry. It was probably borrowed with royal sanction from the Tudor arms.

Outside London

London was not, of course, the only city in medieval England where gold- and silverwares were made. In law, the London guild had jurisdiction over all the provincial goldsmiths, but in practice the local guilds exercised control locally, 'touching' and marking the plate made by their members. That section of the Act of 1300 that read 'and that all the good towns of England, where any goldsmith be dwelling, shall be ordered according to this statute as they of London be; and that one shall come from every good town for all the residue that be dwelling in the same unto London, for to be ascertained of their touch', was somewhat unrealistic in an age when lumbering carriers' wagons might take days to cover a few miles of mired tracks. A statute of 1378, in fact, placed the onus on 'the Mayors and Governors of the cities and boroughs' to carry out the 'assay of the Touch'. In 1423, the Statute of Henry VI appointed York, Newcastle-upon-Tyne, Lincoln, Norwich, Bristol, Salisbury and Coventry to have 'divers touches'. Many other towns, without this express permission from the Crown, marked the plate made by local craftsmen. Silverwares have survived which were marked in Taunton and Exeter, King's Lynn and Leeds, Leicester and

Lewes, Hull and Barnstaple, to name only a few of the towns where at one time or another goldsmiths have worked.

By the middle of the eighteenth century many of the old towns of England had lost their former importance. There was little work for the silversmiths and so the craft died out. At the same time in two newly important and growing towns, Birmingham and Sheffield, manufacturing silversmiths were opening factories and prospering.

These Birmingham and Sheffield smiths soon found that having no assaying and marking facilities locally made life very difficult. The nearest assay offices were those at London and Chester, and it took a long time for goods to travel to and from either of these offices in those days. The silversmiths of Sheffield, the more important silver town of the two, were said to be 'under great difficulties and hardships in the exercise of their trades for want of assayers in a convenient place . . .'. In Birmingham, the great Matthew Boulton was equally concerned. In 1771, several pieces of plate commissioned from him by the Earl of Sherbourne had not only been delayed on the 70-mile journey from Chester, but what was worse, 'the chasing was entirely destroyed by the wilful and careless packing of them at Chester'. This delay and damage, which led to the upsetting of an important customer, was the last straw for Boulton. He drew up and submitted a petition to Parliament on behalf of himself and the other silversmiths of Birmingham. He wanted, he said, to become a great silversmith, but it was useless, unless the powers can be obtained to have a marking hall at Birmingham.

Birmingham and Sheffield got their assay offices under an Act of 1773, despite opposition from the silversmiths in London and the Goldsmiths' Company of London, inspired either by fear for their own trade or concern lest the new provincial offices might not maintain standards. Soon the new offices were working on one or two days a week. The Birmingham Assay Master presided over four or five helpers in the King's Head public house once a week.

Of all the marking halls that have existed at one time or another during the past six and a half centuries, only three now remain in England. These are the offices in London, Birmingham and Sheffield. Scotland has only one assay office now, in Edinburgh (Figure 8.2). The last two offices to close down were Chester, in 1962, and Glasgow, in 1964 (Figure 8.3).

The duty on silver

Besides the four marks to be found on most silverwares made later than the middle of the sixteenth century, a fifth was used between 1784 and 1890 (Figure 8.4). This mark, in the form of a portrait of the reigning sovereign in profile, was struck to indicate that the duty payable under the Act of 1784, which was collectable by the assay offices on behalf of the Excise, had in fact been paid.

At an earlier date, between 1720 and 1758 a duty of 6d. an ounce had been applied to silverwares, but was avoided by many silversmiths. It was this duty-dodging which led, when the tax was reintroduced, to the use of a duty mark being made statutory. It was felt that this mark would make it easier for authority to check up on the duty dodgers. However, a new way was soon discovered of avoiding payment of this unpopular duty, at least on important pieces (see page 143).

The reigning sovereign's head in profile has appeared on silver three times since the repeal of the duty in 1890. The first time was when George V and Queen Mary were celebrating the twenty-fifth anniversary of their accession to the throne (Figure 8.5). To mark this occasion, silverwares made between 1933 and 1935 were stamped with a mark showing the King and Queen in profile. Between 1952 and 1954 gold and

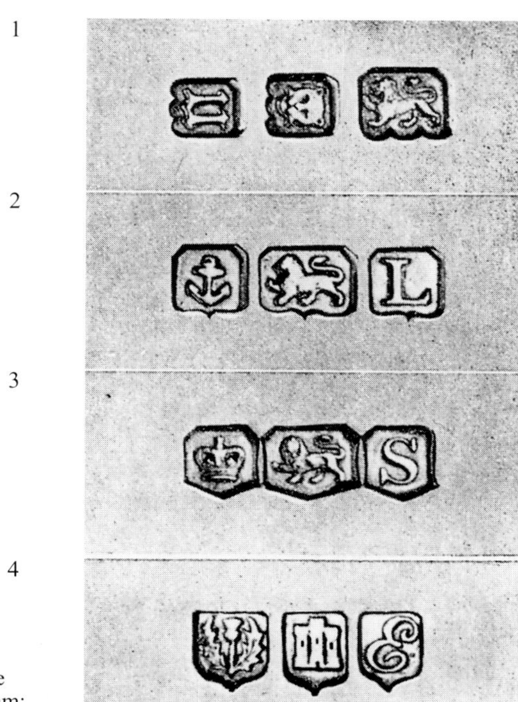

Figure 8.2 Marks found on silver made in the twentieth century: 1 – London; 2 – Birmingham; 3 – Sheffield; 4 – Edinburgh

Figure 8.3 The marks struck in 1935–1936 at the two assay offices which were closed in the 1960s: 1 – Chester; 2 – Glasgow

silverwares were stamped with a Coronation mark showing the profile of Queen Elizabeth II (Figure 8.6). In 1977 the Queen's profile was used again on wares weighing 15 g or over, to mark her Silver Jubilee.

The Britannia standard

Sterling silver (silver which is 925 parts in 1000 pure) has (until 1999) remained the minimum standard which has been legal to offer for sale in this country since 1300, with the exception of a period of 23 years.

Figure 8.4 Duty marks showing the head of the reigning sovereign in profile. Found on silver made between 1784 and 1890

Figure 8.5 The Jubilee mark showing the heads of King George V and Queen Mary, added to the Chester marks for 1935

Figure 8.6 Mark struck at the Sheffield Assay Office (1953) to celebrate the Coronation of Queen Elizabeth II

The great demand for silverwares from the Court of Charles II and the Cavalier courtiers who returned from France and the Low Countries after the Restoration, led many silversmiths to clip the silver shillings of the period. and even to melt them down in order to obtain enough raw material to carry out their commissions. This soon began to have serious repercussions. There was an unofficial devaluation of the shilling which increased the value of a guinea from 20 to 21s, and there was soon a great shortage of silver coins. Charles did nothing. James did nothing. Eventually, in 1697, Dutch William took action. He decided that the melting and the 'wicked and pernicious crime of clipping' had to stop. To discourage these practices he raised the minimum standard for silver plate so that it was higher than the standard for the silver coinage. In future, if the silversmiths continued to clip and melt shillings, they would be put to the considerable trouble of refining their ill-gotten silver if their wares were to pass an assay.

The new minimum standard, set by William III, was called the Britannia standard. Silverwares had to be 958.4 parts in 1000 pure. At the London Assay Office this new standard was marked with a lion's head erased (see Figure 8.7) and a seated figure of Britannia. The wares also bore, of course, the usual maker's mark, consisting of the first two letters of his surname, and the date letter.

Figure 8.7 Marks which indicate Britannia standard silver

The sterling standard was restored as the minimum standard in 1720, but even after that date a few pieces were made to the Britannia standard, and occasionally even today a silversmith will make a special piece to the higher standard and have it marked accordingly.

Unmarked plate

Not all goldsmiths and silversmiths who have worked in England during the past six and a half centuries have been honest. Some avoided the charges for assaying and marking, some evaded the duty, and some were fakers and forgers. Thus. silver with no marks, and silver bearing marks that are not what they purport to be, or do not belong to the piece on which they appear exists.

A lot of silver made during the sixteenth and seventeenth centuries bears no marks at all, or just a maker's mark. The reason is probably that the silversmiths misunderstood, or took advantage of, that part of the Statute of 1575 referring to silver being assayed and marked before being offered for sale. They assumed that this meant that commissioned work was exempted. The result was that by 1643 there was so much unmarked plate in the country that it was given an official price of 8d. an ounce lower than that for marked plate.

False marks

A number of different types of false marks are found on silver, those applied by the duty dodgers among them. With the introduction of a duty mark quite a number of silversmiths when producing a big piece seem to have hit on the idea of sending off a little piece weighing only a few ounces, to the assay office. They paid the small amount of duty and got back the piece complete with the duty mark. They then proceeded to cut out the marks from the small piece and apply them to the big piece, so saving the duty on perhaps a few hundred ounces of silver.

Then there are the marks on faked pieces. Fakes are sometimes made by casting, or electrotyping, from a genuine piece. When this is done the marks are transferred to the fake by the same method used to reproduce the form and decoration of the original. Other fakes are made by more traditional methods and are raised exactly as the piece being imitated would have been raised. If pieces are made in this way, fakers have a choice of two methods of producing a spurious set of hallmarks. They can make a fake set of punches, or take genuine hallmarks from a damaged piece or from a piece of little value of the right date. Forging or transposing hallmarks in this way are, of course, serious offences. They render the perpetrator liable to a maximum penalty of 10 years' imprisonment.

It is equally illegal to sell fakes, forged pieces, or pieces with transferred marks. It is also illegal to sell hallmarked articles that have been altered and not regularized by an assay office. Therefore, anyone selling silverwares must he on the lookout for anything suspicious. One thing that reveals a fake to the expert is decoration that is wrong for the period indicated by the mark. This is more common than might be expected because few fakers seem to have any historical sense. There are also the technical clues, such as signs of soldering round a hallmark, that can often be brought out by breathing on the silver. Blow-holes near marks suggest casting. An exact similarity of the placing of marks on a pair of spoons or casters suggests that one or both may be cast or electrotype copies. Silver was usually marked with four punches individually applied, and so the chance of an exact repetition of the relative positioning of the marks is most unlikely.

In addition to the true fakes, there have been many attempts to deceive the public into believing that they were buying silver when they were not. In writing about pewter it was mentioned that the Goldsmiths' Company was forced to take out an injunction preventing the pewterers from simulating hallmarks on their wares. There have been a number of attempts by manufacturers of electroplated wares to get away with marks which look at first glance exactly like a set of silver marks. Not infrequently American firms have blatantly copied English hallmarks.

Hallmarks and the law

Lawyers faced with the various Acts that used to regulate assaying and hallmarking have described them as 'complicated', 'confusing' and 'archaic'. The Hallmarking Act of 1973, which came into force on January 1, 1975. made things a good deal easier for lawyer and layman alike. This Act repealed the previous Acts, and removed most of the anomalies. An extension of the 1973 Hallmarking Act took place on January 1, 1999, in order to comply with the ruling in the European Court of Justice 1994 Houtwipper case. This required member countries to amend their systems. It meant that Britain needed to accept those marks put on by other European countries which have independent hallmarking systems and satisfactory assaying procedures, and whose marks are intelligible to British consumers, without submitting them for further assay. No further hallmarks are necessary.

These changes included a list of new standards for gold, silver and platinum. Gold standards are now 375, 585, 750, 916, 990 and 999 parts per 1000 fine. Silver standards are now 800, 925, 958 and 999 parts per 1000 fine. Platinum standards are 850, 900, 950 and 999 parts per 1000 fine. The marks have also been divided into those that are compulsory and those that are voluntary. There is now a Convention punch mark for 800 standard quality silver.

During 1999 Birmingham Assay Office assayed wedding rings for one manufacturer with the fineness standard of 990 parts per 1000 fine.

The legislation has no effect, of course, on the validity of the marks struck in the past, and anyone working in the jewellery trade, who will be dealing both with current production and with jewellery and silverwares produced before January 1, 1975, has to be familiar both with the modern and the older legislation. Retail jewellers need not, however, concern themselves with all the intricacies of the law. They must, however, avoid any breach of that section of the law that reads: 'Any person who in the course of a trade or business (*a*) applies to an unhallmarked article a description indicating that it is wholly or partly made of gold, silver or platinum, or (*b*) supplies or offers to supply an unhallmarked article to which such a description is applied, shall be guilty of an offence.'

The law does not require every gold- and silverware to be hallmarked. Certain wares are specifically exempted under the latest Act, though the list of exemptions under this Act is somewhat different from that in force prior to its introduction. The details of the current exemptions are listed in Appendix 2. Today, many manufacturers ignore these exemptions, because they find it to their advantage to have such wares voluntarily assayed and marked, so they can be advertised as hallmarked.

The main tenets of the 1973 Act are that:

1 Platinum, which previously did not have to be hallmarked, now comes within the hallmarking legislation. The legal standard for platinum is 850, 900, 950 and 999 parts in 1000 pure (new standards).
2 Wares of gold and silver below the legal standard or unhallmarked wares, which under the previous legislation could not be offered for sale, may now be legally sold provided they are not described as 'gold' or 'silver'.
3 Changes have been made in the list of exemptions from hallmarking.
4 A Hallmarking Council has been set up, whose principal job is to keep the law continuously under review, and recommend such changes as are desirable.

It is important for the jeweller in dealing with customers' possessions to bear in mind the wording of the law, that the offence is not to possess but to sell or offer for sale in the course of trade or business. There is no reason why customers should not continue to enjoy the use of unmarked wares that they may have inherited or acquired from abroad. If a dealer wished to sell unmarked wares as gold, silver or platinum these would of course, have first to be submitted for assay and marking unless they were in an exempt category. There is the possibility with foreign wares that they might not pass the stringent tests of a modern British assay office. This would not mean, however, as many people believe, that they would be destroyed. Substandard articles that are not new are not 'broken' by the assay office but are returned to their owner unmarked.

Only if a piece bears counterfeit hallmarks is the mere possession of the piece an offence against the law. It is also illegal to deface marks without the consent of an assay office, and the character of wares may not be changed, or additions made, without the prior approval of an assay office. It is also an offence to deal in wares that have been altered without the consent of an assay office.

Solders and adhesives

Specific finenesses are laid down for the solders which it is permissible to use with silver, platinum and the various carat golds. A fineness of 650 parts per 1000 is stipulated for silver, 950 parts per 1000 for platinum, 750 parts per 1000 for 22 and 18 carat gold and so on. From November 1986 it became permissible to use a solder of only 500 parts per 1000 for 14 carat white gold, although for 14 carat yellow gold the acceptable standard remains 585 parts per 1000. From November 1986 non-metallic adhesives became permissible as an alternative to solder.

Silver plate or silver plated

Because of the misuse of these descriptions the Joint Committee of the Assay Offices of Great Britain has issued the following notice:

'For many centuries the words "silver plate" and "gold plate" have been used to denote wares made of silver or gold. These words have been used in that sense over the years in many of the hallmarking statutes. Under the Hallmarking Act 1973 it is an offence (subject to certain exemptions and permitted descriptions) to apply, in the course of trade or business, to an unhallmarked article a description indicating that it is wholly or partly made of gold or silver. "Silver" and "gold" are permitted descriptions if qualified by the word "plated". It follows that base metal wares coated with silver or gold by electro-deposition may not be described as "silver plate" or "gold plate" but may be described as "silver plated" or "gold plated". The words "silver plate" and "gold plate" may only be used to describe wares which can lawfully be described as silver or gold under the Hallmarking Act 1973.'

The marks on gold, silver and platinum

The quality mark on sterling silver

The leopard's head was used originally to denote that both silver and gold were up to standard, and was sometimes called the King's mark. However, the lion passant guardant described as Her Majesty's Lion is found in an indictment of 1597. For many years now all the English assay offices have used either the lion passant or the lion passant guardant to denote that a piece of silver has been assayed and found to be up to the sterling standard. At first the lion passant guardant, which the London office used, was crowned, but the crown disappeared in 1550. From 1821 the London lion no longer looked to the left. It became simply a lion passant.

As reference to *Bradbury's Book of Hallmarks* will show, the shape of the shield containing the lion has changed through the ages, as indeed have the shapes of the shields containing all the assay marks. This is a useful aid when dating pieces with worn marks.

Birmingham used a lion passant guardant as the quality mark on sterling silver from the opening of the office in 1773 until 1875, and then changed to the lion passant. Sheffield used the lion passant guardant until 1975, and so did Chester until the office there closed, though some early Chester pieces are found with the word 'sterling' in a shield. Up to 1839 the lion passant guardant was accompanied by a leopard's head crowned.

The Scottish offices used their own standard marks. Edinburgh used a thistle from 1759 to 1975, but has now adopted a lion rampant. The Glasgow office used a thistle and a lion rampant from 1914 until it closed in 1964. Before that a lion rampant only had been used, and earlier still, before 1800, there had been no separate quality mark at all.

The town mark

The leopard's head has been used by London as its office mark since it ceased to do double duty as an office mark and quality mark in the sixteenth century. At first the leopard was a noble beast bearing a crown, but he lost his crown in 1821. Since then he has looked more like a teddy bear than a beast of the jungle.

Birmingham has used an anchor as its office mark since its establishment. Sheffield had always used a crown on silver until 1975, but has now adopted the York rose previously used only on goldwares. Chester used the arms of the city, a shield bearing three wheat sheaves, and until 1839 a leopard's head was also punched into sterling silver marked at this office – a crowned leopard until 1822, and thereafter uncrowned. Chester's coat of arms were different during the period 1701–1778.

Edinburgh uses a castle as its office mark and has always done so. The distinctive three-turreted castle was introduced in 1617 and has appeared in slightly different forms ever since. Glasgow used its City Arms, a most complicated heraldic device consisting of a tree, a bird, a bell, a fish and a ring.

The date letter

The date letter system, although it provides a unique method of dating antiques, was complicated in the past by three facts. First, rarely were all the letters of the alphabet used, and the various offices were not consistent about which ones they omitted. Then the different offices started the cycles of letters in different years, and introduced their new date letter at differing times of the year. They also used their own styles of lettering in their own styles of shield. The only hope for anyone, without a photographic memory, to find their way around in this gloriously inconsistent world of the date letter on pieces made prior to 1975 is to refer to *Bradbury's Book of Hallmarks*, or the tables in Jackson's *Silver and Gold Marks of England, Scotland and Ireland*. In doing this, care must be taken to check the office mark first, and to see that the letter and the shield really do agree with those in the tables. Positive identification can be difficult if the marks on a piece are badly worn.

None the less, it is perhaps well for those employed in the trade to understand the way in which the various offices organized their date marks system. The following is a simple guide.

London used 20 letters, A to U, omitting the J which in some alphabets could easily be confused with an I. Formerly, London changed from one date letter to another on May 19, the birthdate of St Dunstan, the patron saint of the goldsmiths. In 1660 it was decided to mark the triumphal return of Charles II to London after his exile by changing the date letter on May 29.

Birmingham generally ran a 25-year cycle of letters, starting with A and going right through to Z, but omitting I or J. Sheffield did the same, but though the two offices began marking in the same year the cycles did not coincide. Sheffield's system went haywire for some reason at the start. For example, in 1797 an X followed Z, and then came V, E, N, H, and so on. Chester also had a 25-year cycle and always omitted J.

Edinburgh, except for two periods, used a 25-year cycle from 1681, also leaving out J. Between 1806 and 1832, however, Edinburgh used a 26-letter alphabet. In the period from 1882 to 1906 a 24-letter alphabet was used, with U as well as J being omitted in that series. Glasgow used all 26 letters of the alphabet from 1819 onwards. Its earlier system is very difficult to follow because the office in those days seems to have been particularly addicted to the letter S, and sometimes used it in different forms year after year. At the time this office closed it was using a Celtic alphabet with only 18 capital letters. This cycle had started in 1949.

London changed its letter in May. Birmingham's and Chester's letter was altered on July 1. Sheffield changed its letter on the day following the office's annual general meeting, which is held in July. Edinburgh introduced the new letter on the day following the third Thursday in October, and Glasgow used to make the change on the first Monday in July.

Since January 1975 this chaotic situation has been rationalized. Now all the offices using the same letter in the same year change the letter on January 1

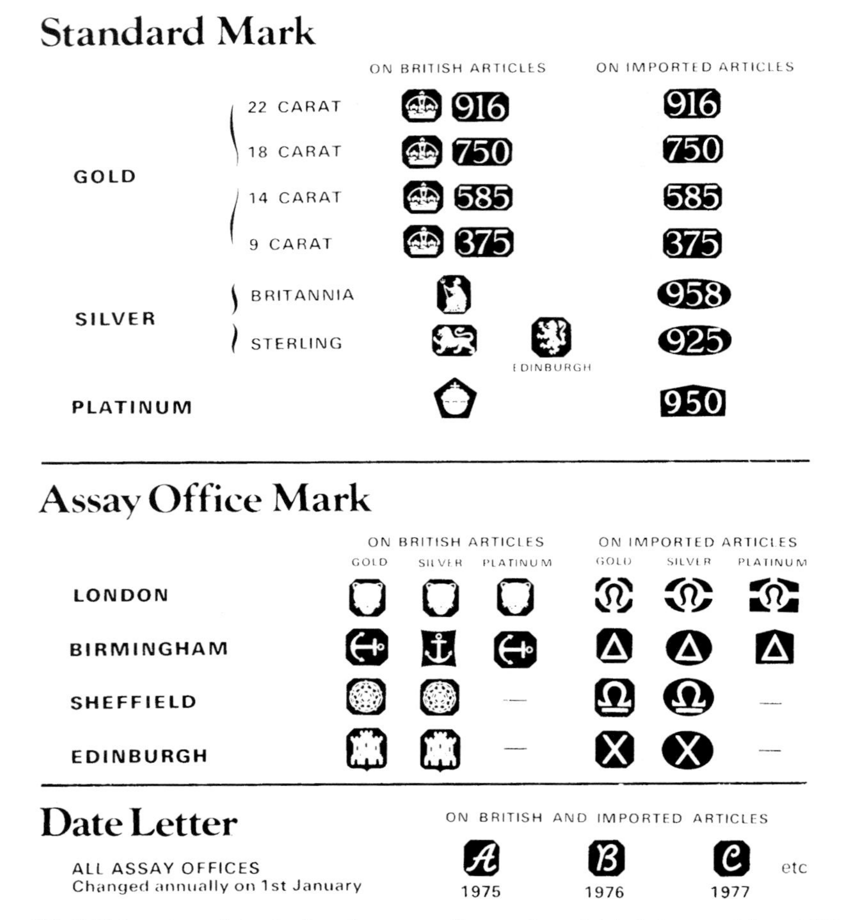

Figure 8.8 British system of standard marks, assay office marks and date letters used from 1975 to 1999

each year. The offices all began a new cycle in 1975 using a script capital alphabet (see Figure 8.8).

The sponsor's (maker's) mark

As a glance through Jackson's invaluable guide shows, the goldsmiths in the early days of marking put some very attractive marks on their wares before sending them to the Hall to be assayed. But nowadays the maker's mark, renamed the sponsor's mark under the new legislation, is more mundane. Gone are the crossbows and the fish, the lambs' heads and the eagles. Today, initial letters are used – those of the firm's name, or those of the individual smith submitting the ware.

The sponsor's mark is registered with the office, and is stamped on a plate by the 'maker' so that the office can check who has sent in a particular piece.

The term 'maker's mark' was obviously a misnomer. Anyone can register a mark,

Plate 1 Pink YAG (yttrium aluminium garnet) spectrum

Plate 2 Pink CZ (cubic zirconia) spectrum

Plate 3 Natural zircon spectrum

Plate 4 Quartz citrine

Plate 5 Zircon

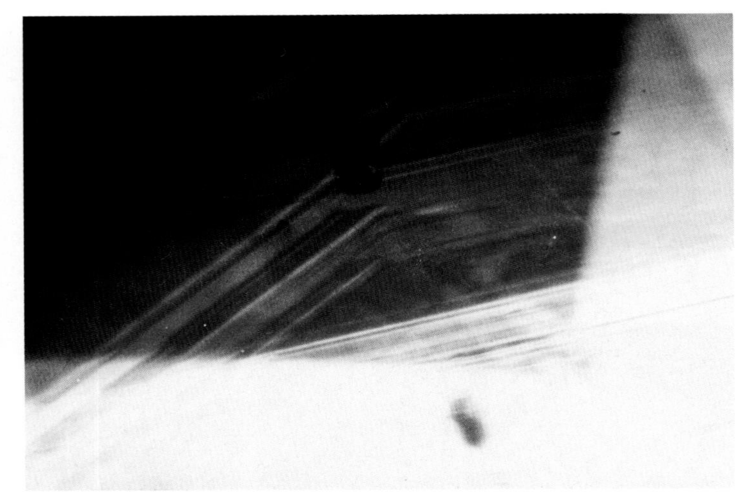

Plate 6 Natural sapphire

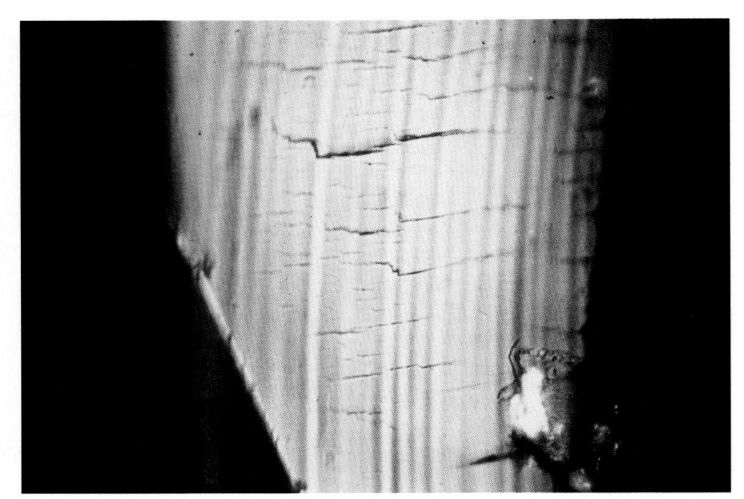

Plate 7 Synthetic sapphire

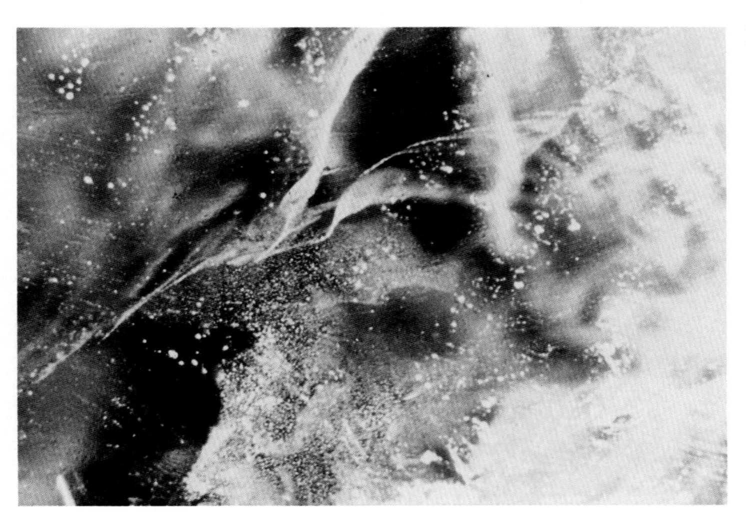

Plate 8 Synthetic emerald

and wholesalers and retailers often have wares that have been made for them marked with their own mark by the manufacturer. It is not unusual for a firm of manufacturing silvermiths to put its own name on only 50 per cent of the silver it produces, while putting the marks of its customers on the rest.

The marks on Britannia silver

The Britannia standard, the higher standard introduced to discourage the clipping and melting of the silver coinage, was the minimum legal standard from 1697 until 1720. After 1720 Britain reverted to sterling as the legal minimum, but silver to the higher standard continued to be made, and indeed is still made to this day. Silversmiths, when they are making a reproduction of a Queen Anne piece, for instance, sometimes use Britannia silver so that the marks, except for the date letter, will be similar to those on the original.

From the beginning it was necessary to differentiate between Britannia standard and sterling standard silver by the punching of distinguishing marks. So, in 1697, a new mark, showing the seated figure of Britannia with her shield and her spear, was struck on English silver for the first time and has been used ever since by the London Assay Office to indicate that silver has been assayed and found to be at least 95.84 per cent pure. When offices were opened in Birmingham and Sheffield they also adopted the Britannia mark as a quality mark for the higher standard, and so did Chester up to the time the office closed. Edinburgh added the Britannia to its usual marks as did Glasgow when that office was operating.

To further distinguish Britannia from sterling, London used a different town mark for Britannia, substituting a lion's head erased, that is to say a lion with a ragged neck, for the leopard's head. Chester used the lion's head erased and the Britannia mark, but retained its normal town mark, the wheat sheaves in a shield from 1778 until it closed in 1962.

From January 1975 the seated Britannia has been used as the standard mark by all assay offices, and each uses its normal town mark (Figure 8.8). The lion's head erased mark, formerly used by the London office, has been dropped.

The duty mark

Most gold- and silverwares made between 1784 and 1890 should bear a mark depicting a portrait in profile of the head of reigning sovereigns. The presence of the profile of George III, George IV, William IV or Queen Victoria, indicates that the duty payable under the Act of 1784 had been paid. Edinburgh, however, stamped no duty mark on goldwares below the 18 carat standard.

There exist some pieces marked in Sheffield bearing two portraits instead of one. This was because when the duty was raised those in charge of that assay office decided, for a six-month period in 1797, to use a double mark to show that the higher duty had been paid. After this little burst of Sheffield individualism the office reverted once more to normal practice.

The marks on gold

The hallmarking of goldwares was complicated because there were four legal standards, 9, 14, 18 and 22 carat gold (Figures 8.9, 8.10, 8.11). The two higher standards were marked until 1975 in a different way from the lower. The 18 and 22 carat gold-

Figure 8.9 Standard marks used by various assay offices for 22 carat gold before 1975

Figure 8.10 Standard marks used by all assay offices for 14 and 9 carat gold before 1975

Figure 8.11 Marks for the discontinued 15 and 12 carat standards

wares assayed and found to be up to standard were marked with a number indicating the carat quality and with an additional mark.

The English offices used a crown with the figures 18 or 22. The crowns used by the various offices were slightly different in detail one from another. The Scottish offices did not use a crown. Edinburgh used a thistle, and Glasgow used a thistle together with a lion rampant. The use of the standard mark for 18 carat was authorized by an Act of 1798, and for 22 carat by an Act of 1844. The assay offices added the usual date letter and town mark to these marks, and, of course, makers had to stamp their marks on the piece before submitting it.

The Sheffield Assay Office, which started to mark gold only in 1904, realized that the use of its normal town mark, a crown, on goldwares would be confused with the quality mark, so it used the Yorkshire rose as a town mark on gold (Figure 8.12).

The lower standards of gold were also marked with two quality marks – a figure in a shield denoting the standard of the gold, and a second shield bearing a set of figures indicating the parts per thousand of gold in the alloy. The figures ·585 appear on 14 carat gold, and the figures ·375 on 9 carat gold.

At varying periods gold of other standards has been legal. From 1854 to 1932, for instance, both 12 and 15 carat were legal. They were marked in the same way as 14 and 9 carat before 1975, but with the appropriate carat and parts per thousand figures in the two shields.

From 1477 to 1575 the minimum legal standard was 18 carat. Then for 223 years to 1798 it was 22 carat. From 1798 to 1854 the minimum standard was lowered to 18 carat again.

Figure 8.12 The rose, the alternative Sheffield town mark used on goldwares only, before 1975

Figure 8.13 Mark (not strictly a hallmark) used from 1942 to 1952 on 9 carat plain wedding rings weighing not more than 2 dwt

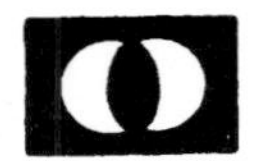

It was in 1854, because of a growing demand from the watch manufacturers, that gold of a standard lower than 18 carat was first allowed. That year saw the introduction of 9 carat as the minimum standard for goldwares permissible in this country. Since that date it has been legal to sell 9 carat goldwares, but until 1975 it was illegal to sell goldwares of a quality lower than 9 carat, whether described as gold or not (Figure 8.13).

During the period 1975–1998 all the offices used the same two standard marks on goldware – a crown in a square with cut corners, and a set of figures in an oblong with cut corners denoting the gold content of the piece in parts per thousand (see Figure 8.8).

The marks on platinum

With the advent of platinum hallmarking a new quality mark had to be devised, and this mark depicted from 1975 to 1998 an orb (see Figure 8.14). The mark was punched by the various offices together with the office mark, the date letter and the sponsor's mark.

Figure 8.14 Marks used on platinum by the London Assay Office in 1975. The lion passant, Britannia, the crown and the orb are now optional marks

Summary of the hallmarking changes introduced on January 1, 1999

(This section has been reproduced with the permission of the British Hallmarking Council)

Figure 8.15 shows some examples of the types of marks which are being stamped on articles of gold, silver and platinum in the UK.

Modifications to the Hallmarking Act, effective from January 1, 1999, have changed the way articles made of precious metal are hallmarked. The UK hallmark now comprises a minimum of three compulsory symbols as follows (see Figure 8.16):

Sponsor's or maker's mark indicates the maker or sponsor of the article. In Britain, this mark consists of at least two letters within a shield, and no two marks are the same.

Metal and fineness (purity) mark indicates the precious metal content of the article, and that it is not less than the fineness indicated. Since 1999, all finenesses are indicated by a millesimal number (i.e. 375 is 9 ct). This number is contained in a shield depicting the precious metal.

Assay office mark indicates the particular assay office at which the article was tested and marked. There are now four British assay offices – London, Birmingham, Sheffield and Edinburgh.

ASSAY OFFICE MARK
The mark of the Assay Office where the piece was tested

DATE MARK
A letter representing the year in which the piece was hallmarked

COMMEMORATIVE MARK
Struck on the occasion of a special event

Figure 8.15 UK hallmarks

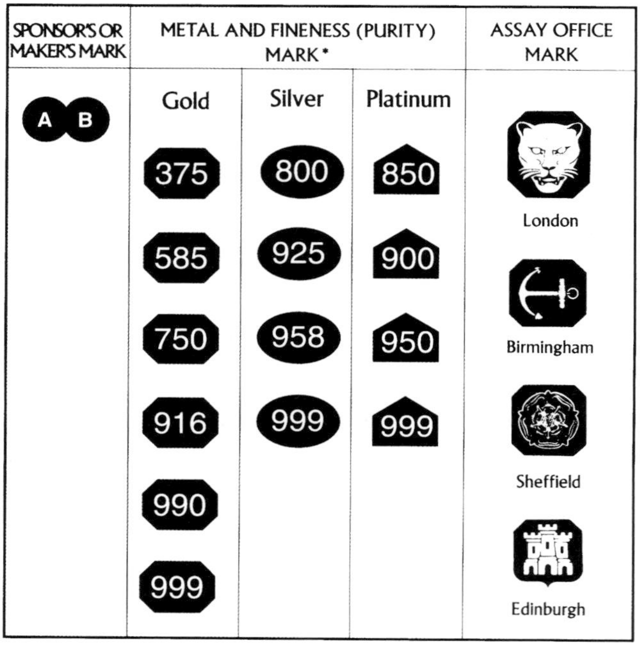

SPONSOR'S OR MAKER'S MARK	METAL AND FINENESS (PURITY) MARK*			ASSAY OFFICE MARK
	Gold	Silver	Platinum	
A B	375	800	850	London
	585	925	900	Birmingham
	750	958	950	Sheffield
	916	999	999	Edinburgh
	990			
	999			

*The Hallmark guarantees that the purity of the metal is at least that indicated by the Fineness Number.

Figure 8.16 Compulsory marks

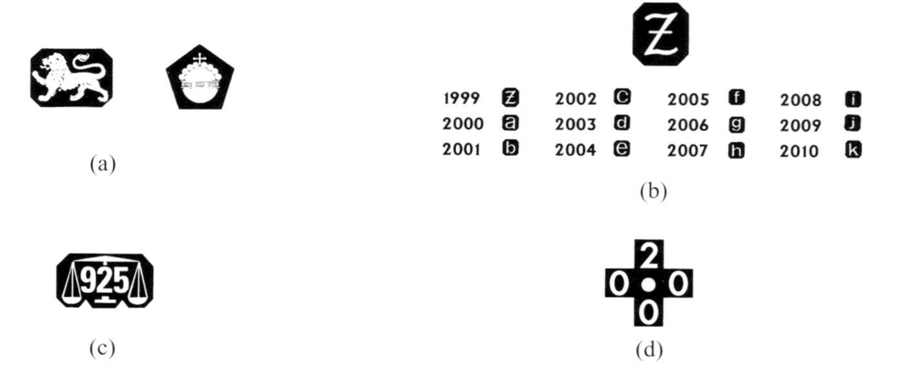

Figure 8.17 Voluntary marks: (a) traditional fineness (purity) mark; (b) common control mark; (c) date mark; (d) commemorative mark

925

FINENESS (PURITY) MARK

Figure 8.18 Convention mark

Voluntary marks may also be found on articles of precious metal (Figure 8.17). The traditional fineness (purity mark) was used to indicate silver and platinum finenesses prior to 1999. The **common control mark** is used by countries which are signatories to the International Convention on Hallmarks. The date mark, a letter indicating the year of hallmarking, was compulsory before 1999, and can be applied voluntarily in addition to the compulsory marks. One example of a commemorative mark is the Millennium Mark, which will be applied to precious metals by the four UK Assay offices during 1999 and 2000.

The UK has been a signatory to the International Convention on Hallmarks since 1972. This means that UK assay offices can strike the Convention Hallmark, which will then be recognized by all member countries in the International Convention. Conversely, Convention Hallmarks from other member countries are legally recognized in the UK. Articles bearing the Convention Hallmark do not have to be re-hallmarked in the UK. An example of a Convention Hallmark is given in Figure 8.18.

The assay office marks of member countries of the Convention are illustrated in Figure 8.19. The shield design around the Assay Office Mark sometimes varies according to whether the article is gold, silver or platinum. The key mark to look for is the Common Control Mark. The three other marks must also be present.

Historic UK hallmarks are illustrated in Figures 8.20 and 8.21.

Marks stamped on imported wares

All platinum, gold and silverwares imported into Britain are subject to the same legal requirements as British-made articles. Hallmarking of foreign articles began in 1842. From 1883 until 1904 the letter 'F' for foreign was added in an oval. Following the passing of the Foreign Plate Act of 1904, the various offices began to use their own special marks for foreign wares (see Figure 8.22). These were placed in different-shaped surrounds depending on the metal to be marked. Gold was marked with an oblong shield with cut corners, silver with an oval, and platinum with a pentagonal surround shaped like the façade of a building.

At first London used a Phoebus, Sheffield two arrows crossed and Glasgow a bishop's mitre, but after two years London and Sheffield adopted their present marks. Today, London uses the sign of the constellation of Leo, Sheffield the constellation of Libra, Birmingham an equilateral triangle and Edinburgh the Cross of St Andrew. Glasgow used two letter 'F's on top of each other, and Chester used an acorn and oak leaves.

The quality marks placed on foreign wares consist of figures indicating the various standards of gold and silver (see Figure 8.8). To these two marks are added the date letter and the sponsor's mark.

Jewellery made from a combination of platinum and gold has become popular all over the world and has led the trade in the UK to seek to have the 1973 legislation

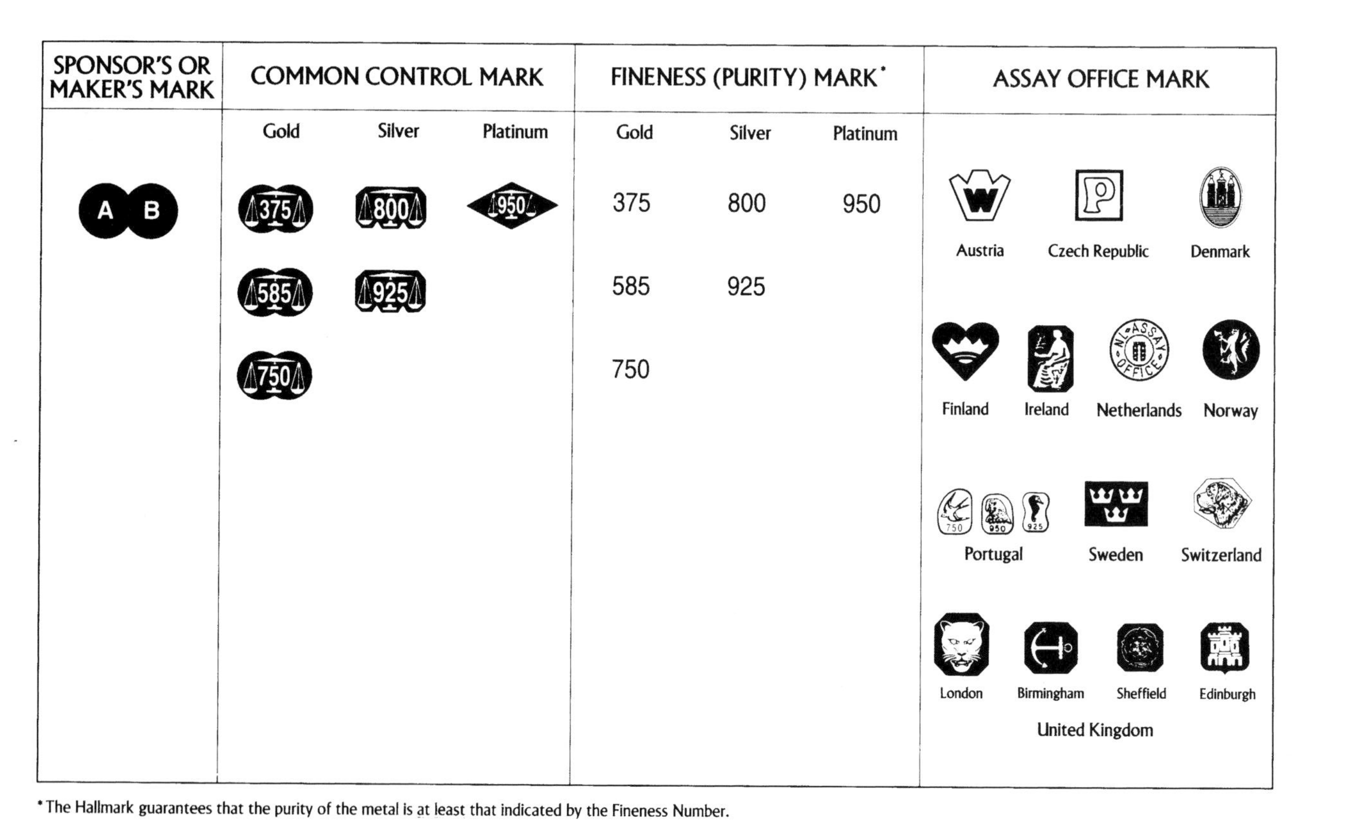

Figure 8.19 Assay Office Marks of member countries of the Convention

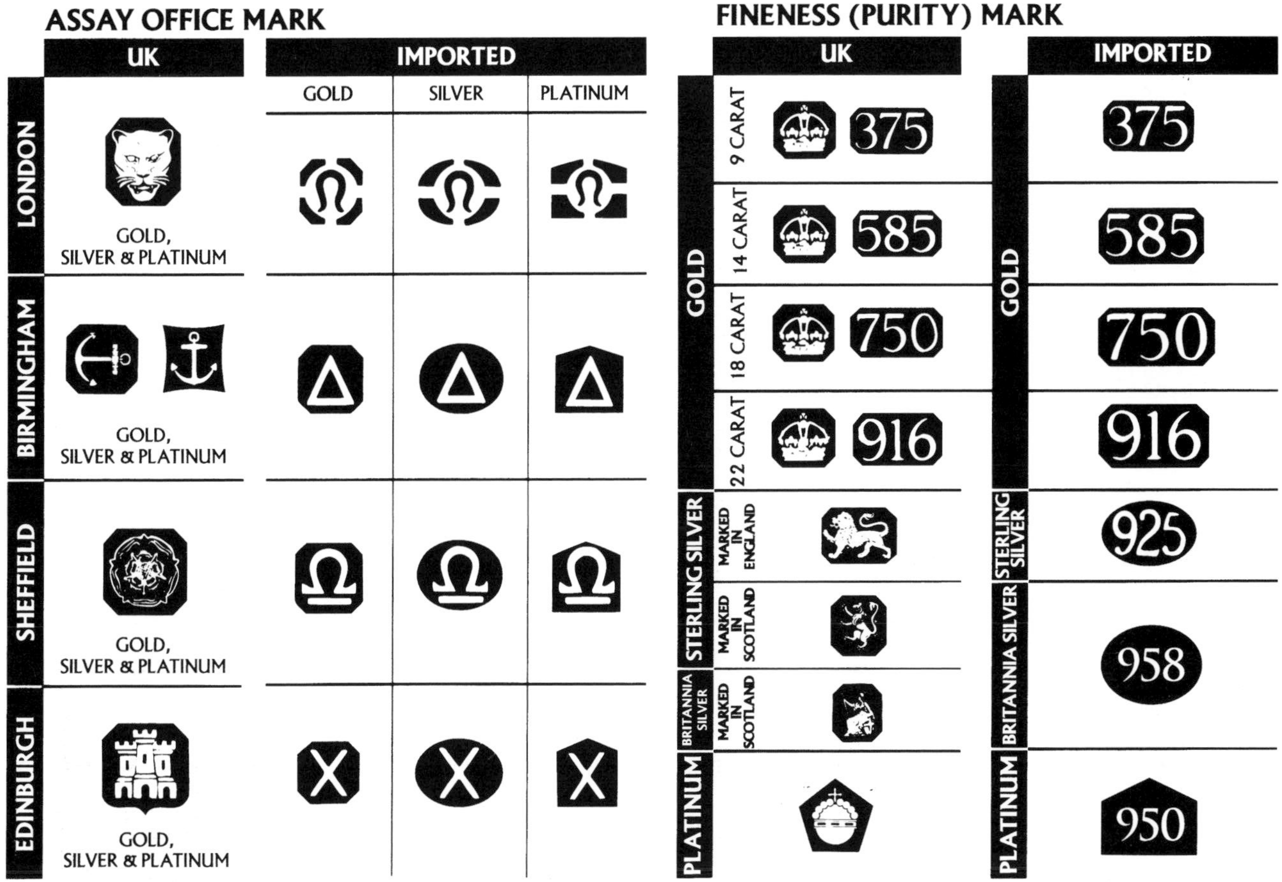

ASSAY OFFICE MARK

	UK	IMPORTED GOLD	IMPORTED SILVER	IMPORTED PLATINUM
LONDON	GOLD, SILVER & PLATINUM			
BIRMINGHAM	GOLD, SILVER & PLATINUM			
SHEFFIELD	GOLD, SILVER & PLATINUM			
EDINBURGH	GOLD, SILVER & PLATINUM			

FINENESS (PURITY) MARK

UK			IMPORTED	
GOLD	9 CARAT	375	GOLD	375
GOLD	14 CARAT	585	GOLD	585
GOLD	18 CARAT	750	GOLD	750
GOLD	22 CARAT	916	GOLD	916
STERLING SILVER	MARKED IN ENGLAND		STERLING SILVER	925
STERLING SILVER	MARKED IN SCOTLAND		BRITANNIA SILVER	958
BRITANNIA SILVER	MARKED IN SCOTLAND			
PLATINUM			PLATINUM	950

Figure 8.20 UK hallmarks, 1975–1998

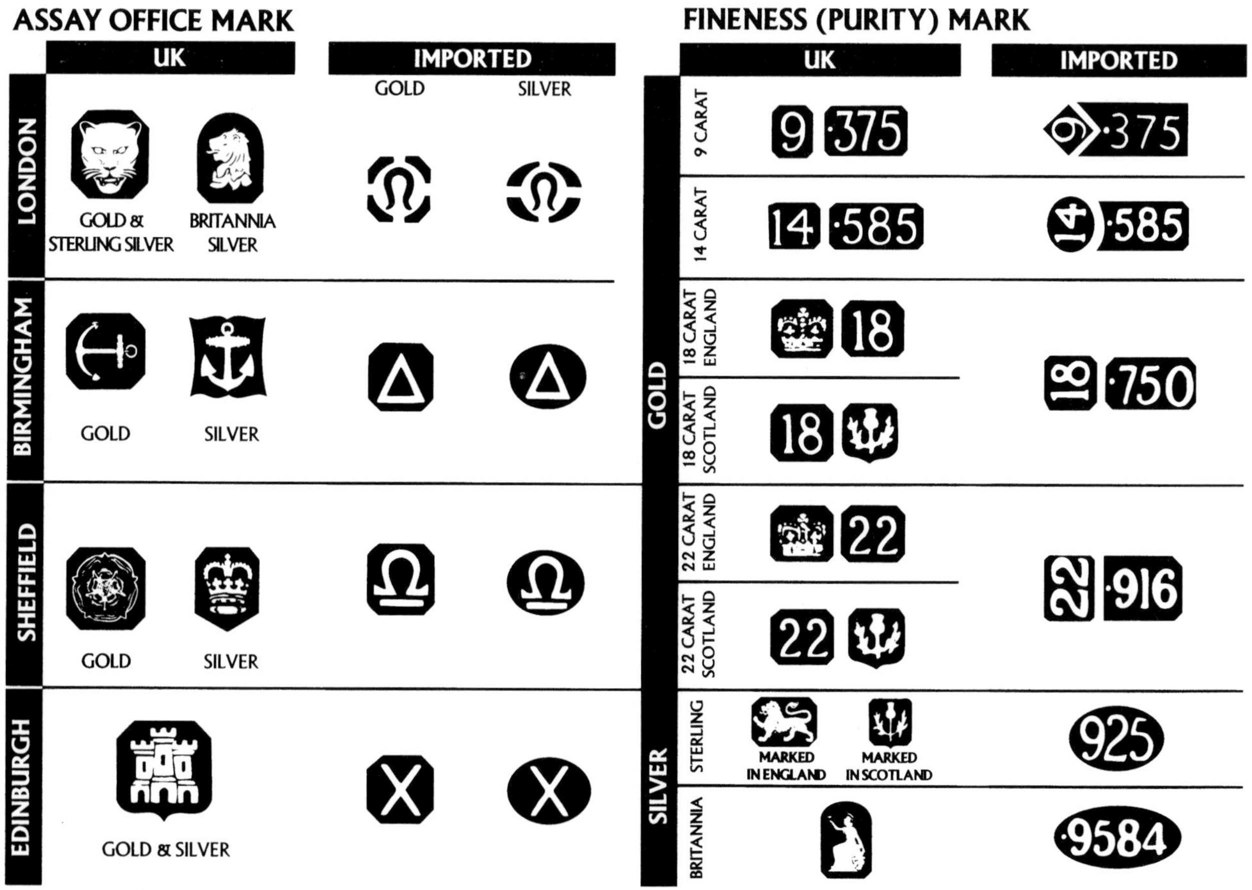

ASSAY OFFICE MARK

UK | IMPORTED (GOLD, SILVER)

LONDON: GOLD & STERLING SILVER; BRITANNIA SILVER

BIRMINGHAM: GOLD; SILVER

SHEFFIELD: GOLD; SILVER

EDINBURGH: GOLD & SILVER

FINENESS (PURITY) MARK

UK | IMPORTED

GOLD

9 CARAT: 9 ·375 | 9 ·375

14 CARAT: 14 ·585 | 14 ·585

18 CARAT ENGLAND: 18 | 18 ·750

18 CARAT SCOTLAND: 18

22 CARAT ENGLAND: 22 | 22 ·916

22 CARAT SCOTLAND: 22

SILVER

STERLING: MARKED IN ENGLAND; MARKED IN SCOTLAND | 925

BRITANNIA: | ·9584

Figure 8.21 UK hallmarks before 1975

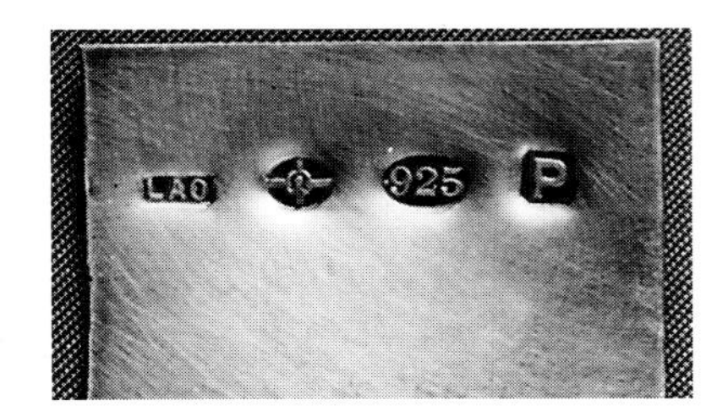

Figure 8.22 Marks struck by the London Assay Office on a piece of imported sterling silver

amended. This was carried through in 1975. Prior to this the legislation was based on the concept that platinum was more precious than gold and that if the two metals were used in conjunction the piece would be marked as though it were fabricated from the less precious of the two metals, i.e. the gold. Only if the platinum content exceeded 50 per cent by weight would the piece be considered to be an article of platinum. The manufacturers found this to be too restrictive.

Under the amended legislation any amount of gold and platinum can be used in the making of a piece of jewellery, providing that the gold is yellow in colour and of the 18 carat standard. Such pieces will be marked for both metals whenever it is practical to do so. If the platinum in the piece represents more than 50 per cent of the total, the platinum will be marked with a full set of platinum marks and the gold with the relevant quality mark only. If the gold represents more than 50 per cent by weight then this will be marked with the full marks for 18 carat gold and the platinum will be marked with the platinum quality mark only.

Irish marks

The Dublin Assay Office was set up in 1637 as a result of a Royal Charter of Charles 1. In 1637 a crowned harp was added by this office to the maker's mark, and in 1638 a voluntary date letter was introduced. After the passing of the Tillage Act of 1729 the Hibernia mark (Figure 8.23), not dissimilar to the Britannia mark on English silver, was used to denote that the duty levied under the Act had been paid. From 1807 the English practice of using a profile of the reigning sovereign was followed in Dublin, but the Hibernia mark was retained as a town mark.

Before 1784, 22 carat was the legal gold standard in Ireland, and was marked with the crowned harp. From 1784, 20 carat and 18 carat gold became legal. Twenty carat gold was marked with a plume of three feathers and 18 carat with a unicorn's head erased. In 1854, 15, 12 and 9 carat gold were permitted, and were marked by the maker with a number to indicate the carat standard of wares (Figure 8.24). From 1935

Figure 8.23 The Hibernia mark

22 carat

20 carat

18 carat

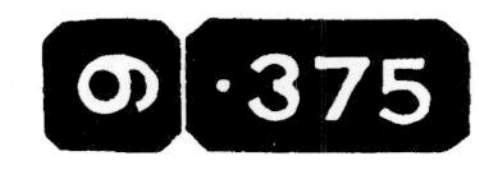

Figure 8.24 Marks struck by the Dublin Assay Office on 22, 20, 18, 14 and 9 carat gold

the 15 and 12 carat standards were replaced by the 14 carat standard, leaving the Republic of Ireland with five standards for gold – 22, 20, 18, 14 and 9.

The Irish Free State Order of 1923 ordained that 'plate manufactured in the Irish Free State should not be deemed to be British wrought plate within the meaning of the Customs Acts'. And so Irish plate (unless Convention marked, to 1998) entering Britain has to be marked just as other imported plate. Also British plate entering Southern Ireland has to be submitted to the Dublin Assay Office for assaying and marking.

Sometimes Irish silver with marks other than those punched by the Dublin office is found. Cork, Galway, Limerick and Youghal marked their own plate in the seventeenth century.

Hallmarks of other European countries

The following is a brief summary of the hallmarking systems of France, Belgium, Germany, Holland, Italy, Portugal and Denmark.

France

Among European countries, France has the oldest and most complicated system of hallmarking. The French guilds started marking gold and silver in the thirteenth century. Every town had its own marks, and the mint master responsible for marking stamped his initials in the town marks. These are often quoted in sales catalogues. In the reign of Louis XIV a system of tax marks was introduced. There were two tax marks, the charge mark stamped on the piece in its unfinished state by the Warden of the Guild, and the decharge mark stamped into the finished piece by the tax collector to show that the duty had been paid.

Current marks on gold

The French mark three qualities of gold: 900/1000, 840/1000, 750/1000. Since 1919 an eagle's head has been used for all qualities, but the shape of the shield varies and the figures 1, 2 or 3 within the shield identify the quality of the gold. Wares which are not tested by cupellation but by the touchstone are marked with an eagle's head with no surrounding shield and no quality number. Goldwares are also stamped with a maker's mark. French manufacturers stamp their wares with their initials within a diamond. Imported wares have the importer's initials within an oval. Wares below the French standard intended for export are stamped with an obus, which looks a little like a space capsule.

Chains

What complicates the system is that some types of goldwares bear different marks from those already mentioned. The system used to mark chain necklaces and bracelets is almost incomprehensible to anyone but a French bureaucrat. The idea is to indicate the weight of the chain. The eagle's head is used in combination with the head of a rhinoceros. By their position on the clasp or on the outer link of the chain it is possible to determine what the weight of the chain should be, and whether some of the links have subsequently been removed.

Imports and exports

On imported wares which conform to the French standard, including chains, a weevil in an oval replaces the eagle's head. An owl in an oval is stamped on all imported watches and on wares which come from countries whose legal standards are not the same as the French ones. French gold wares intended for export are stamped with the head of Mercury and with a figure indicating the quality. Such a piece can only be sold in France if it is stamped with a return mark in the form of a hare's head to show that tax has been paid on it. A system of counter-marking exists in France to make forgery more difficult; pieces are marked over an anvil covered with bands of insects that indent the metal on the opposite side to the hallmark.

Current marks on silver

The French have two standards for silver, 950/1000 and 800/1000, which they mark with the head of Minerva in an octagonal shield and with a crab mark, respectively. The Mercury mark in a six-sided shield is used for exports and the weevil in an oblong for imports.

Marks on platinum

In France, platinum of the standard 950/1000 is marked with the head of a dog. Platinum wares intended for export bear the head of a girl and imports the head of a bearded man.

Office marks

All the different French offices have distinguishing marks. Some like Marseilles use a letter, in this case an 'M', others like Amiens use a symbol, in this case a star.

Belgium

Hallmarks were first used in the thirteenth century when this area was under French rule, and the various towns continued to strike their guild marks on gold- and silver-wares as was done in France. With the unification of the Netherlands in the sixteenth century, a unified system was introduced. This lasted until the French restored their system again in 1795 when they reoccupied the Netherlands. After the French withdrawal in 1814 the Netherlands again introduced their own system.

Current marks

Hallmarking has been voluntary in Belgium since 1869, and in that year new marks were introduced to indicate the different standards. For gold of the standards 800/1000 and 750/1000, two different devices, both containing stars, are used, the first containing the figure 1, and the second the figure 2.

For silver, for which there are also two standards, 925/1000 and 835/1000, rose marks are used. The higher standard is marked with a rose within an octagon containing the figure 1. The lower standard is marked with a rose within an octagon containing a Roman two. Platinum is marked with a crown mark.

If the maker or importer chooses not to have his wares officially marked, he must still place his own mark in a barrel-shaped shield on the pieces together with a standard mark (see Figure 8.25).

Germany

There is no hallmarking system in Germany. Government inspection is said to ensure that manufacturers' wares are of the standard stamped on them. The minimum accepted standard for gold jewellery is 8 carat and for silverwares the minimum standard is 800 parts per thousand.

Holland

Makers' marks were stamped into Dutch gold- and silverwares as early as 1386, but no hallmarking system proper existed until the formation of the Republic of the Seven Provinces, which passed hallmarking laws in 1663. However, as in Belgium, the French imposed their system between 1795 and 1814.

Gold marks

The current system has been in force since 1953. The Dutch prior to 1987 recognized three standards of gold: 833/1000, marked with a rampant lion in an octagon; 750/1000, marked with a lion passant in an octagon holding a sword; and 585/1000, marked with a leaf in an oval. On smaller goldwares different stamps were used (see Figure 8.18). The assay offices also stamp goldwares with the head of a lion in a circle with a capital letter to indicate the office. Holland also has a date-letter system which changes every year.

Current silver marks

From 1953 to 1987 two standards of silver were permitted in Holland; 925/1000, marked with a rampant lion in a shield and the figure 1, and 835/1000, a lion passant in a six-sided surround accompanied by the figure II. The assay office's mark for silver was a helmeted head.

Platinum

Platinum is marked with a standard mark indicating the 950/1000 official standard. There are also special import marks which are obligatory.

New hallmarks

The changes in the hallmarking legislation which came into force on March 1, 1981 included the introduction of an 800 standard for silver for which a lion couchant mark is used. Articles which are too small to be tested will be struck with a ZIII mark. The 22 carat standard, which was abolished in 1951, has been reintroduced because 22 carat gold is being used increasingly for jewellery in Holland.

The hallmark used to mark 22 carat jewellery is in the form of a crocus flower; a rampant lion mark is used for larger 22 carat gold articles. The minimum legal standard for gold in Holland remains at 14 carat. Aside from these changes the marks used on gold and silver in Holland remain the same as before.

In the past medals have been exempt from hallmarking in Holland, but because both the trade and the public have expressed themselves in favour of having medals hallmarked these will no longer be exempt. Also under the new legislation the hallmarking of platinum and gold articles weighing less than 0.5 grams and silver articles weighing less than 1 gram is optional.

Italy

Like Germany, Italy at one time consisted of a collection of city states, each of which had its own system of marking. In Naples, gold and silver was marked as early as the fifteenth century. Parts of Italy came under French domination during the nineteenth century and were subject to French hallmarking regulations. In 1934, hallmarking became compulsory throughout the country. After a period of grace of four years, during which time goods in stock were marked with a single mark consisting of a radiating star, all goods have to bear two marks, a maker's mark consisting of a three-digit number assigned to the maker, followed by the initials of the province in which the goods are made. Before 1946 goods were also stamped with the old Roman fasces symbol, a bunch of twigs surrounding an axe. A second mark consisted of a standard mark expressing the quality of the metal in thousandths.

Current marks

In 1968 the form of the maker's mark was changed and a five-point star incorporated with it. A third mark was also added, a Perseus mark applied to works optionally assayed at the request of the manufacturer. Otherwise, marks are applied by the manufacturer and are subject to checking by the state. The standards current in Italy for goldwares are 750/1000, 585/1000, 500/1000 and 333/1000. A standard of 753/1000 is also permitted for goldwares made by lost-wax casting. There are three standards for silver, 925/1000, 835/1000 and 800/1000, and a single standard for platinum of 950/1000.

Denmark

Wares made from precious metals have been hallmarked in Denmark since the beginning of the seventeenth century. The first marks to be used were town marks, like the three towers used by Copenhagen. After 1679 the assay master also stamped his initials on the wares.

Current marks

The present voluntary system was introduced in 1893 and modified in 1961. Two standards are recognized for silver, 830/1000 and 925/1000, with a tolerance of 2 parts per thousand. The minimum quality for gold is 585/1000, or 14 carat, and imported wares must be of a minimum standard of 15 carat. A standard of 950/1000 is laid down for platinum. If a ware is submitted for assay and marking, it is stamped with the old three towers mark of Copenhagen, otherwise all articles of gold, silver or platinum must bear a marker's mark registered by the State and the quality of the metal expressed in parts per thousand. From the annual total of 600 000 to 900 000 wares produced in, or imported into Denmark, government inspectors check the high proportion of 12 000 to 14 000 to discourage abuse of the regulations.

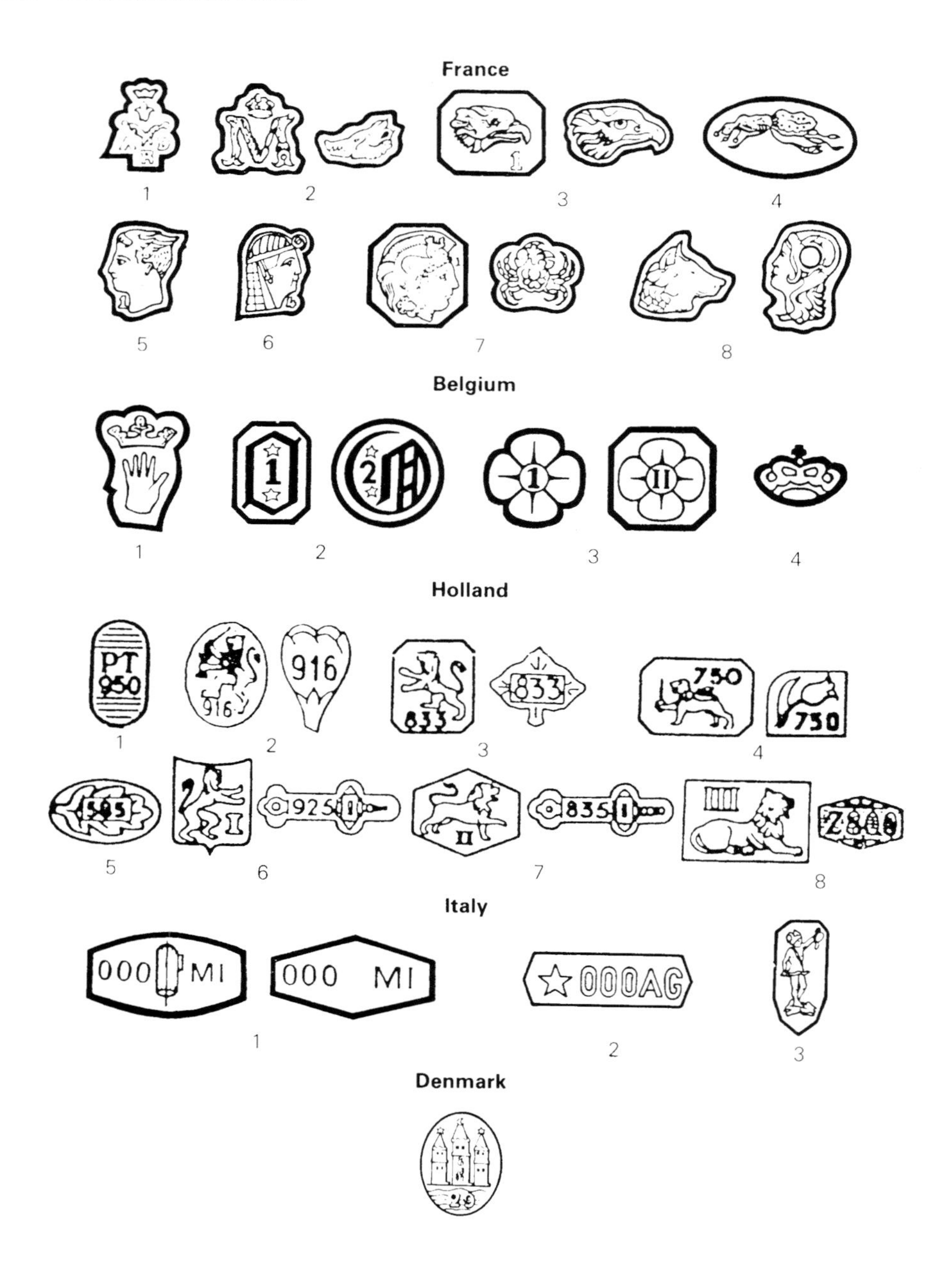

Figure 8.25 European hallmarks

France

1 Early Paris town mark incorporating the mint master's initials.
2 Charge and decharge marks used in Toulouse from 1780 to 1789.
3 Current standard marks for goldwares. *Left*: the number 1 incorporated in this mark indicates the highest standard of 900 parts per 1000. *Right*: The head without a number indicates that the piece was tested by the touchstone method.
4 Marks used on imported wares. The weevil in an oval is used for gold. For silver the weevil is contained in an oblong.
5 The head of Mercury used on goldwares produced for export.
6 The mark used on exported watch cases.
7 The current silver marks. The Minerva's head incorporates a number indicating the standard. The crab is used on wares tested only by the touchstone method.
8 Marks used on platinum. The dog is used on wares designed for home consumption. The girl's head is used on wares designed for export.

Belgium

1 Old town mark of Antwerp.
2 State marks. *Left*: gold of the 800/1000 quality. *Right*: 750/1000 quality.
3 State marks on silver. *Left*: 925/1000 silver. *Right*: 835/1000 silver.
4 State mark for platinum.

Holland

1 Platinum 950/1000.
2 Gold 22 carat 916/1000.
3 Gold 20 carat 833/1000.
4 Gold 18 carat 750/1000.
5 Gold 14 carat 585/1000.
6 Silver 925/1000.
7 Silver 835/1000.
8 Silver 800/1000.

The marks shown left in the cases of 22, 20 and 18 carat gold, and 925, 835 and 800 silver, apply to larger items. Those on the right apply to smaller items.

Italy

1 *Left*: makers' mark used from 1934 to 1938. The three zeros represent the three-letter number allocated by the state to the manufacturer. *Right*: after 1938 the fasces was removed.
2 The current maker's mark.
3 The optional state mark.

Denmark

The optional official guarantee mark is the historical three tower mark, once the town mark of Copenhagen. The date is incorporated in the mark, the last two figures of the year being used. The mark shown would have been stamped on pieces assayed in 1929.

Portugal

The Hallmarking International Convention Amendment Order 1987, which recognizes that the Portuguese hallmarks have changed to a swallow for gold, a sea-horse for silver and a parrot's head for platinum, came into force in the UK on December 3, 1987. The marks incorporate the precious metal content of the alloy in Arabic numerals. The shape of the mark identifies the office which applied for the mark.

International agreement

It has been recognized for some time that a common marking system would be of benefit to international trade. The first steps were taken in 1972 to produce a system of assaying and hallmarking that would be generally acceptable internationally. The former EFTA countries signed an agreement in that year known as 'The Convention on the Control of Articles of Precious Metals'. The Convention is open to any country to join provided it has the necessary assay office facilities. The Convention has been ratified by Austria, Eire, Finland, Norway, Portugal, Sweden, Switzerland and the United Kingdom. Under the Convention, an article bearing certain specified marks (namely a sponsor's mark, fineness mark, common control mark and assay office mark) is accepted in any of the above countries without any further assay or marking. The common control mark and assay office marks must be applied at an authorized assay office under agreed rules.

The standards recognized under the Convention are 750 (18 carat), 585 (14 carat) and 375 (9 carat) for gold; 925, 830 and 800 for silver; and 950 for platinum. 830 standard silver is not a legal standard in the UK. 830 Convention marks do exist and may be used on exports to convention or other countries which accept this standard. The standard 800 parts per 1000 fine silver is now a legal standard in the UK (since January 1999).

Examples of the new common control marks which are used under the Convention are shown in Figure 8.26.

Figure 8.27 shows the assay office marks that are used under the Convention by other signatory countries. In the UK the normal assay office marks are used on articles of the legal standards.

Assay office procedure

The assay was previously known as the 'touch'. This referred to the method of testing gold and silver by rubbing it on a touchstone – a black siliceous stone or an earthenware block. The streak left by rubbing the metal on the stone was compared with the streak made by gold or silver of a known standard. (See Appendix 4 for details of the touchstone method.) This method is still used by many Continental assay offices, and

Figure 8.26 Common control marks under the Convention on the Control of Articles of Precious Metals. On the left is the mark for 18 carat gold wares, right, for sterling silver articles; and centre, for platinum wares. Countries may refuse to import articles which are of a lower fineness than its national minimum standards

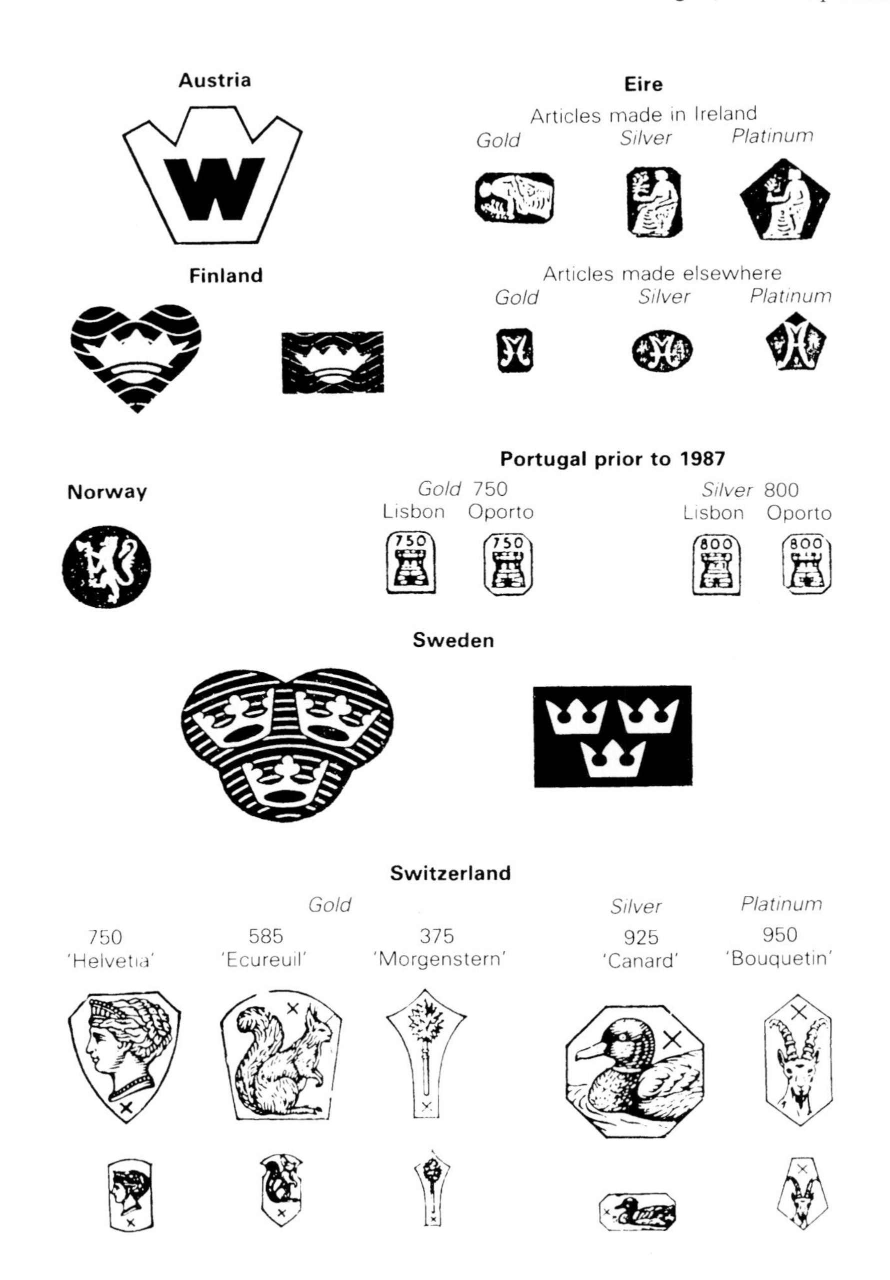

Figure 8.27 Foreign assay office marks used with Convention Hallmarks

though it is a surprisingly accurate technique when used by an expert, it is far less accurate than the cupellation method for gold, and the titration method for silver, used by British assay offices.

The London Assay Office abandoned the touchstone method at a very early date, though the term 'touch' continued in use for some time. In an entry in his diary for May 19, 1669, Samuel Pepys described a visit to the Royal Mint. He recalls how he 'after dinner went to the Assay Office and there saw the manner for assaying gold and silver'. The technique of assaying gold which he witnessed and described differs little from that used to this day. He wrote that the method '. . . if it be for gold is by taking an equal weight of that and of silver . . . this they do wrap up within lead.' Then they put the samples 'into little earthenware cups made of stuff like tobacco pipes and put them into a burning hot furnace, where after a while, the whole body is melted, and at last the whole body of the lead . . . is sunk into the body of the cup, which carried away all the copper and the dross with it, and left the pure gold and silver embodied together. . . . ' The silver was then dissolved in acid leaving pure gold.

If, however, the method used is basically the same as that of Pepys' day, the modern assay office is a very different place from the one which he visited and described. It is a combination of well-equipped laboratory and workshop. When wares are delivered they are weighed; the accurate weighing of wares sent in for testing and marking ensures that the correct amount of gold or silver submitted by the manufacturer or retailer is returned to him – an important consideration when precious metals are concerned.

After weighing, the wares go to the drawing department, where the drawer takes scrapings from different parts of each piece. If a coffee-pot is being sampled, for instance, drawings are taken from the body, the spout, the handle fittings and the lid, to make sure that all the silver and all the solder used are up to the sterling standard.

This drawing is a highly skilled operation. For one thing drawers have to be careful not to mar the work. If a piece of Continental manufacture, which has been submitted in the finished state is being sampled, it may not only be delicate, but also highly finished or expensively textured. So skilled are most drawers that they can not only sample delicate wares without damage, but they can often spot base metal or substandard metal just by the feel of it under the scraper. This feel for metal is perhaps understandable when it is realized just how much work passes across the drawer's bench in the course of a year. In 1987, the four assay offices tested and marked 21.5 million gold, silver or platinum articles.

After sampling, the samples are weighed so that a known amount of metal is sent forward for testing. The scrapings of gold are wrapped in small sheets of lead with pieces of fine silver and placed in cupels, cup-shaped depressions in a refractory block. This block is then put in a furnace, where the lead and the base metal oxidize at a temperature of 1100°C in a current of air. The oxides are absorbed in the pores of the refractory materials. After removal from the furnace, an alloy of gold and silver in the form of little beads remains in the depressions in the block. Each bead is flattened, annealed, rolled into a thin strip and placed in a small platinum cup and immersed in boiling nitric acid. The silver dissolves in the nitric acid and the liquid is poured off leaving pure gold in an amorphous state behind. The amorphous gold is then annealed and weighed. The weight of the pure gold is compared with the weight of the sample scrapings, and the percentage of pure gold in the sample can then be calculated.

The Birmingham method for assaying silver is now potentiometric titration (Figure 8.28). After the silver is accurately weighed the sample is dissolved in nitric acid and

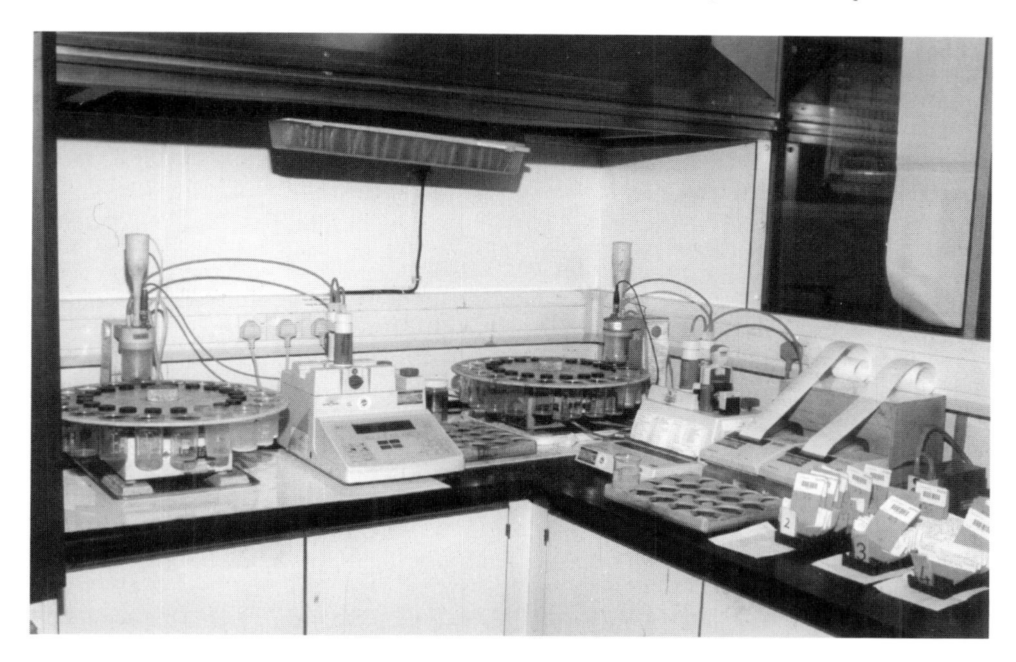

Figure 8.28 Potentiometric titration method

the silver is determined directly by potentiometric titration. Here potassium chloride solution is added to the silver solution via an automatic burette. A silver electrode detects the change in electrical conductivity of the solution as the potassium chloride is added until it finds the end point of the titration when all the silver has been precipitated as silver chloride. A calculation is performed automatically and the silver content of the sample in parts per thousand is printed out. There are now four legal silver standards in the UK: 800, 925, 958 or 999 parts per thousand.

After testing, the wares that have passed the assay are marked. If they are wares made in series, such as gold lockets, they are marked in a press with tools specially made for the job. Individual pieces are marked by hand; the punches are struck with a hammer to drive them into the metal. The assay offices have hundreds of shaped beds on which to rest the wares so that they will not be damaged during stamping, and recently the use of plastic beds that can be formed to the exact contours of the piece means that it is possible to mark almost any silver- or goldware, no matter how small or thin, without damage.

Birmingham Assay Office now has the largest number of laser hallmarking machines in full operation in the UK. The first machine was installed in 1997. It enables the operative to control the depth of cut, by use of a computer system. The size, place and depth of the hallmark can now be carried out without bruising or damaging the item. This method of marking is increasingly used, especially on finished items, notably watches (Figure 8.29).

In addition to the day-to-day work of assaying and marking, the assay offices also do valuable research work. Experiments carried out to discover the best way to assay platinum in anticipation of any future legislation resulted in the development of the technique now in daily use. Spectrographic analysis has also been used at the London

Figure 8.29 Demonstration of laser hallmarked goods to the Lord Mayor of Birmingham and the Assay Office Master, Michael Allchin

Assay Office to test antique silver. Apparently silver alloys have varied down the ages, and it is possible to date an alloy by analysing its constituents. This confirms that the silver used and the marks on a piece belong to the same period. The technique has been of great assistance to the Goldsmiths' Company's Antique Plate Committee, an advisory committee of experts, who, over a 20 year period up to 1983, had examined 17 000 pieces of plate submitted to them because their authenticity was in question.

Assaying platinum

The new hallmarking legislation in force since January 1975 made it necessary for the assay offices to evolve a technique for assaying platinum. The technique that they use is called atomic absorption spectrometry. Scrapings are taken from the platinum jewellery submitted for assaying and marking in the same way as for gold- or silverwares. Much smaller samples are taken of the platinum, only 10 milligrams as compared with 100 to 250 milligrams for goldwares. An electronic microbalance is employed for weighing these samples.

After weighing, the scrapings are dissolved in *aqua regia* (a mixture of hydrochloric acid and nitric acid). The solution is diluted to 100 ml and introduced into an air/acetylene flame in the atomic absorption apparatus. The platinum is converted into free atoms in the flame. Radiation from a special lamp which has a platinum cathode

Figure 8.30 Beds used by the Birmingham Assay Office to support pieces when they are being struck with the marks

is passed through the flame. Part of this radiation is absorbed by the platinum atoms in proportion to their number. The intensity of the incident radiation is measured and recorded. From this measurement the platinum content of the sample can be calculated.

Chapter 9

The history of watches and clocks

Developments down the years

The ancient Egyptians probably made the first time-measuring devices. Observi that the sun threw lengthening and shortening shadows on its journey across the s from dawn to dusk, they built sun clocks. It has been suggested that the famo needles, such as Cleopatra's needle now standing on the Thames embankment, wer the gnomons of enormous Egyptians sundials, casting their shadows on huge dials These first sundials would have been the forerunners of the more compact sundial (Figure 9. 1) which were still in use well into the eighteenth century to check the time-keeping of mechanical clocks.

The Egyptians also devised other ways of measuring the passing of time. Amon these were the water clocks called clepsydra. In their simplest form clepsydra consisted of vessels with holes in their bases and a scale marked on their interiors. As water trickled out of the holes, the level in the vessels fell, and by reading off the level against scales the owners knew how much time had passed since their clocks were filled up. There is an alabaster cast of one of these primitive clepsydra in the British Museum, the original of which is said to date from between 1415 and 1380 BC. It looks like a giant flower-pot decorated on the outside with hieroglyphs, the ancient Egyptian picture writing.

Figure 9.1 Copy of a tenth century portable sundial (by courtesy of the Science Museum)

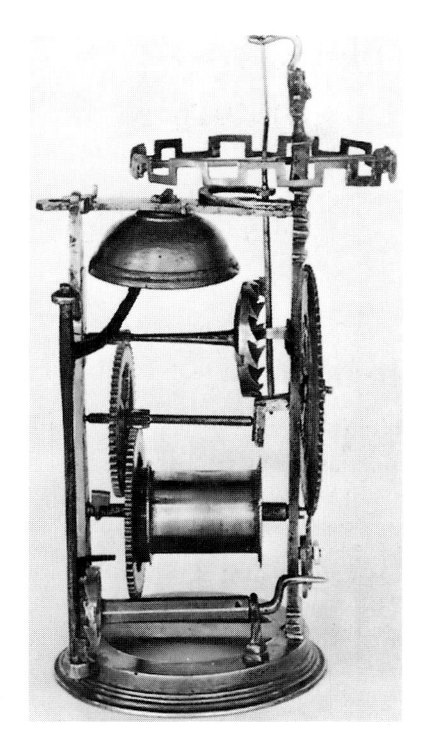

Figure 9.2 Alarm clock fitted with a verge escapement. Said to be a fifteenth century monastic alarm clock

Later civilizations made much more sophisticated clepsydra. In Alexandria in the last century BC, clepsydra were being built with a wheel train, a pointer and a column dial. Highly developed clepsydra must have some claim to be deemed the first mechanical clocks. Most horologists, however, consider the first mechanical time-keepers to have been those clocks, built in the last quarter of the thirteenth century, which were fitted with a mechanical escapement. No one is certain who invented the first of these mechanical clocks, but the inventor was most likely a monk, for in those days all learning was enclosed within the walls of the monasteries. The first clock may not have been a clock as we know it, with hands and a dial, but probably just a mechanism that rang a bell at the appropriate time to call the inventor's fellow monks to attend matins, vespers and the other canonical hours.

The mechanical clock

When a dial was added to this mechanism, the main elements of the weight-driven clock, as we know it today, came into existence (Figure 9.2). All the horological inventions affecting the weight-driven clock made in all the centuries since have only been improvements. Fundamentally, the weight-driven clock remains just as it was at the beginning. There was a source of power, a rock suspended from a rope coiled round a pulley. There was an escapement, the purpose of which was to check the fall of the rock and to stop all the power from escaping when the rock plummeted to the end of its rope, and to divide the power up into regular impulses. The earliest type of escapement was known as the 'verge escapement'. It consisted of a crown-shaped

Figure 9.3 A foliot from the Salisbury Cathedral clock said to have been made in 1386 (by courtesy of the Keystone Press Agency Ltd.)

wheel and two pallets set at right angles to the wheel. These pallets engaged and released the wheel alternately, so halting and releasing the weight on its downward journey.

The pallets of the verge escapement were attached to an upright rod. On top of this rod and above the rest of the clock mechanism, was fixed either a metal bar or a wheel. This was called the foliot. As the pallets of the verge wagged back and forth so did the foliot (Figure 9.3). The foliot was in fact the equivalent of the modern balance wheel found in most mechanical watches today. Its back and forth swing, like the back and forth swing of the watch balance, beat out the regular rhythm on which timekeeping depends.

This regular beat was transferred to a set of wheels and toothed pinions, which drove the single hand of the clock round and round the dial as the hours passed.

The timekeeping of those early clocks was by no means accurate by present-day standards, but seconds and even minutes were not important at that time. The early clockmakers put no minute marks on their dials, but only calibrated them for hours and quarter hours. This practice continued until the latter half of the seventeenth century. Any clock which stylistically belongs to a period earlier than this that has either a dial calibrated in minutes, or two hands, should be viewed with suspicion.

A clock with the falling weight as a source of power was all very well to install in the turret of a castle or a church, or to mount permanently on the wall of a great hall, but it was not easy to move about. Not until the fifteenth century did it become possible to build portable timekeepers other than portable sundials. A new source of power a good deal less cumbersome than the weight and line had to be devised. At some time during the second half of the fifteenth century some unknown person discovered this alternative source of power. It was discovered that if a ribbon of steel was coiled it would naturally seek to unwind itself. This uncoiling action, which the late T. P. Camerer Cuss described as 'the pent up power of a ribbon of steel', provides enough energy to drive a clock (Figure 9.4). It was, of course, necessary to have an escapement to prevent the spring from uncoiling in a fraction of a second, and at the same time to set the balance beating a steady rhythm.

Many of the early spring-wound clocks were in the shape of drums with the dial on the top. It was inevitable that sooner or later someone should have had the idea of making a little drum clock small enough to wear on the person.

Figure 9.4 A sixteenth century example of the spring powered drum clock with the dial on top and a single hand

The first watches

There has been considerable controversy about who can be considered the inventor of the watch. The controversy has never been finally settled, but Peter Henlein, a Nuremberg watchmaker, is thought to have a strong claim. He made 'horologia' from a 'trifling amount of iron . . . which both indicate time and strike for forty hours . . . even if they are carried on the bosom or in the purse' (Figure 9.5). Henlein's little horologia date from early in the sixteenth century.

Many of the early watches were made in fanciful shapes – crosses, balls, shells and skulls and they were very poor timekeepers. One reason for the great inaccuracy of the early spring-driven timekeepers was that the power output of a fully wound spring was far greater than the power output from one that was almost uncoiled. This variation in power output had a serious effect on the performance of the verge escapement.

During the fifteenth century the clockmakers and watchmakers devised two mech-

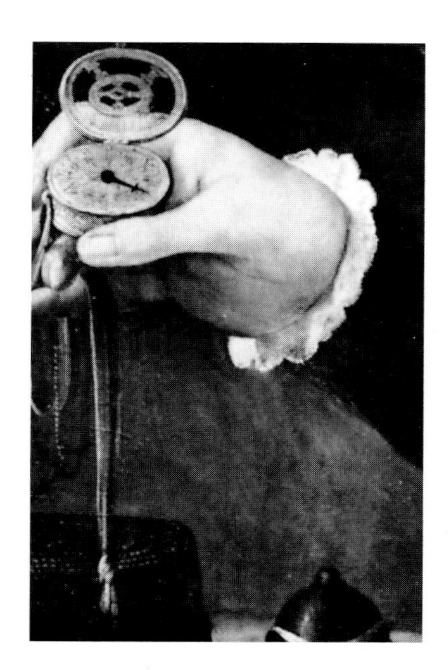

Figure 9.5 Detail from a portrait painted in Germany in 1560, which shows one of the first watches ever made

Figure 9.6 A watch given by King Charles I to Lord Panmuir and typical of watches made at this period

anisms designed to overcome this problem of power loss. One of these, a German invention, consisted of a spring acting on a snail-shaped cam. This was called the stackfreed. It has been described by one modern writer as 'a most brutal arrangement', and it was shortlived. The other device, a drawing of which was made by Leonardo da Vinci, was much more satisfactory. It became a standard component of good-quality spring-driven clocks and watches up to the time when the verge escapement was finally dispensed with in the nineteenth century.

This power compensation device was called a fusée. It consisted of a grooved cone, which was in effect a stack of graduated pulleys. A gut line, later a fine chain, was wound on to the barrel containing the coiled spring, the mainspring, that powered the clockwork. The gut line was then attached to the fusée. As the mainspring unwound, the gut passed on to bigger and bigger grooves on the fusée and this provided ever-increasing leverage to compensate for the ever-diminishing power of the spring.

The seventeenth century was to witness a horological revolution. At the beginning of the century the watch was more of a toy than a timekeeper. A typical watch of the first quarter of the century, the kind being produced by David Ramsey, one of the best watchmakers of the period, was very small, perhaps 2½ in. across (Figure 9.6). It might have an engraved metal case, or the case might be made of two hollowed-out pieces of rock crystal held in gilt-brass rims. The dial would probably be made of silver and engraved, or it might be decorated with champlevé enamel. On the dial would be a tiny chapter-ring with rather stumpy Roman numerals incised in it. The single hand would be made of steel and would be dart-shaped. The movement would almost certainly be fitted with a fusée and chain, and the verge escapement would be controlled by a wheel foliot.

By the second quarter of the seventeenth century, such makers as Edward East were turning out watches that looked much more like the pocket watches of later periods. These watches were bigger than those made earlier in the century. The flattened spherical cases were sometimes decorated with leather and sometimes with biblical or mythological pictures painted in enamel. The movements of these watches, though

Figure 9.7 Two clocks illustrating the development of the English brass lantern clock from the Continental iron clock

bigger and more robust than those used by Ramsey for his little watches, were still poor timekeepers.

Seventeenth-century clocks

During the seventeenth century a clock, probably Continental in origin, made its appearance in England. It was called the lantern clock, because it looked like a brass lantern (Figure 9.7). This clock was produced for over a century, and consisted of a brass box with turned brass pillars at the corners. These pillars terminated in baluster-shaped finials, and ended below the case in little turned feet. Above the brass box, which had hinged sides to give access to the movement, were two intersecting arches. Another baluster finial rose from the junction of these arches, and an alarm bell was suspended below it. Above the case, between the arches, were decorative frets. The dial usually overlapped the case, and it tended to get bigger as the century progressed. Also the Roman numerals on the chapter-ring became progressively taller and more graceful. These clocks, which were hung on the wall, were from 6 to 16 in. tall (Figure 9.8). Their movements, which were controlled by a verge and wheel foliot, were powered by a weight that dangled below the clock on the end of a rope. The other end of the rope was coiled round a spiked wheel in the movement. A second, and smaller weight, powered the alarm mechanism. The alarm setting was in the form of a small calibrated disc in the centre of the dial.

The long-case clock was born in the middle years of the seventeenth century. Wooden-cased wall clocks were already being produced and it was perhaps only to be expected that, sooner or later, someone would feel that a weight-driven clock would look much tidier if the weights were enclosed. So the clock became a piece of furniture that could stand on the floor. All that was necessary was to add what amounted to a narrow cupboard, fitted with a door to give access to the weights, below the pedimented wooden case of the period. These first long-case clocks were small (about 5 ft high), slender, simple and classically proportioned. Examples by such makers as Ahasuerus Fromanteel and Edward East are very desirable collectors' items today (Figure 9.9).

Figure 9.8 A pendulum version of the lantern clock made about 1680

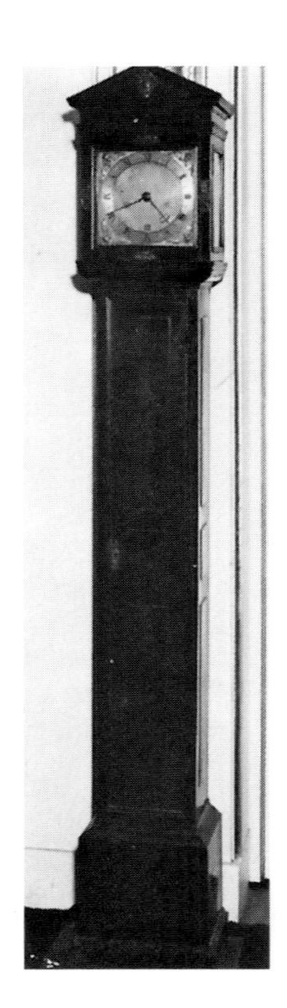

Figure 9.9 A long-case clock made by Ahasuerus Fromanteel about 1665

The pendulum

Though undoubtedly better timekeepers than the watches of the period, clocks made during the first 60 years of the seventeenth century including the lanterns, the wooden-cased wall-clocks and the long-case clocks, were still far from good time-keepers. Then about 1657 a clock with a new regulator made its appearance in Holland. It was almost certainly the astronomer, Galileo Galilei, who invented this new regulator, the now familiar pendulum. It is said that when he was a young man, he was sitting one day in Pisa Cathedral when he noticed that one of the lamps was swinging. He became intrigued and began to measure the time of the swings by comparing them with his pulse beats. He found that whether the swings were long or short, they occupied the same amount of time. The swinging lamp was in fact, he realized, a natural timekeeper, it was as the horologists say 'isochronous'. Galileo eventually made a pendulum, which like the suspended lamp was nothing more or less than a swinging weight. He used his pendulum in connection with his astronomical observations, and conceived the idea of fitting a pendulum to a clock. Indeed he left a drawing of a pendulum clock, but died before he actually made one. In the end it was the Dutch physicist Christiaan Huygens who became known as the inventor of the pendulum clock, though he was not the actual maker of the clock. It was a fellow Dutchman, Samuel Coster, who made the clock to Huygens' specification (Figure 9.10).

It was soon realized that the pendulum deserved a better escapement than the old verge, the action of which was interfered with by the train. In 1671 William Clement invented a much superior escapement, called the anchor escapement because it looked like the business end of an anchor. This escapement was much less subject to outside

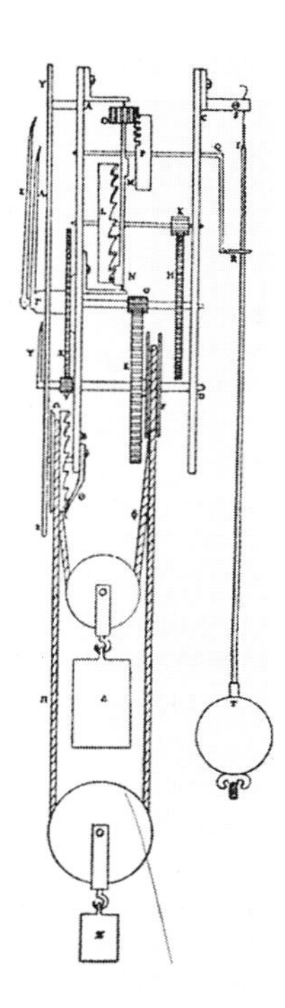

Figure 9.10 Drawing of the first pendulum clock made in Holland by Samuel Coster to a design by Christiaan Huygens (by courtesy of the Science Museum)

Figure 9.11 A Thomas Tompion spring wound bracket clock movement with fusée

influences than the verge, and has been the standard escapement for pendulum clocks ever since. In the next two centuries, between two and three hundred escapements for watches and clocks were invented. Some, like George Graham's famous dead-beat escapement, were successful, but the majority made no lasting contribution to the science of horology and therefore passed into oblivion.

The invention of the pendulum led very quickly to the addition of minute hands to clocks. It then became logical to place five one-minute marks between the hour marks on the chapter-ring, in place of the four quarter marks that had been there for the previous 300 years or more.

The invention of the pendulum and the anchor escapement coincided with the emergence of some of the finest clockmakers who have ever lived. The greatest name among these makers was Thomas Tompion (Figure 9.11). His fame was so great at home and abroad, that some of his Continental competitors paid him the backhanded compliment of engraving his name on the dials of clocks far inferior to his. It is a tribute to the craftsmanship of this period that so many of the clocks made in the last quarter of the seventeenth century have survived to our own day. Many of them still keep excellent time nearly 300 years after they were made. The two typical clocks of this period were the bracket clock and the long-case. The long-case had grown bigger to accommodate a long swinging pendulum. Its case now was enriched with all the fashionable decorative elements of the day (Figure 9.12). Gilded classical pilasters or twisted columns flanked the dial, beautifully matched walnut veneers, marquetry, or chinoiserie lacquer-work enriched the case. In the corners of the dials were little applied brass decorative frets called spandrels, often in the form of plump-cheeked cherubs, and the elegant Roman numerals on the applied chapter-rings were cut into the brass and filled with pitch.

The bracket clock, so called because it was designed to stand on a bracket fixed to the wall of a room, was a spring-wound clock with a short bob-pendulum as a regulator. Some of these bracket clocks were made with travelling cases, so that their

Figure 9.12 A Daniel Quare long-case clock of about 1695, illustrating how the long-case developed into a substantial piece of furniture

owners could carry them down to their country estates when the London season came to an end (Figures 9.13(a) and (b)). These were the forerunners of the famous carriage clocks of the nineteenth century.

The cases of the bracket clocks had an elegant simplicity. They were made of ebony or ebonized wood. Above the dial, which was similar to that used on the long cases, was a domed moulding carrying a hinged brass handle. Usually there were a few pierced brass enrichments, but nothing else to detract from the classical proportions.

The balance spring

Within a few years of the application of the pendulum to the clock, Huygens also devised a method of rendering the old wheel foliot, the balance wheel as it is known today, isochronous. He attached a coiled spring to his balance wheel, which pulsed like a heart when the balance received its impulse, tightening and then uncoiling again.

That irascible figure of the seventeenth century scientific scene, Dr Robert Hooke, argued and raged by turn. He claimed that he had anticipated Huygens' invention of the balance spring by some years. The validity of Hooke's claim to this invention is still a matter of controversy, but most of the evidence seems to be in Huygens' favour.

The balance spring did for the watch what the pendulum had done for the clock. It made possible an accuracy previously undreamed of, and put a minute hand and minute marks on the dial. Soon every man who could afford one was carrying a watch in the pocket of his fashionable embroidered waistcoat.

Because they were the prerogative of the rich, watches were usually heavily ornamented in the style of the period. The repoussé-chasing and the florid piercing, similar

Figure 9.13(a) A Thomas Tompion bracket clock – typical example of about 1690, with winding squares and cherub spandrels in the dial corners

Figure 9.13(b) The back of a movement showing the short pendulum used with a verge escapement, and the richly engraved back-plate

to that found on Caroline silverware, were applied to the 18 carat gold or silver watch cases. These creations of the goldsmiths' art, carried out in soft metals, were very liable to sustain damage, and the fine detail of the chasing soon became blurred from daily wear. So it became customary to provide an outer case for the protection of the ornate inner case (Figures 9.14(a) and (b)). The outer of these so-called pair-cases was more workmanlike than the inner. Often it was made of unadorned metal, or sometimes it was covered with leather.

Figure 9.14(a) An eighteenth century pair-case watch by George Graham showing a richly chased gold case and a simpler outer case to protect it

Figure 9.14(b) A decorative outer case which would have a third case, possibly covered with leather or shagreen for further protection

Temperature and friction

As so often happens in any mechanical science, the solution of one problem tends to draw attention to others. It soon became apparent that though the pendulum and the balance spring brought vast improvements in timekeeping, the full potential of these inventions could not be realized until the errors due to changes of temperature and friction had been greatly reduced. The problem of friction in watch bearings was tackled very early in the eighteenth century. In 1704, a Swiss and two Frenchmen, all living in London, made a joint application for a patent. In this application they stated that as a result of 'great charge and continual labour', they had acquired 'an art of working precious or more common stones . . . so that they may be employed and made use of in clockwork or watchwork . . . not for ornament only but as an internal or useful part of the engine itself.' These three men, Nicholas Facio, and Peter and Jacob Debaufre, met opposition to their application from the Clockmakers' Company. In fact they shared the fate of so many horological inventors, and became the centre of a long drawn-out controversy. There now seems little doubt that they were indeed the inventors of those watch jewels, which provide hard, and relatively friction-free, cups for the pivot of the balance wheel and the train wheels to run in.

Temperature compensation was not so easily achieved. It was not until the end of the nineteenth century that this bogy was finally laid. The horologists of the eighteenth century did, however, come to terms with the problem. Their work in this field contributed much to the science of navigation, and this in turn laid the foundations of Britain's great programme of imperialist expansion into territories far across the seas.

The main impetus for the eighteenth century experiments in temperature compensation was an Act of Parliament of 1714, an Act which was recommended to the government of the day by, among others, Sir Isaac Newton. The Act aimed to foster the building of a timekeeper that would make it possible for the captains of the ships, upon which Britain depended so much for her wealth and her security, to navigate with much greater accuracy than in the past. A prize of £10 000 was offered for an instrument that would have an error equivalent to not more than one degree of longitude on a voyage to the West Indies. Such a voyage across the Atlantic would of course mean that this timekeeper would be subjected to considerable changes of temperature. Therefore, temperature compensation was one of the first problems that any inventor who set out to win the prize would have to tackle.

In 1725 John Harrison, a carpenter's son, devised a pendulum that provided temperature compensation by making use of the fact that brass and steel have different coefficients of expansion. This 'grid-iron pendulum', as it was called, consisted of bars of the two metals lying parallel to one another, five of steel and four of brass, set up in such a way that the expansion of one set of metal bars was offset by the expansion of the other. No matter how hot or how cold it became the pendulum remained the same length. This pendulum was used in high-class long-case clocks and regulators for two centuries. Of course, as Harrison realized, you could not use a pendulum clock on the heaving quarter-deck of a ship. He went on, however, to apply the same bimetallic principle to the balance of a spring-driven chronometer.

Harrison discovered that the balance spring of a watch loses some of its elasticity when heated, and that if the spring is shortened the lost elasticity is restored. Therefore he attached a bimetallic strip to a block holding the curb-pins that determine the effective length of the balance spring. When the temperature rose, the brass on one side of the strip expanded more than the steel on the other side. This caused the strip to bow, draw the curb-pins further along the balance spring, and so shorten the spring. This simple device was fitted to Harrison's famous Number 4 chronometer, which on the voyage of the *Deptford* from Spithead to Jamaica in 1761, lost only five seconds.

In 1767, the French watchmaker, Pierre le Roy, achieved compensation by making a bimetallic balance wheel, the rim of which was cut on opposite sides. Changes of temperature altered the form of the balance wheel, and this compensated for the changes of elasticity of the balance spring. This system was used on good watches until the problem of temperature was solved finally by the Swiss, Dr Charles Edward Guillaume in 1896. He evolved two nickel-steel alloys, one of which he called Invar, short for invariable. As its name suggests it expanded only a negligible amount when heated. It was therefore an ideal metal from which to make balance wheels. Guillaume's other alloy, called Elinvar, does not lose elasticity to any appreciable extent within the range of temperatures that a watch is likely to encounter. It was just what the manufacturers of balance springs had been looking for.

Lever escapement

The second half of the eighteenth century saw the invention of a new watch escapement. Of the hundreds invented during the century, the one which Thomas Mudge used for a watch made for King George III to give to Queen Charlotte was by far the most important. About 80 years passed before another watch with this lever escapement was made, but during the last century tens of millions of watches have been fitted with this escapement every year.

Just what the detached lever escapement achieved was summed up by the late T. P. Camerer Cuss in his book *The Story of Watches*. 'Every escapement up to 1759', he wrote, 'had been very much under the influence of the watch train . . . Mudge's lever freed and detached the balance for the greater part of its swing. In fact it is only for a brief moment when it unlocks the train and receives the impulse that the balance has any connection with the rest of the watch.' (See Appendix 3 for full description of the lever escapement.)

Three other important horological inventions also date from the second half of the eighteenth century: the shock-absorber; keyless work; and the self-winding mechanism. The great Swiss watchmaker Breguet was almost certainly the first to fit a shock-absorber to a watch. It was called the parachute, and worked on the same principle as the shock-absorber still used in mechanical watches. The end stone for the balance staff was mounted on springs which gave when the watch received a jolt.

Keyless work

It was well into the nineteenth century before keyless work became a standard part of every watch, and owners no longer had to carry a little key on their watch chains that went into a hole in the back of the watch case or through the dial and fitted over the winding square. The first keyless mechanism was applied as early as 1752 when the French maker, Pierre Caron, made a watch for Madame de Pompadour. He fitted this with a little projecting hook that its owner could draw round the dial with her nail when she wanted to wind it. It was 68 years later before a London watchmaker, Thomas Prest, improved on Caron's idea. In 1820 he took out a patent for the pendant winding system.

Self-winding mechanism

In the meantime, the eighteenth century horologists had tried to get rid of winding altogether. However, it is still not really known who first fitted a self-winding mechanism to a watch. Was it Breguet, a Swiss working in Paris; was it Abraham Louis Perrelet, another Swiss working in Switzerland, or was it a third Swiss, Louis Recordon, who worked in London? All three men made self-winding watches at much the same time, about 1780, and they all worked on much the same principle – the pedometer principle (Figure 9.15). Like the pedometer, which measures the distance covered by a walker, these watches were fitted with a pivoted weight that swung when the wearer moved. In the self-winding watches this swing of the weight activated a simple train of gear-wheels and pinions that wound the mainspring barrel. This was yet another invention that lay dormant for a long time before its full potentialities were realized.

Fashion and electricity

Nineteenth century clocks changed little in their essentials between the early years of the eighteenth century and the middle years of the nineteenth. Their appearance

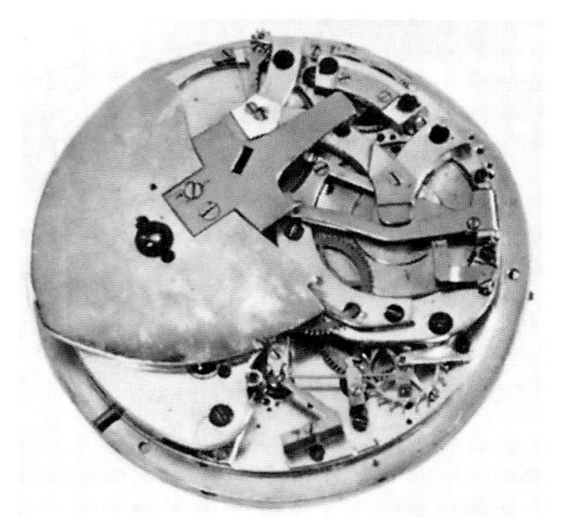

Figure 9.15 A Breguet self-winding watch (*c*. 1780) with heart-shaped weight

changed, of course, as fashions changed. Round dials replaced square ones in the eighteenth century, and nineteenth century romanticism was reflected in the transformation of the clock case from a sensible wooden box to a confection of marble and bronze (Figure 9.16). In this period, too, many experimental clocks were produced incorporating more or less successful new escapements. None of these, however, enjoyed more than a brief popularity.

Figure 9.16 A Benjamin Vulliamy clock made at the beginning of the nineteenth century

In 1840, a Scotsman, Alexander Bain, took an important step into the future when he applied electricity to a clock. His patent No. 8783, of October 1840, was not fully exploited until the middle years of the twentieth century, when it was to become increasingly apparent that the days of the spring-wound clock were numbered, and that we were entering the age of electric and electronic timekeeping.

Few horological inventions have not been challenged and, as has been seen, some of them have been acrimoniously disputed. Bain's invention was no exception. Alexander Bain had arrived in London from his native Scotland in 1837 'to seek employment as a journeyman clockmaker'. It was in the same year that Cook and Wheatstone took out their first patent for the electric telegraph, and Wheatstone had certainly, at some stage in his work on the telegraph, considered applying electricity to timekeeping. On August 1st, Sir Charles Wheatstone was introduced to Bain, and on August 18th he bought apparatus for the telegraph from Bain. In November, Wheatstone read a paper to the Royal Society and demonstrated an electric clock. Mutual accusations of stolen ideas followed, but whatever the truth of the controversy there is no doubt that Bain got his patent registered first.

In the first electric clocks, an iron pendulum bob was attracted and repelled by electromagnets, or the pendulum bob was made into a magnet which was attracted and repelled by electrically impulsed coils. In the earliest clocks the electric supply was provided by batteries. Later the mains electrical supply was called into service to power a synchronous electric clock, which, incidentally, the purists say is not a clock at all because it has no escapement.

During the period when electric timekeeping was being developed there had been an important breakthrough in mechanical horology. An Englishman, John Harwood, revived the old pedometer system of self-winding. In 1922 he applied it to the wrist watch which had begun to replace the pocket watch in the first decade of the twentieth century (Figure 9.17). Harwood's invention was taken up by a Swiss factory and his watches enjoyed a brief popularity before the firm went bankrupt in the early thirties. The self-winding watch, which subsequently became so popular, is merely an improvement on Harwood's patent.

Electrical discoveries

Two nineteenth-century discoveries eventually revolutionized timekeeping, but there was to be a considerable time lag before the full potential of these discoveries was

Figure 9.17 Drawing of the movement of the John Harwood first self-winding wrist watch

exploited. What made it possible for Bain to make the first electric clock was the invention of the battery. It was found that electric current could be generated by chemical reaction. This was achieved by placing two metal plates called electrodes in a substance that would set up the chemical reaction. This substance, known as the electrolyte, was originally damp earth and the batteries were therefore known as earth batteries. The latest button-sized silver oxide cells used today to power watches, work on exactly the same principle, though of course more sophisticated electrolytes are used.

The other nineteenth-century discovery which was to make such a dramatic impact on the jewellery trade, occurred in 1880 when Pierre and Marie Curie observed the piezo-electric effect in quartz. They found that if an electric current was passed through a slice of rock-crystal it would vibrate at a regular rate. Like the swaying lamp in Pisa Cathedral which Galileo had observed, quartz is isochronous when electrically impulsed. Whereas the controllers that evolved from Galileo's observations swung slowly, the oscillation of quartz is relatively fast, and it has been found that the faster the controller in a timekeeper oscillates, the greater is the potential accuracy of that timekeeper.

The first quartz clock

It was not until the 1930s that the Curies' discovery was applied to timekeeping, when a Canadian, W. A. Marrison, working for Bell Telephone Laboratories in the USA constructed the first quartz clock for use in the telecommunications industry. That first quartz clock occupied a good sized room and its complex circuits were contained in a series of tall metal cabinets. This clock proved to be far more accurate than any timekeeper that had ever been built. It was only later superseded for use in telecommunications and in observatories when controllers with an even faster rate of oscillation were devised. The first of these used an ammonia molecule but later ones employed a caesium atom. The resulting clocks therefore became known as atomic clocks and they had a theoretical error of only one second in a thousand years.

Miniaturization

Although these inventions had little immediate relevance to the horological industry, they did direct the attention of the boffins in the back rooms of the clock factories to the new technology. The invention in 1948 of the transistor to replace the cumbersome valve, by three American physicists working for Bell Telephone Laboratories, John Bardeen, Walter H. Bratain and William Stockley, made it possible for electronics engineers to miniaturize the circuit of the quartz clock.

This led to the launch at the 1960 Basle Fair of what was described as the first domestic quartz clock. This clock consisted of a number of circuit boards incorporating transistors and had an analogue readout. Shortly afterwards the first solid-state quartz domestic clock made its appearance in Switzerland, employing valves which contained a stack of filaments in the form of numbers that lit up when electronically impulsed. See Figure 9.18.

Early electric clocks

Electric clocks powered either from the mains or by batteries were on the market for many years. The most commercially successful and, under normal circumstances, far

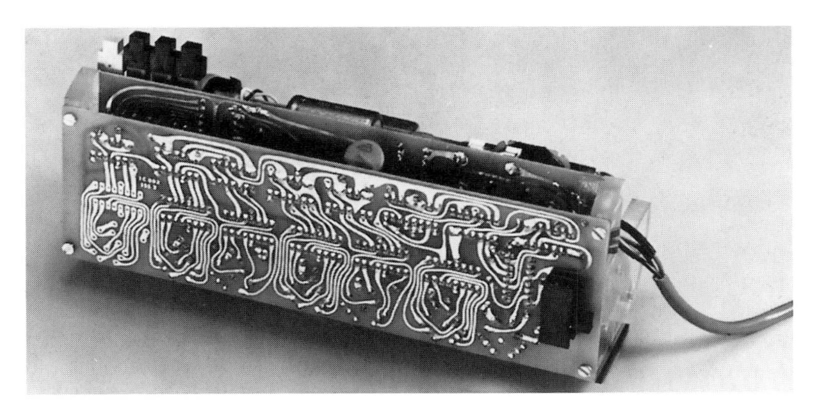

Figure 9.18 Inside the case of a British-made mains-powered quartz digital clock. The digital display was provided by valves containing a stack of numbers which light up successively. Shown here are the printed circuit and the push buttons for setting the valves to time. This model showed hours, minutes and seconds

and away the most reliable of the pre-war electric clocks were the synchronous clocks. These became very popular in the 1930s, particularly the alarm versions. They were not really timekeepers, but slave clocks which employed the alternations in the mains supply as a time standard. So long as the mains frequency remained stable, which was usually the case, these clocks were very accurate indeed. However, during the 1939–1945 war and its aftermath, the mains supply was often interrupted or the frequency reduced. The reputation for reliability which these clocks had gained for themselves was undermined and the public never regained its confidence in them.

The possibilities of the battery clock, which Bain had demonstrated, led to the production of many experimental battery clocks over the years. A few of them came on to the market and had a limited commercial success. Perhaps the best known of these were the Eureka clock invented by T. B. Powers in 1906 and the Bulle clock, invented by M. T. Favre-Bulle in collaboration with Professor Marcel Moulon. The principles on which the Bulle clock operated were those which Bain had invented, but the contact system was improved and various other modifications were incorporated. The mechanical contact system employed in the Bulle clock in fact proved troublesome. Contact troubles also proved to be the main drawback to the new generation of battery electric clocks which came on the market after the last war. These contacts were improved as time went on, but as long as mechanical contacts were employed dirt and chemical deposits could lead to failure.

Those battery clocks, which were imported into the UK in their thousands through a loophole in the licensing regulations, offered electric timekeeping without the need for unsightly leads or special plugs. However, they were not outstandingly accurate. Perhaps their most important contribution was that they introduced the public at large to the concept of a clock that didn't need regular winding. Up to this time only mains-powered clocks and the remarkable Atmos, which used changes in temperature to wind the mainspring, had offered this boon and the Atmos was a clock for the wealthy few.

Eventually the application of the transistor to the battery clock, which made it possible to replace the mechanical with an electronic contact, resulted in greater reliability. The timekeeping of these later clocks was acceptable, but it was never outstanding, no better than a good quality mechanical clock. These battery clocks were, however, relatively cheap and they sold in very large numbers.

Electric watches

The production of button-sized batteries for use in hearing aids in the early 1950s, immediately led horologists to think in terms of an electric watch. It took some years, however, for this dream to become a reality. The now defunct Lip Company in France and Hamilton in the USA jointly researched the project and eventually launched the first electric watch in 1957. This watch used an electromagnetic make-and-break system. This was the system which Bain had invented, but in the watch it was applied to the balance wheel instead of to a pendulum.

The first electric watches were large and ugly and didn't work very well. After some initial interest they found few customers. A number of Swiss manufacturers subsequently produced improved versions over the next decade or so, but the electric watch was never a great commercial success, with one exception, a very sophisticated version called the Accutron.

The sonic watch

This watch came on to the market in 1960, and was created for Bulova by a Swiss electronics engineer, Max Hetzel. He employed a tuning fork with a natural frequency of 360 cycles per second as his time standard. It is the natural frequency of a tuning fork when struck that makes it so useful to musicians for tuning their instruments and it was this association with sound frequencies which led to watches and clocks incorporating a tuning fork to be called 'sonic'. Instead of activating his tuning fork by striking it, Max Hetzel impulsed it electrically by using two coils wound round its tines. He then converted the resulting vibrations into a mechanical drive to the train wheels by means of a small jewel-tipped spring attached to one of the tines. This spring moved an index wheel, only 1 inch in diameter, but with 360 teeth cut round its edge, once every second. The coils were wound with 30 000 feet of wire one fifth the thickness of human hair.

This watch proved to be 99.9977 per cent accurate. This enabled the manufacturers to guarantee its accuracy, something that no watch manufacturer had up to that time been able to do. This guarantee of timekeeping to within one minute a month and the revolutionary concept on which the watch was based caught the public's interest and the Accutron sold well. Nor were the public disappointed. The watch proved to be reliable as well as accurate though its timekeeping could be upset by shocks. Max Hetzel went on to improve upon his original design, but by the time that he had completed his work on his improved sonic system the era of the electro-mechanical watch was over.

The first quartz watches

If, however, the electro-mechanical watches, with the exception of the Accutron were anything but a commercial success, they did lay the foundations for the electronic revolution to come. In the 1960s people in the industry were beginning to speculate about the possibility that one day it would be possible to miniaturize the circuits employed in quartz clocks and produce a quartz watch. Indeed electronics engineers in both

Switzerland and Japan were already working towards this end, the production of the watch of the future.

In 1968 both the Japanese and the Swiss were able to announce that they had produced prototype quartz watches. When these prototypes were submitted to the Neuchâtel Observatory for testing they proved to have an error of only two seconds a day. They were, however, clumsy, inelegant watches compared to the mechanical watches being produced at the time. Of course they represented an outstanding scientific achievement, but nobody in the trade saw them as a challenge to the traditional mechanical watch.

The Japanese, however, who had the advantage of a vigorous electronics industry, continued to develop the new concept. The Swiss on the other hand, with their vast investment in the old technology, though they continued to fund research, were somewhat half-hearted in their attitude to the quartz watch. They didn't really want it to have a future and at first it seemed that their wish might be granted.

Solid state

To electronics engineers those first analogue quartz watches were an anachronism. Why not, they argued, get rid of those expensive and potentially troublesome gear trains and produce a watch that was completely electronic. This thinking led, only four years after the first quartz watches made their appearance, to the production of a solid-state quartz watch, that is a watch with no moving parts. Launched by Pulsar in 1972, this new watch employed light-emitting diodes to provide a digital readout, which led the watch to be christened an LED. These diodes were a miniaturized form of those valves with their stacks of numbers used for the readouts of the earlier solid-state quartz clocks. The LED had one major drawback, however: the diodes consumed a lot of power. To conserve the battery therefore, the LED watch was designed so that the diodes were activated for only a second or two when the wearer wanted to read the time. This meant that the wearer had to press a button to find out what time it was. In the event these first digital quartz watches had little more commercial success than the first generation of quartz analogues had done. Button pressing at that time seemed to be an unacceptable disadvantage.

Chips

The early 1970s witnessed a new invention that eventually changed all this. A group of electronics engineers in the development department of Texas Instruments invented the integrated circuit, soon to became known more familiarly as the 'chip', in the course of their work on the space programme and on computer development. The chip is a thin wafer of silicon, often no larger than a cornflake, on which a circuit incorporating large numbers of transistors, resistors and capacitors is imprinted by a complex photographic process. What the chip achieves is perhaps best illustrated by the fact that one chip can do the work of the room full of circuits that Marrison employed for the first quartz clock.

To produce these chips in quantity called for a vast capital investment, but once a production unit had been set up it would turn out chips by the million, far more than the electronics industry could at that time absorb. So the producers began to look round for other outlets. Could they find employment for their chips in the area of consumer durables, they wondered. The first outlet they found was in the production of pocket calculators, then very much in their infancy. Soon the chip made it possible to reduce the price of the pocket calculator dramatically and over the next few years untold millions of them were sold. But still the production lines were pouring out chips so fast that even this rapidly growing new industry could not absorb them all.

It was then that someone had the idea of applying a chip to the solid-state quartz watch, believing that at the right price this too might take off as the pocket calculator had done. Before long half a dozen major American electronics firms had moved into the watch business and greatly to the surprise of the traditional watch industry the public did in fact buy these new press-button LEDs. They bought millions of them. It seems that the fact that for the first time in hundreds of years a watch looked different and employed a space age technology blinded people to its disadvantages. But not only were the public being offered a new look, they were being promised an unbelievable new accuracy, a watch that would not have an error of more than a minute or two a year. Before long they were also being offered a bewildering variety of additional functions. Complex LEDs with day-date calendars, alarms, and incorporating split-second chronographs were soon coming on to the market. There seemed to be no limit to what these new watches could do.

The retail jewellery trade resisted the new watches, feeling not only that they were an unknown and untried commodity, but also because they were reluctant to deal with new and equally unknown suppliers. They were relieved that in the end the LED turned out to be a nine-day wonder. The decline and fall of the LED came about for a variety of reasons. The most important of these was probably that the product's inherent shortcomings soon began to reveal themselves. Many of the watches that came on the market proved to be unreliable, battery life often proved in practice to be unacceptably short and once the novelty wore off the public found that it wasn't much fun to have to press a button in order to find out what time it was. Also the big electronics companies and the smaller assemblers, who had jumped on the bandwagon, discovered that the watch business was no easy area in which to make a fortune. Instead of profits, they found they were recording big losses and all but a handful of these firms either moved out of the watch business or disappeared altogether. In many instances this led to the disappearance of an after-sales service and to the final disillusionment of the public. The final nail in the coffin of the LED was the appearance of a better alternative, a solid-state watch that gave a continuous readout, was nothing like so battery hungry and which ultimately proved much more reliable.

Liquid crystals

While the LED boom was in progress, both the Swiss and the Japanese were working on a new readout known as the liquid crystal display, or the LCD. These readouts consisted of organic crystals held between two sheets of glass. The crystals consisted of a series of segments, arranged in the form of a figure 8. Various combinations of these segments would form any numeral from 0 to 9. The segments were activated electronically by means of a chip, resulting in the required ones becoming opaque.

The first LCDs were difficult to read and their life expectancy was as little as a year. Before long, however, both the readability was improved and their life greatly extended. Even so, they remained not all that easy to read in marginal light conditions and or course, unlike the LEDs, they could not be read in the dark. This problem was solved either by providing a small back light to the dial or by the use of a radioactive substance that illuminated the dial. The light, if frequently used, drained the battery. The chemical system, while it did not affect battery life, was outlawed in many countries and was abandoned.

Like the LED before it, the LCD soon became available with a multitude of ancillary functions, some watches having as many as fifty, including up and down 100ths of a second chronographs, time in two time zones, musical alarms and so on. It seemed for a long time that this new wave of digital watches would not only change the public's idea of what a watch ought to look like in the space age, but also the structure of the watch manufacturing industry. Anyone with access to a supply of the components could now produce watches independent of a labour pool skilled in the traditional craft of watchmaking. Hundreds of new firms sprang up in the Far East and put watches on the market at ever lower prices. This eventually persuaded most of the traditional manufacturers who had taken up the LCD to loose interest in it. Today, though large quantities of LCDs are still being produced, the vast majority of them are being sold outside the jewellery trade.

Back to hands

Fortunately for the future both of the traditional watch manufacturers and for retail jewellers, the public eventually made two important discoveries. Most of those who bought complicated LCDs found that once they had mastered the logic, by studying the not always very lucid instruction manuals, and had played with their watches for a time, they made little use of all those much lauded additional functions.

What most people want their watches for is to tell them the time and perhaps the date – and when it came to telling the time the digital readout was not very informative. The time, 8.37 for example, is not particularly meaningful. The information we

Figure 9.19 Inexpensive quartz clock movement made in Germany and used by a number of clock manufacturers

want, when we look at our watches, includes such facts as how long is it since 8 o'clock or more importantly, how long there is to go before say 8.45, when our train or our plane leaves or when we have to turn up for an appointment. It is possible, of course, to work out this information from the time displayed on a digital readout, but it calls for a little calculation. The relationship of time present to time past and time future, on the other hand, is there on a conventional dial for all to see. We are assured by the advocates of the digital readout that the younger generation will become used to this method of telling time and this may well prove to be the case though there has recently been a marked decline in the popularity of these watches.

Many members of the public found that they just didn't like digitals and felt much more at home with the time-honoured system of hands traversing a dial. A further drawback to the digital watch is that it is very limiting as far as design is concerned. Once the novelty wore off we began to realize that not only did all LCDs look much alike, but they were not particularly attractive. Slowly but surely the analogue quartz watch, so long neglected, began to make a comeback and, as these watches became slimmer and smaller and the case and dial designers were able to give them style and elegance, it seemed that this after all was the most acceptable form of quartz watch. It was as though tradition had tamed the new technology. Now it was necessary to look for the word 'quartz' on the dial to know whether a watch was controlled by a sliver of synthetic rock crystal or by a balance and escapement. The LCD is far from dead, but the majority of those being sold today probably sell for less than £20. In the price brackets above this, the analogue quartz watch is taking a bigger share of the market every year.

While this revolution had been taking place in the watch trade, a quieter but no less devastating revolution had taken place in the clock industry. Sonic clocks employing a tuning fork system enjoyed a brief popularity, but this was soon followed by the development of inexpensive, accurate and reliable quartz clock movements by a handful of German clock manufacturers. These not only ousted the electric clock movement but also the mechanical clock movement as well, except at the top end of the market. Today these inexpensive quartz movements are fitted to every type of clock from miniature alarms to expensive brass-cased mantle clocks. Even anniversary clocks are nowadays fitted with quartz movements and the revolving controller with its brass balls, once the bane of the workshops, is now no more than a decorative feature. These clocks work on the same principle as the quartz watch, with a chip, a stepping motor and a modified train. The only major difference is that most of them use a quartz which has a megahertz frequency.

Chapter 10

Watches and clocks continued

Despite the advent of the quartz watch, large numbers of mechanical watches continue to be made and sold to the public, particularly in the third world. The last few decades have seen a number of important developments both in the design and construction of mechanical watches. The high cost of producing mechanical movements, compared with solid state modules, has forced the watchmaking industry to evolve so-called 'economical' movements and to introduce increasingly more sophisticated mass-production techniques in both the ebauches and the assembly factories. The developments in production technology are also applied to quartz analogues of course.

Automation

The ebauches, the so called 'rough' movements consist of plates and bridges, wheels and pinions and screws, which together with the balance and escapement and the springs, are assembled to produce mechanical watch movements. These parts have been mass-produced for decades. The automatic machines which turn up parts from steel rods, performing a pre-determined series of machining processes as one tool and then another comes into operation, date back to the beginning of the twentieth century (Figure 10.1). The transfer machines which today carry out all the machining and drilling necessary to turn brass sheet into plates and bridges are, however, a much more recent development. Relative newcomers, too, are the carousels. These are fascinating miniature roundabouts, which revolve and vibrate. They supply the components with which they have been loaded to the tools. When correctly orientated, they allow sub-assemblies to be carried out completely automatically (see Figure 10.2).

Many of the machines seen nowadays in the ebauches factories are computer programmed. Balances are also poised on computerized machines. The computer measures and records every dimension of the balance and then instructs the tools to remove a little metal here and a little there from the rim of the wheel so that it will run perfectly true. Even the working drawings which are supplied to the various workshops, which once took skilled draftsmen many hours to produce, are now produced by a computer in minutes. Today a complete new calibre can be designed by a computer into which the necessary information has been fed.

Indeed, one of the things which always impresses the visitor to a modern ebauches factory is just how few people there are to be seen. Row upon row of machines, hundreds of them, working away, produce millions upon millions of components, seemingly of their own volition. The aim of all this automation is to reduce the number of skilled workers to a minimum, so controlling overheads and keeping prices competitive.

The application of automation to the assembly of movements has been a much

Figure 10.1 The five cutters of an automatic machine cutting the rod to make an arbor

slower process. Partly this was because the assembly factories each produced so many different calibres and this diversity did not lend itself to modern mass-production techniques. Another factor was that in the past the assembly factories tended to be small, under capitalized, family businesses which did not have the resources to either develop or install the necessary plant. More recently the coming together of these factories into more powerful groups resulted in finance being available for rationalization. This has led to the widespread adoption of the modern technology which had been pioneered by the larger or more forward looking companies both in Switzerland and Japan.

Thirty years ago watches were still being assembled, virtually by hand, by teams of highly skilled watchmakers, whose apprenticeship occupied many years. However, using such methods the factories found that they were unable to exploit the greatly increased demand for their products in a world where more and more people wanted to own a watch. To achieve increased production it was necessary to dilute the labour force by making use of semi-skilled workers. To this end the assembly had to be broken down into a series of relatively simple tasks that workers, assisted by ingenious jigs and power tools, could carry out after a relatively short period of training.

So the production line made its appearance in some watch factories and watches began to be assembled in much the same way as motor cars and refrigerators. To facilitate this rationalization a number of ingenious assembly systems were devised in the 1960s and 1970s. The first of these was known as the Lanco bicycle chain system (Figure 10.3). The basis of this system was a unit consisting of 50 to 100 movement holders mounted on cogs linked by a continuous chain. The whole unit was covered by a protective plastic lid designed to exclude dust. This unit was positioned in front

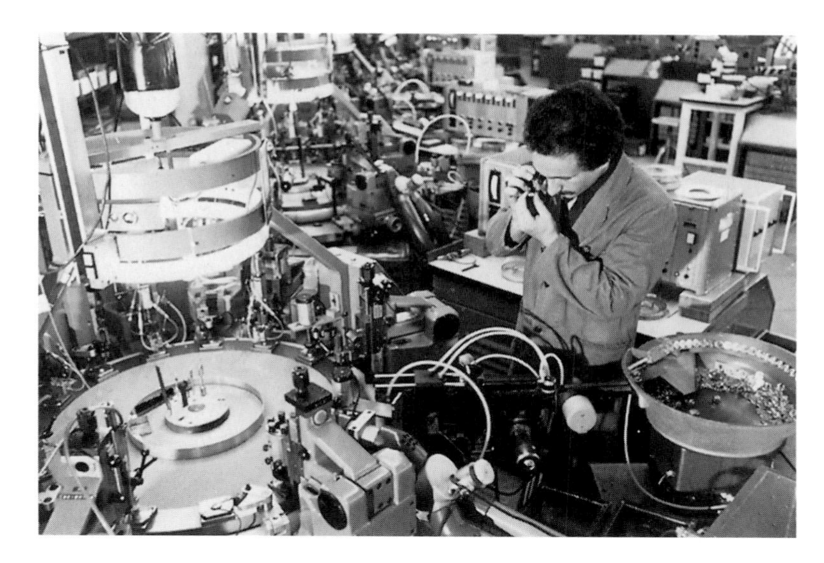

Figure 10.2 An automatic unit in a Swiss ebauches factory, which was fed by a carousel (right) and carried out nine operations on plates or bridges. The operative in the picture is an inspector. All these machines were programmed to process the components without human intervention

of an operative on the line who would pull a lever which brought the first of the fifty or so movements into a position where they could carry out the operation for which they were responsible. Having done this they brought the next movement forward until eventually they had completed their work on all the movements in the unit. The unit then passed to the next operative on the line and so on until all the operations were completed and all the movements were assembled, oiled and ready for testing. At intervals along the line inspectors checked the work as it was carried out in order to reduce the number of completed movements that needed to be rejected or expensively rectified.

Providing inspection standards were maintained at a high level it was found that watches put together in this way were in no way less satisfactory than those produced by traditional methods. It was subsequently found, however, that any failure to maintain quality standards could result in a quite unacceptable rate of returns under guarantee. Under the old system the odd stopper might get through. Under the new system there was always a danger that these might pour into the market place in their thousands and, in the late 1970s, this was exactly what happened. Like all systems, this one was only as good as the people who operated it.

A number of variations on the Lanco system were subsequently introduced, but all of them worked on similar principles. Over the years, however, more and more automated tools were introduced in to the lines in the more progressive factories. Eventually, first of all in the giant Japanese factories, robots took over from people.

Not only are watches now invariably produced on these lines except in a few prestigious Geneva factories, but quartz analogues also take advantage of the economies which such mass production techniques provide. The quartz, the chip and the stepping-motor constitute a sub-assembly, replacing the balance and escapement sub-assembly. The mainspring is replaced by a battery, but otherwise there is no

Figure 10.3 A production line in a Swiss factory employing the Lanco assembly system

essential difference in the assembly of an analogue watch whether it has a quartz or a mechanical controller.

The introduction of the assembly line, if it was to achieve the increased output and the desired economies, had to go hand-in-hand with a reduction in the number of calibres the individual factories offered to their customers. In the past a single factory might assemble 20 or even 30 different calibres. With the new system it was not feasible for a factory to produce more than two or three calibres. It is costly to rejig a line and to retrain the operatives to work on a different movement from the one on which they are accustomed to work. Although it has become common practice for the different factories to assemble movements for one another so that each can offer the trade a wider choice, the number of calibres produced by the whole industry has been greatly reduced compared with thirty years ago. Now the majority of quartz movements in Switzerland are centrally produced by the former producers of mechanical ebauches.

Designed for economy

The pressure on the producer of popularly priced mechanical watches to keep prices down also led to the rationalization of the movements themselves. Perhaps the simplest illustration of this is the employment of standard sized screws. At one time as many as twenty or more different screws might be found in a single movement. Subsequently only two or three different sizes were used, resulting in considerable

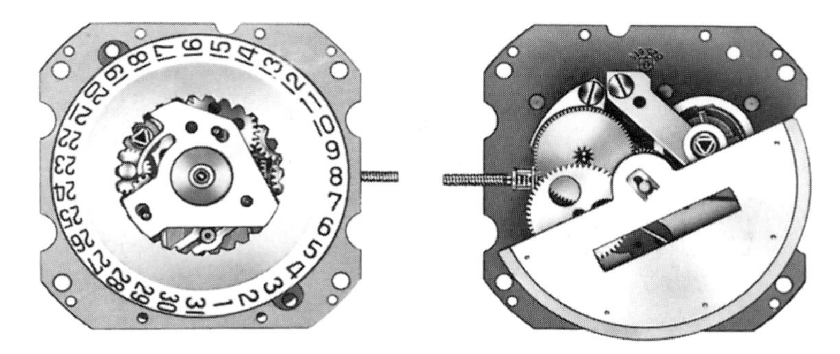

Figure 10.4 One of the economical calibres, which is self-winding. Parts were reduced to a minimum. Screws were standardized and the calibre designed for automated assembly

economies both at the ebauches and assembly stages of production. The number of components in some of the movements were also reduced in the course of designing the 'economical' movements, while great attention has been given by the designers to ease of assembly by the use of automated tools (Figure 10.4).

Mechanicals for the few

If the mechanical watch may be said to be fighting for its survival in the mass market, many people in the industry believe that at the top end of the market the mechanical watch has an assured niche for the foreseeable future. Indeed the mechanical watch is acquiring a snob value among the rich and famous. This probably lies at the root of the revival of the skeleton watch, which allows the owner to study the working of its wheels and pinions. Producing fine skeleton watches is a minor art. The craftsman reduces the plates and the bridges to produce an intricate design of engraved scrolls while at the same time retaining adequate strength in the frames to resist the considerable forces to which they will be subjected. The classic complicated watches are still being made by the Geneva houses. Chiming repeaters and perpetual calendar watches are made for the fortunate few with no thought of expense. There are obviously people in the world who still appreciate such outstanding examples of engineering in miniature, and prefer them to quartz watches, which though offering a far greater number of functions and far greater efficiency as timekeepers, are by comparison perhaps a little soulless.

Mechanical knowledge

There are indeed still some people who find a mechanical watch, which they have to wind and which ticks away on their wrist, companionable, seeming to have a life of its own. So long as there are people who feel like this the mechanical watch is far from moribund. Therefore for a long time to come anyone working in a jeweller's shop is going to have to understand the principles on which mechanical watches work, what different types are available and what their merits are. They must retain an interest in the old technology as well as embracing the new.

The principles on which a jewelled lever watch functions are very simple, though the designing of an efficient mechanical movement calls for mechanical and mathe-

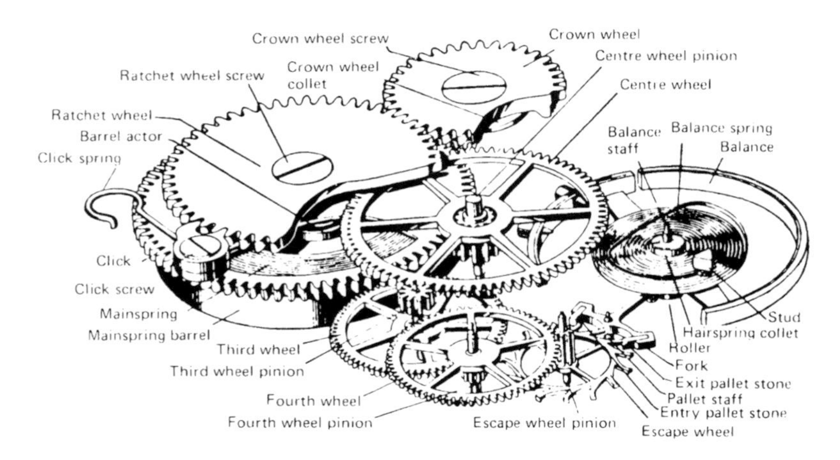

Figure 10.5 Diagram of an older type of jewelled-lever movement with screwed balance

matical skills of a high order. Perhaps the easiest way to appreciate the modern watch is to consider it in terms of those basic elements of which any timekeeper is comprised – the power source, the controller and those elements which enable the owner of the watch to tell the time (Figure 10.5).

The power source

The power source of a modern watch consists of a coiled alloy spring contained in a disc-shaped brass box called the barrel. In a manual watch the power is, as it were, turned on by the wearer winding up this spring by rotating the winding crown. This activates a series of gear wheels, known as the motion work, which turn the barrel and so coil the spring. The winding crown is, of course, also used, in the pulled-out position, to set the hands by means of another set of gears.

At one time mainsprings used to break; indeed this was so common that it was usual for mainspring breakage to be excluded from the guarantees issued with watches. Nowadays, however, a mainspring breakage is a rare occurrence because so-called 'unbreakable' alloy mainsprings are universally used. In fast-beat watches, which will be described later in this chapter, a more powerful spring was required than for a watch with a standard beat of 18 000 per hour. One watch company, Longines, introduced a fast-beat with two mainsprings in two separate barrels working in tandem a few years ago. This arrangement allowed them to provide the extra power required and still produce a very thin movement.

A watch used to stop while it was being wound, but for many years now all watches have been fitted with what are known as 'going barrels'. This means that the spring is wound in the same direction as that in which it drives the watch and so the driving continues during winding. This system is also known as 'maintaining power'. Many members of the public believe that they can damage their watch by overwinding. Though it is conceivable that someone may, by applying elephantine force to the winding crown, wreck the motion work, it is not possible to overwind a modern watch. Many people make a habit of winding their watch once a day, but in fact it is probably better to give the crown a few turns from time to time during the day because this results in keeping the spring more or less fully wound so that it gives out a more or less constant torque.

Controlling the power

In the same way that the weight used to power a weight-driven clock would expend all its energy in plummeting earthwards if it were not controlled, so the pent-up power of a coiled spring would be expended in a fraction of a second if it were allowed its freedom. One of the functions of the escapement in a watch is to control the release of power from the spring. The type of escapement used in virtually all modern watches is the jewelled-lever escapement. This consists of a lever, not unlike an anchor in appearance. The flukes of the lever consist of two slabs of synthetic ruby, with the working faces cut to extremely fine tolerances. These synthetic rubies are called the pallets and they release and check the progress of the escape wheel by acting on the boot-shaped teeth which have been cut round its perimeter. When a tooth is released a little power is allowed to escape from the mainspring.

Many other types of escapement have been invented over the centuries. Prior to the appearance of inexpensive digital quartz watches, cheap mechanical watches fitted with a pin-lever escapement were made in their millions. In this escapement two steel pins served as pallets. The trouble was that these were more likely to suffer wear than ruby pallets, causing the timekeeping of the watch to suffer.

Providing a time standard

Control of the power output is not the only function of the escapement. The escapement works on a reciprocal basis in conjunction with the balance. The balance is a wheel with a heavy rim controlled by a fine spring known as the 'hairspring', or more often these days as the 'balance spring'. This spring is attached at one end to the axle of the balance wheel, the balance pivot as it is called. The other end is attached to a regulator which enables the spring to be either lengthened or shortened in order to adjust the timekeeping of the watch. The operation of the balance and escapement, which have been described as the heart of the watch, are described in detail in Appendix 3. They are briefly summarized here.

The balance is impulsed by the power released from the mainspring. The impulse is conveyed to the mainspring by the escape wheel and the fork. This is achieved by the slot in the top of the fork acting on a synthetic ruby pin on the hub of the balance wheel. This impulse causes the balance wheel to swing, which serves to coil the balance spring. The balance spring now becomes a souce of power it its turn, causing the balance wheel to swing in the reverse direction. This winding and unwinding produces a rocking action in the lever which causes it to release and lock the escape wheel by turns, switching on and off the power supply. Because the balance spring, like the pendulum, is isochronous, it causes the balance to oscillate at a regular rate, impulsing the escapement in a standard watch every fifth of a second. By this means time is divided into intervals of ⅕th of a second, which provides the watch with what is known as a time standard.

The readout

These fifth-of-a-second impulses are conveyed to that set of brass wheels and steel pinions known as the train, which gears them down and turns the hour, minute and seconds hands by way of the sleeved cannon pinion in the centre of the dial. This gear train is sometimes referred to as the 'motion work'. The mathematics and geometry involved in the design of the teeth is very complicated.

This is a very simplified account of the working of a lever watch, but the sales person really only needs to understand the principles involved in order to have an appreciation of what they are selling. What must also be appreciated, however, is that the care with which the various components are made, finished and assembled has a direct influence on the performance of a watch and that the time taken to produce a fine watch as well as the quality of the components used is reflected in the price.

Jewels and shock proofing

Two other features which are standard in a modern mechanical watch need some comment, however, if for no other reason than they probably elicit more questions from the public than any other feature of the movement. These are the jewels and the shock absorber. It hardly needs to be said nowadays that the jewels used in watches are cut from synthetic corundum, but perhaps there are still a few people who think of them as real jewels and equate the value of the watch with the number of these gems present in the movement. There is, however, an optimum number of jewels and additional ones are unnecessary, or produce only a negligible improvement in performance.

A fully-jewelled standard hand-wound movement requires 15 jewels, 17 if it has a centre seconds hand. There are two pallet jewels and the ruby pin on the balance. Then the balance wheel has four jewels, two jewelled bearings and two endstones which are disc shaped and are located below the bearings to reduce friction and keep out dust. The other eight jewels serve as bearings for the pinions of the escape wheel and the train wheels. Sometimes in expensive watches the bearings in which the escape wheel and the train wheels run are provided with endstones, but this is unusual. On some self-winding watches the wheels of the self-winding mechanism also run in jewelled bearings. Thus there may be more than 15 jewels in a watch, all of which serve a purpose even if they are only marginally functional. If a watch is advertised as having so many jewels these must, however, all function to some degree. A law case some years ago established the precedent that to put in jewels for the sake of making a watch seem more desirable was calculated to deceive the public and was therefore illegal.

The jewels that are used for bearings are designed so that they will hold and retain the oil which reduces friction and wear. This oil does, however, tend to dry up in time or to become contaminated by any dust that enters the case. Dirty oil is one of the major causes of malfunction in mechanical watches. It results in the watch stopping or running irregularly. This is why all watches with mechanical components need regular cleaning and oiling. The period of time which can safely be allowed to elapse between services varies. Very small women's calibres may need to be serviced once a year. A standard 10½ ligne will probably perform satisfactorily if it is serviced at two or even three year intervals, while a waterproof watch the case of which is resistant to dust as well as to water can go for even longer periods without attention.

But the conditions under which a watch has been kept before it was sold to the customer are important too. Many watch windows are like ovens and a lot of returns under guarantee have resulted from the oil having dried up in this unsuitable atmosphere. This fact has led jewellers to seek window lighting for their watch windows that generated less heat. To a great extent, modern lighting has helped to solve this problem. Periods during which sales have fallen off also tend to produce a high level of returns under guarantee, because in these circumstances watches tend to be stored for long periods in hot dry warehouses and showrooms.

Shock proofing is a feature of virtually all watches nowadays. The Incabloc system, which is the most widely used, is really only a more sophisticated and refined version of the shock absorber invented by Breguet in the eighteenth century. Its purpose is to protect the balance staff, which because of the weight of the balance wheel is liable to fracture if a watch is dropped. The Incabloc consists of two conical jewels in metal sleeves which are located in conical holes. Below these jewels endstones are located and held in position by lyre-shaped springs. The springs act like buffers, absorbing any thrust. The cone-shaped arrangement ensures that the jewels are returned to their proper alignment.

Increasing the beat

The normal mechanical watch beats 18 000 times in the course of an hour. Pursuing the theory that the faster the time standard of a timekeeper the more accurate it is likely to be, a number of high-beat calibres were designed in the past. Beats double the normal, that is of the order of 36 000 an hour have been used, as well as a number of intermediate frequencies. All these fast-beats have proved themselves capable of producing a much steadier rate than the standard watch, partly because the higher beat reduces positional error, though even their performance pales, of course, in comparison with a quartz watch. But for the appearance on the scene of the quartz watch the fast-beat would probably have become the standard watch of the future. The initial problem of providing the necessary extra power was soon overcome and so too were some initial lubrication problems. The fast-beats that were produced proved to be both reliable and accurate.

Self-winding watches

Had the quartz watch not been invented, probably every watch produced would eventually have been a self-winding watch with the exception of a small number of dress watches. It is difficult, in fact, to understand why automatics took so long to gain acceptance. They have much to recommend them. Perhaps the initial resistance on the part of the public was due to the fact that the early models tended to be rather fat and that some of them were not particularly efficient and failed to build up an adequate power reserve. Even after slim and small automatics (see Figure 10.6). became commonplace and built up enough power to run for a couple of days after a few hours

Figure 10.6 Typical automatic movement incorporating a two-way winding system with the weight swinging through 360 degrees

wear, sales of these watches continued to represent only a very small proportion of total watch sales until the middle of the 1970s. Perhaps the public could not see that it was worth paying a high purchase price and higher service charges just to be relieved of the insignificant task of daily winding, a task to which they had anyway become habituated.

Once, of course, anyone has worn an automatic they would be unwilling to go back to manual winding again, just as anyone who has worn a quartz watch would be unwilling to put up with the vagaries of a mechanical watch. What, of course, was not generally appreciated was that self-winding watches have other advantages besides not needing to be wound. They are potentially more accurate than manually wound watches, because their mainsprings tend to remain more or less fully wound for most of the time. At one time variations in mainspring torque had a considerable influence on timekeeping. Up to the middle of the nineteenth century, any good-quality watch was fitted with a fusée to compensate for variations in the power output of the mainspring as it wound down. Improvement in movement design led the variation in rate due to this cause to be reduced to an acceptable level, but the problem still existed.

The ability of a watch to keep time is limited by a number of inherent factors such as positional error and this problem of torque. The maintenance of power at a relatively stable level in a self-winding watch leads to better timekeeping. If self-winding is combined with a higher beat, optimum accuracy for a mechanical watch will be approached. Another not-insignificant inbuilt advantage of the self-winding watch is the fact that, because the crown is used only infrequently, wear is less likely to occur. Wear to the stem and O-ring can result in the case becoming less dust resistant, or if the watch in question is a waterproof, less water resistant.

Almost all the self-winding watches that have been produced in recent years have a semi-circular rotor with a heavy rim pivoted on the centre of the back plate of the movement. This swings through 360 degrees when the wearer of the watch moves their arm. An ingenious set of gears, located under the bridge that supports the other end of the weight pivot, allows both clockwise and anti-clockwise swings to wind the mainspring barrel. A clutch is also provided which disconnects the gearing when the mainspring is fully wound.

Calendars

The most popular additional function nowadays on both mechanical and quartz watches is a calendar either one giving just the date, or the day and the date. The mechanism (Figure 10.7) required to provide the date is simply a disc with the figures 1 to 31 printed round its circumference. This is driven by an additional wheel, the 24-hour wheel, which in its turn is driven by the hour wheel of the train. The provision of day indication, which is shown next to the date in a slot in the dial requires a second disc printed with the days of the week. Usually nowadays there is a choice of two or three languages on this disc which makes it more economical to produce calendar watches for world markets.

In the early days, the setting of these calendars was very laborious and required the owner to turn the hands of the watch through 24 hours to advance the calendar by one day. The date on later watches could be changed in seconds because an intermediate position of the winding crown allows the date to be flicked over independently of the hands. If the watch is a day-date, the day is changed by moving the crown clockwise and the date by moving it anti-clockwise (see Figure 10.8). Calendars on LCD watches are invariably programmed for a period longer than the anticipated life of the watch.

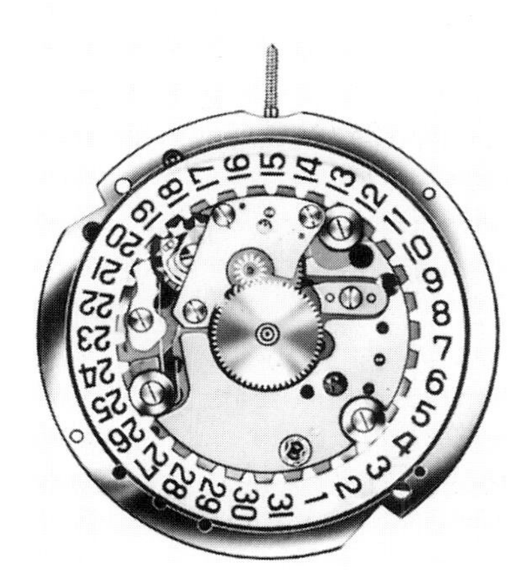

Figure 10.7 The calendar mechanism added to many modern watches is relatively simple and does not add greatly to the cost

Setting is therefore only necessary when a battery is changed because the programme automatically takes account of 28, 30 and 31-day months and even leap years.

Anti-magnetic

Because virtually all quality mechanical watches are nowadays fitted with alloy balance springs they are not affected by normal magnetic fields. In the past the fact that balance springs could become magnetized and the coils adhere to one another did cause problems.

Sizes

The diameter of a watch movement is sometimes expressed in ligne. One ligne is the equivalent of 2.26 millimetres. Thus a 10 ligne movement has a diameter of 22.6 millimetres. However, nowadays the diameter and height are more often simply expressed in millimetres.

Figure 10.8 The two discs for the date and the day used in a day-date watch. As can be seen there is a choice of two languages for the day

Quartz watches

Though the mechanical watch is just about with us, the quartz watch now has the lion's share of the market except in the underdeveloped countries. Everyone working in the retail jewellery trade therefore needs to understand at least the rudiments of quartz watch technology. If the production of quartz analogues involves many of the processes described in connection with mechanical watches, the production of the electronic components introduces us to another world, while the production of solid-state watches involves processes completely alien to the watch industry.

Making chips

Making the chips, which have been described as the brain of the quartz watch, is carried out in conditions and by techniques more reminiscent of the laboratory than the workshop. Production takes place in conditions of scrupulous cleanliness in a controlled atmosphere behind glass panels by operatives dressed in special clothing. No visitor is allowed to penetrate these sealed compartments. The processes which are involved in creating these miraculous chips and which produce a complex and powerful circuit on to a miniscule slice of silica are highly technical. Thin sheets of silica are first produced and then processed so that they will react to a photographic process. The required circuit is then transferred to the silica by a technique known as photo-resist. This imprints a number of circuits upon the silica. The number depends on the size of the chip. The sheet is then chopped up into the familiar little squares each with its transistors, resistors and capacitors linked to form an integrated circuit. The necessary leads are then added and the chip is ready for testing.

Only a relatively small percentage of the chips produced on any production line are up to specification; an acceptable average is about 30 per cent. The highly technical nature of the processes involved in this production is indicated by the illustration in Figure 10.9. The circuitry is incorporated in a slice of silica only perhaps $\frac{9}{16}$ in. by $\frac{3}{16}$ in. for a man's watch and one not much more than half that size for a woman's watch.

The quartz

The dimensioning of the quartz and mounting it in a tiny vacuumized compartment and the production of LCD readouts are equally technical and demand the same scrupulous attention to the conditions of manufacture. The sub-assembly of the electronic components of an analogue and the assembly of a digital module are carried out by soldering. This work, like the assembly of mechanical parts, has now been automated. One computer-controlled unit now carries out a number of soldering operations to a programme without human intervention.

The analogue takes over

It now seems certain that the analogue quartz watch rather than the digital will turn out to be the watch of the future although complex digitals continue to be sold in large numbers, mainly low-priced models through outlets other than the jeweller. It is significant that even in the catalogues the analogue tends to oust the digital, which at one time took pride of place on these colourful pages.

The watch trade views this development as a return to normality. The analogue is a far less alien product than the digital. It looks more like a watch. It lends itself

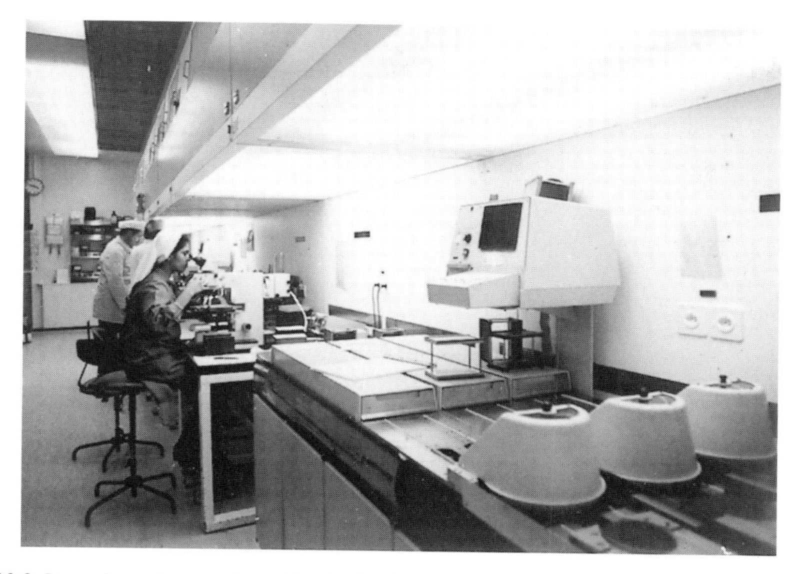

Figure 10.9 Part of a unit producing chips in the Ebauches SA factory at Marin in Switzerland. Here a photo-sensitive coating is being applied to the silicon before the circuit is photographed on to it

better to styling and tends to sell on its design. For this reason it is easier for a watch salesman to sell. It is certainly easier to sell-up these watches on the basis of such readily recognizable features as the elegance of its case, the quality of the dial and so on, particularly now that these watches have become so slender, unbelievably so, in some instances! (See Figures 10.12 and 10.13.) Indeed the quartz analogue differs from the mechanical watch only in having certain virtues inherent in its technology. Its time-

Figure 10.10 A semi-automatic assembly system for quartz watches, which have a synthetic overmoulded grid instead of conventional plates

Figure 10.11 The electronic components of a watch nowadays are assembled on these semi-automatic machines. They pass through the line attached to thin film and are separated into individual pieces when the assembly has been completed.

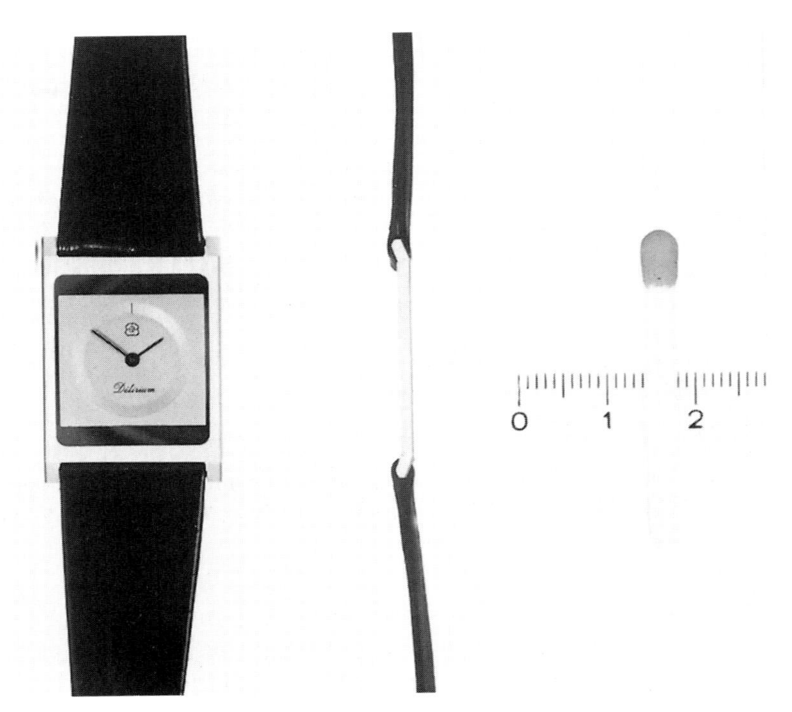

Figure 10.12 An ultra-slim quartz watch

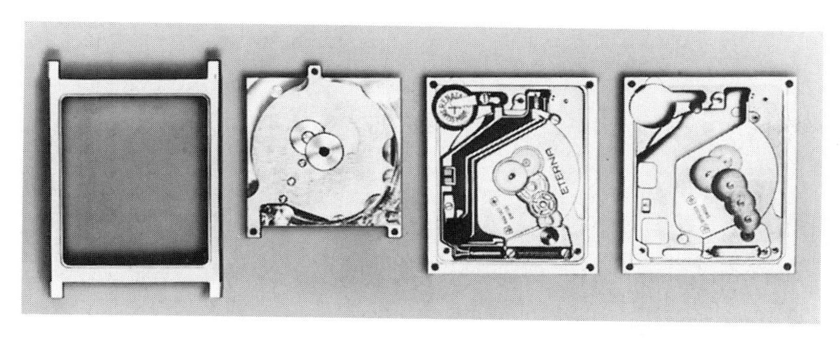

Figure 10.13 The result in Figure 10.12 has been achieved by re-designing the movement, using the dial and the back of the case as the frame and dispensing with conventional plates and bridges. The train wheels are located under the dial but the quartz, the chip, the stepping motor and the battery are displaced down the sides of the movement. This technology was subsequently used in the development of the Swatch movement

keeping far surpasses that of any mechanical watch. Though it is perhaps dangerous to dwell too much on precision for fear of turning customers into seconds watchers, accuracy is a virtue in any timekeeper. More important virtues are perhaps the fact that quartz watches are, by-and-large, trouble free, reliable and dependable.

Most jewellers record return rates for quartz analogues in the region of 0.5 per cent compared with perhaps 10 per cent for medium priced mechanicals, though some brands may achieve more acceptable figures. From both the jeweller's and the customer's point of view this reliability is perhaps the best reason there is for selling and for buying quartz. More meaningful to the average watch owner than the fact that his watch has an error of only two minutes a year is the knowledge that when he looks at his watch he knows that it is telling the right time, that at worst it is a few seconds adrift. This is true whether the watch is worn regularly or laid aside for a period. Women, in particular, because they tend to take their watch off from time to time or wear different watches for different occasions, find this a great boon. Perhaps the only snag to the quartz watch is the need to replace the battery. However, with battery life extended to from two to five years, depending on the design of the watch, and likely to be further extended in the future this is much less of a drawback than it was initially.

There are always going to be people who will want the facilities offered by a complicated LCD, so the jeweller cannot really ignore the digital. Digitals do indeed have certain virtues. Having the calendar programmed for a period of years, for instance, is an advantage and there are occasions when it is useful to have an alarm on your wrist or a timer handy.

There are currently three types of quartz watch of which the salesperson needs to have an understanding. These are the analogue, the LCD and the combined analogue and LCD.

Inside an analogue

In an analogue quartz watch the power source is a battery which produces electrical energy as a result of a chemical reaction between the electrodes and the electrolite. The controller in a quartz watch consists of the quartz and the chip working together.

Figure 10.14 A modern Swiss analogue movement. The chip can be seen to the left of the stem with the quartz in its vacuumized container to the left of this. The long coil of the stepping motor is at the bottom left and the battery bottom right

These replace the balance and escapement of a mechanical watch. Like the balance and the escapement, the quartz and the chip have reciprocal roles. The chip activates the quartz and gears down the quartz frequency to a single impulse every second. This impulses a miniature electric motor, the so-called stepping motor, which translates electrical impulses into mechanical impulses and sets the gear train which drives the hour, minute and seconds hand, in motion. Such a movement can be seen in Figure 10.14.

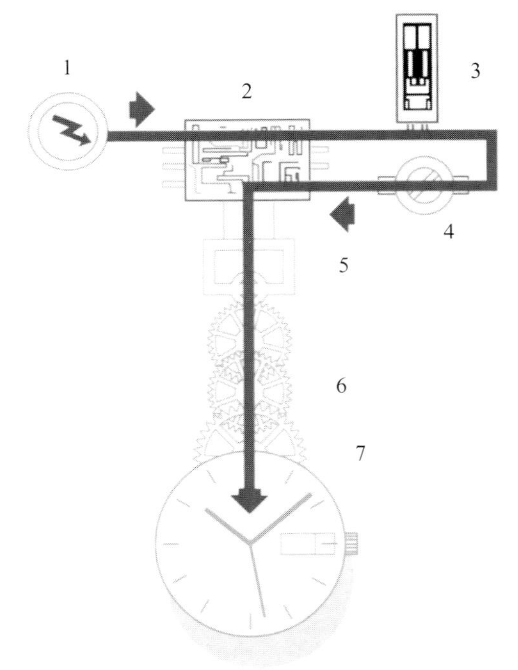

Figure 10.15 Although the construction of different makes of analogue quartz watches can vary, they are all based on the same working principles described as follows. The battery (1) supplies power to the electronic circuit. The integrated circuit (2) makes the quartz (3) vibrate; at each vibration the quartz sends a signal back to the integrated circuit in the form of an impulse. The integrated circuit reduces the frequency and amplifies the signals so that just one impulse is transmitted to the stepping motor (5). The trimmer (4) is used for regulating the frequency of the quartz if necessary (used less often in quartz watches now). Impulses received from the integrated circuit activate a rotor which is part of the stepping motor (5) and is geared to the train (6) which drives the hands (7) and when included, the date and day of the week displays (courtesy of Rotary Watches)

These are the underlying principles on which a quartz timekeeper works. Looking at the various elements in more detail let us first consider the quartz.

The quartz

The quartz in today's quartz watches consists of a bar, or more often a U-shaped or tuning-fork shaped, paper-thin sliver of synthetic rock crystal. A block of this synthetic material is reduced to slices by sawing with diamond saws. The slices are then cut up to produce the quartzes, which are dimensioned to very fine tolerances. These quartzes are given a coating of silver or gold, leads are attached and they are then mounted on suspension springs, to absorb shocks which would affect their rate. They are finally encapsulated in a little brass box, which is subsequently vacuumized.

To begin with, the quartzes that were used had the relatively low frequency of 8192 Hz. Some makers in search of the ultimate in timekeeping, employed a lenticular quartz with a frequency of 2 359 296 or 1 548 288 Hz. However, these high frequencies required more complex circuits and did not produce any great advantage. Today megahertz frequencies are used only for clocks. For virtually all watches a standard frequency of 32 768 Hz is now employed, and this provides all the accuracy anyone could expect of a watch.

The performance of a quartz crystal alters slightly with age and initially quartz watches needed trimming after a period of use. Today the quartz is artificially aged in the course of manufacture. Although ageing continues after this, its effect on rate is so slight as to be acceptable to most people.

The chip

The chip or integrated circuit has two functions. It converts the power from the battery into impulses which cause the quartz to vibrate. It also takes those 32 768 vibrations per second and reduces them, dividing by two and then by two again, to one pulse per second. For this reason the integrated circuit is known as a binary circuit. In some very small watches that can accommodate only a very small battery, a pulse once every five seconds is employed to conserve power. The minute transistors, resistors and capacitors implanted in the chip have been described as electronic gears. They do indeed perform the same function as mechanical gearing by gradually reducing the rate of impulse. Some of the slightly sub-standard chips that come off the production line have, over the years, been bought in and used by the manufacturers of cheap quartz watches. These chips, though not up to specification do more often than not work quite satisfactorily, but there use does go some way to explaining the high failure rate sometimes experienced with cheap modules. A quartz watch, like a mechanical watch, is the sum of its parts and a reputable manufacturer with a brand name to protect will only employ grade A components, whether balances or chips.

The stepping motor

The stepping motor which activates the train wheels of a quartz analogue is perhaps best described as 'an electro-magnetic device'. Its function is to translate electric impulses into mechanical impulses. It consists of a magnet, activated by a coil, which turns a rotor on the make-and-break principle. Considering how small they are, these motors have proved to be extremely reliable.

The trimmer

The only other component in the electronic part of a quartz analogue is the ceramic trimmer which takes the place of the index on a mechanical watch in that it enables the beat to be adjusted. This can readily be done by a watchmaker with the necessary training and the necessary instruments.

Some watches are built on a modular principle and any of the electronic components that prove faulty can be replaced. An increasing number of watches produced these days employ electronic units assembled in batches on automated equipment and enclosed in a protective plastic overlay. In this type of watch a fault in the electronics would entail the replacement of the entire unit. Because of the economies achieved by mass-production the difference in cost would probably be minimal.

The train

There is one essential difference between the train in a quartz analogue and the train in a mechanical watch. Whereas the mainspring in the mechanical watch exerts considerable force, the force exerted by the stepping motor is negligible. However, dust still remains an enemy of the train. Even a very tiny particle of dust can impede a quartz watch train, a particle which the force in a mechanical train might easily push aside. It follows that the assembly of quartz watches must be carried out in dust-free conditions and that the cases of quartz watches need to be well made and dust proof. Also, although it is not always appreciated, quartz watch trains need oil and so do the bearings of the stepping motor. Because this oil can dry out or be contaminated by dust, quartz analogues may require cleaning and oiling at regular intervals.

The battery

Not the least important component of the quartz watch is the battery. Indeed, as the battery manufacturers are not slow to point out, without their product the quartz watch will not function and its performance is entirely dependent on those tiny cells which they have expensively researched. This research has led to great improvements in watch batteries over the years. As any retailer knows, batteries can still be a problem area, however. Not the least of these problems is the diversity of types and sizes which the retailer has to stock and the related problem of shelf life.

The shelf-life problem means that retailers cannot be too specific in talking about battery life because they have no way of knowing how much time has passed since the battery came off the production line. Today a battery can be expected to keep a watch going for upwards of two years. By redesigning their movements some manufacturers have been able to advertise a five-year battery life. Some battery firms do feel, though, that there is a danger that a battery which is kept in a watch for five years could leak and that even over a two-year period, the possibility that a battery will leak does exist. That this happens very infrequently these days is proved by the fact that it is the normal practice of the battery manufacturers to take responsibility for damage caused by one of their products. Although modern batteries are unlikely to leak when they are operational the danger is much greater when they are expended. It is essential to inform the public in some way that they must replace an expended battery as soon as possible. Anyone who has seen what a used up battery can do to the casing of a torch can imagine what a leaking battery could do to the electronics of a quartz watch.

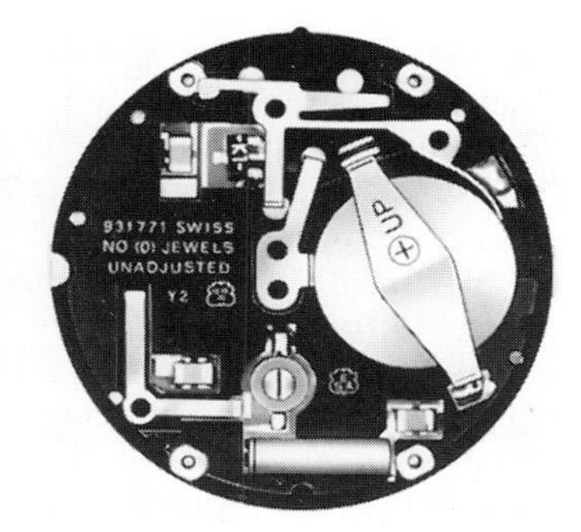

Figure 10.16 This solid state watch has an LCD analogue readout for the time, the hands are simulated by liquid crystals arranged in a circle. The LCD digital dial provides a readout for the additional functions which include alarm, chronograph and time in a second zone

Inside an LCD

The main difference between an LCD and an analogue is that a second chip replaces the train and a digital display replaces the conventional hands and dial. The LCD (Figure 10.16), therefore, has no moving parts which need oiling and could wear out. The second chip has the job of passing messages to the liquid crystals and is more complex than the binary chips used to reduce frequency. In some LCD watches, more complicated chips called microprocessors are used. These are very powerful circuits and make it possible to provide an almost infinite number of functions.

As for the liquid crystal itself, the main developments have been the use of different backgrounds to increase readability. Some readouts have employed larger digits with the same end in view. Seiko at one time introduced a different system called 'dot-matrix'. In these readouts the crystals are arranged not in the form of figures of eight, but in the form of a matrix of vertical and horizontal squares. This arrangement allows not only numbers but words and even diagrams to be built up in response to electronic signals. The display can either be stationary or can move across the screen like the messages on electronic information readouts in airports and other public buildings.

LCD analogues

Mention must be made also of the so-called LCD analogues. These are solid state watches which have liquid crystals simulating the hands and the cyphers of a conventional dial (see Figure 10.16), and are read in the same way as an analogue dial. These watches have been heralded as the watches of the future, combining as they do the advantages of solid-state and analogue display. However, so far their rather unattractive appearance and the difficulty of reading the time have mitigated against their wide acceptance. The inherent problems may, however, be overcome ultimately, as most technical problems seem to be these days.

The best of both worlds

These watches must not be confused with the combined analogue and digital watches

Figure 10.17 Watch combining a conventional quartz analogue and a solid state digital giving additional functions

which have been produced in Switzerland and Japan (Figure 10.17). These combine the two systems. They have a stepping motor and a train driving hands round a conventional dial. A second chip controls an LCD readout and gives a variety of additional functions such as time in another zone, date, alarm and chronograph. To begin with the analogue dials were very small and the LCD readout tended to predominate. Models in which the LCD has a subsidiary role were later introduced and this is a more logical arrangement.

Cases, dials and hands

The case, the dial and the hands of a watch are hardly less important than the movement. Indeed they are likely to be the determining factor in the choice by the customer of one watch rather than another. It is an old adage that 'it is the dial that sells the watch', and this still remains true today. Besides the influence which a watch case has on the saleability of a watch, it also has a very important job to do in protecting the movement or module throughout its life from knocks, dust and, if the case is designed to be waterproof, from the corrosive presence of water. Cases nowadays are made from gold, silver, stainless steel or brass, which is usually gold plated. A handful of manufacturers have also produced limited ranges of platinum watches. Some plastic cases have been used and these have proved popular for LCD timers and chronographs. Plastic cases are a feature of Swatch and many other low-priced fashion watches.

The components of a standard case consist of a ring of metal with four ear-pieces, or lugs, into which the spring lug-pins, which secure the strap or bracelet, are fitted. Then there is a domed disc which forms the back of the case and a narrow rim, called the bezel, which secures the crystal that covers the dial. These components are either machined up from sheet metal or first formed in a press and then machined to the close tolerance laid down by the case makers. They are then polished. Some of the surfaces of high quality cases are diamond-milled to give them a brilliant finish. Besides these three-piece cases, so called one-piece cases are produced in which the ring and back are all of a piece. Such cases are used for waterproof watches and obviate the problem of sealing the back of the case satisfactorily.

Waterproof cases

Waterproof cases, or 'water resistant' cases as they are often described these days, have seals round the circumference of the back and glands where the winding stem enters the case. In contrast the back of a normal case just snaps on and the O-ring on the winding stem is only designed to keep out dust. It may well be necessary to replace the seals of a waterproof case when the watch is being serviced and the case should, of course be tested before it is handed back to the customer. Customers should be informed if an extra charge is made for this. If they demur then they should be warned in writing that the watch can no longer be deemed waterproof. A number of claims against the jeweller resulting from watches admitting water after they have been serviced have been made in recent years and such a written declaration would probably serve to protect the jeweller from prosecution.

The crystal is an integral part of the case. Though plastic crystals are still used, most good watches are today fitted with mineral crystals which are much more scratch resistant and greatly enhance the appearance of the watch. Some expensive watches are fitted with synthetic colourless sapphire crystals.

Fashions in cases and dials

The design of cases has become subject to the whims of fashion. One year they may be round, another year square, and in recent years shaped cases have been in fashion, squares or oblongs with rounded corners. Sometimes bezels are narrow and self-effacing, sometimes broad and decorative. At some periods masculine styles predominate, at others very small or very thin watches are in vogue.

Dials too are subject to changing fashions. Sometimes full-figured white dials with Roman or Arabic numerals are in fashion. At other times the figures disappear altogether and we have what are known as 'sans heure' dials, or dials with only a single cypher at 12 o'clock. Then there are cypher dials with only four cyphers at the quarters and full cypher dials with a full complement of hour markers and minute marks interspersed between them. The showroom of a dial manufacturer looks like a library, with fat volumes lining the shelves and every page of every volume has perhaps ten sample dials attached to it. These shelves hold tens of thousands of designs. Dials may be made of gold or silver or base metal, which may be gilded or coloured. They may be enamelled white or overlaid with a thin slice of one of the decorative minerals such as malachite or lapis lazuli. Figures may be painted on or applied by silk-screen printing, cyphers may be raised in the press or applied. On very high grade gold dials each gold cypher will be individually made and finished and then riveted on to the dial. At the other end of the price spectrum are the popularly priced fashion watches with their colourful printed dials.

Like all branches of watchmaking, the making of fine cases and dials calls for great skill. Artistry and attention to detail are important factors in determining the selling price of a watch. One watch may cost many times more than another watch with an identical movement because of the workmanship and the materials employed in the making of the case and dial. Therefore, the salesperson needs to have an appreciation of the arts of case and dial making and be able to convey their appreciation to the customer.

Clocks today

The majority of clocks sold today are quartz clocks. The inexpensive quartz movements produced initially in Germany and more recently in the Far East, have superseded not only battery movements, but also spring-wound movements in the low to medium price brackets. Indeed today even expensive clocks are often fitted with these movements because these are so very accurate and reliable. Now that prices have come down most of the alarm clocks being sold are also quartz controlled these days. Quartz clocks costing only a few pounds have an accuracy comparable to that achieved in the past only by the finest marine chronometers. By far the majority of these clocks are analogues. After some initial interest, digital clocks have not turned out to be popular with the public and the only clocks with LCD readouts which are now sold in any quantity are tiny miniature travel alarms and clock radios.

A place for mechanicals

In the higher price brackets there is still, however, a considerable demand for mechanical clocks, invariably reproduction designs. Consumer research has established that so far as clocks are concerned women's influence is paramount and women have decided ideas about the kind of clocks they want to have in their homes. They like pretty clocks rather than functional ones and they prefer nostalgic rather than modern designs. These attitudes are clearly reflected in the ranges which manufacturers and importers offer to the retailer. The reproduction carriage clock with its elegant brass and glass panelled case and white enamelled dial, seems to epitomize women's taste in clocks. No doubt this explains the continuing demand for these. The revival of interest in long-cased clocks is another facet of this interest in the traditional. It also emphasizes the fact that the public see clocks as furniture, an element of interior decoration, rather than as instruments for telling the time.

Chapter 11

Gemmological instruments and their use

Because of the increase in the latter part of the 20th century in the treatment of diamonds and coloured gemstones it is vital for the professional jeweller in the 21st century to master the basic skills of gem identification using at least the basic gemmological instruments. The vast improvement in the recent techniques used for the manufacture of the synthetic gemstones has also made it very difficult in some cases to identify the natural material from the synthetic. The Trade Descriptions Act and the Sale of Goods Act mean that the correct trade description must be applied (either verbally or written) to merchandise when offered for sale. For example, it is illegal to describe a gemstone as natural when in fact it has been made synthetically. The terms applied must be valid and technically correct. There has been in the last few years a growth in the number of jewellers who undertake valuations either for insurance or probate, mainly as a direct result of insurance companies having to validate claims made by their clients with regards to the loss of jewellery. The principles of carrying out valuations on jewellery, silver and other items have already been set out in Chapter 4. Suffice it to say any person who acts in this field of professionalism must be able to use gemmological instruments. The ability to identify the imitation and synthetic gemstones from natural materials is of great importance not only when selling stock but also when taking in repairs, purchasing second-hand goods and assessing a customer's jewellery for the purposes of insurance valuation.

10× LENS

There are a small number of gem materials that can be detected with the aid of a 10× lens (Figure 11.1) due to their optical and physical properties. An example is a cultured pearl necklet. It is sometimes possible to view the inner mother of pearl nucleus by examining down the drill hole of the larger central beads. There is usually a dividing line of materials at this point, the one being the nacre and the other the nucleus. Neither a natural pearl nor a glass imitation pearl would display a mother of pearl core. There are several tests for cultured pearls one using an anglepoise lamp and the other making use of UV light. The former is carried out by holding the pearls across the lamp hood while the light is on and looking for the tell-tale darker ball-like bead. It is also possible with a little experience to assess the thickness of the nacre around the nucleus.

A number of gemstones possess what can only be called typical natural birthmarks which can be identified using a 10× lens. Amethyst, sapphires, peridot and some emeralds, particularly those from Columbia, often exhibit such natural birthmarks. When these gemstones are made synthetically it must be remembered that while their

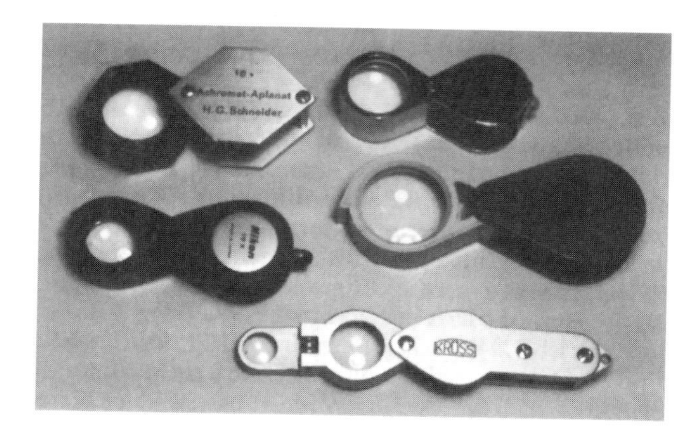

Figure 11.1 A range of hand lenses

physical and optical properties are identical to those found in the natural stones, their internal growth development differs greatly. The reference here to the physical and optical properties covers such things as the hardness of the gemstone on Mohs' scale of hardness, refractive index, chemical composition, specific gravity, crystal system and lustre. Inclusions are nature's birthmarks and these should always be included in any listing of the physical and optical properties of a gemstone. This chapter is devoted to a simple step-by-step guide to the critical gemmological expressions used to describe the results we see when an unknown gemstone is inspected using the appropriate gemmological instrument or instruments.

When purchasing a 10× lens it is false economy to purchase a cheap version. A good lens will make it easier to study the tiny inclusion in a natural ruby or peridot. A quality lens will have no or very little distortion of the field of view at the outer edge whereas the cheaper versions will give a problem in this region of view. It is not recommended that you simply purchase a powerful 10× lens and use it for everything because this could result in eyestrain. A less powerful lens is often adequate to identify any external fault on a gemstone such as cracks, join lines at the girdle on a doublet (although the join is not always at the girdle area) and the facets on a brilliant cut stone. Some older worn natural zircons often show 'paper wear' on the crown facets which can often be detected with the lower powered lens. It is important, however, when using a lens to have a really good bright light source otherwise the effectiveness of the instrument will be minimal. It is not advisable to start the examination of a piece of jewellery or gem set jewellery without first of all making sure that both the lens and the item under examination have been cleaned. There have been numerous factual records of instances when an item of jewellery has been hastily examined and thought to possess natural typical inclusions. These on further examination and cleaning of the item has proved to be nothing of the kind.

Many complaints by customers regarding their repair, or recent purchase, could have been avoided if the item had first been carefully examined with a 10× lens, noting any defects and then dealing with their notation on the spot. Useful as a 10× lens may be, it is an unwritten rule practised by most gemmologists that one should not give a positive identification of the nature of the unknown gemstone based on one test alone. A minimum of three tests is recommended.

It would be as well to include the fact that it is possible to purchase a hand lens or, as they are sometimes called in the trade, loupes. These have a higher magnification than a 10× and include 15× and 20×. Such instruments are invaluable to the jeweller and the manufacturer of jewellery when carrying out quality control examinations, particularly of diamonds for clarity grading and the other gemstones for internal and external defects that directly affect the quality value of the item. Incidentally members of the retail jewellery trade have always been encouraged to use a 10× lens at the counter at the point of sale or when purchasing jewellery from the public.

Refractometer

The refractive index reading or readings of an unknown gemstone can provide the key factor in identification. This is because most natural gemstones belong to one of the seven crystal systems: cubic, hexagonal, tetragonal, trigonal, monoclinic, triclinic and orthorhombic. Light travels through a gemstone in a given manner which is dependent upon the crystal system to which a particular gemstone belongs. Gemstones belonging to the cubic crystal system include diamond, spinel and garnet. In these gemstones light travels into it as one single ray, goes through it as one single ray and leaves it as one single ray. This in gemmological terms is called single refraction. So a natural spinel would be said to be singly refractive and gives a single reading on the refractometer (Figure 11.2). All the other gemstones which belong to the other six crystal systems are said to be doubly refractive, because when a ray of light enters them it is broken up into two separated rays which travel in slightly different

Figure 11.2 Rayner Dialdex Refractometer with sodium filter, polarizing filter and refractive index liquid

directions and at different speeds through the gemstone. The actual amount of separation of these rays and their speed of travel through the stone will be particular and constant to a given gem material. This means that a ruby will have two refractive index readings on the scale of the refractometer between 1.76 and 1.78. No other natural red gemstone will give the same readings on the refractometer. It is true to say that the synthetic ruby and the synthetic sapphire will indeed possess the same refractometer readings but close microscope examination of their inclusions will provide the evidence to identify them as synthetic and not natural gemstones. Gemstones belonging to these systems include quartz, ruby and sapphire, emerald and aquamarine, tourmaline, zircon, peridot and topaz. The specific reading for each of these gemstones can be found in any good textbook on practical gem testing. The difference between the highest and lowest readings taken is called birefringence. The refractometer in many instances when used correctly can provide an indicator as to the nature of the unknown gemstone simply by recording its refractive index. It will not as already explained identify the natural from the synthetic, this requires the use of the microscope. Perhaps the exception would be spinel, where the synthetic Verneuil spinel usually has a fairly constant refractive index of 1.727, whereas most natural spinels have a refractive index close to 1.718. However, as in so many cases in gemmology, there are exceptions, and natural spinels have been found with a much higher refractive index. Yet another reason why it is as well to do a second or even third test, using a different instrument, on a stone. Synthetic spinels do have a slightly higher refractive index readings than natural ones due to the additional amount of alumina added to the basic chemical composition of the gemstone in the process of its production.

To use the refractometer effectively it is necessary to use a sodium light source or the equivalent. It is possible indeed to use a white light source provided that a yellow filter is placed over the light window or eyepiece. The contact liquid for the refractometer is usually supplied with the instrument but can always be purchased separately. Caution must be used when using this liquid because it is not only expensive but toxic. It should not be inhaled, and one should always wash ones hands after using it. Normal liquids currently available have a refractive index of 1.79, and this represents the upper limit of the refractive index of the stones that can be tested on the refractometer; thus diamond (R.I. 2.42) and many of the diamond simulants have too high a refractive index to be tested.

Steps in the use of the refractometer.

1. Clean the surface of the gemstone and the table of the refractometer.
2. Apply a small amount of the refractive index liquid to the centre of the table of the refractometer.
3. Place the gemstone table carefully on the prism of the refractometer so that the contact liquid makes the two surfaces come into close contact with each other.
4. If white light is being used then employ a sodium filter. In cases where the readings are taken in white light without a filter the result is a dark edge with a spectrum of colours. Take the reading at the yellow/green end of the spectrum.
5. Place the polarizing filter in place on the eyepiece, making sure it is the correct way round. It is advisable to check the condition of the filter for cracks or damage because if it is damaged it will directly affect one's ability to read the refractive index shadow edge on the instrument.
6. Focus the eyepiece of the instrument to give the sharpest of images when the

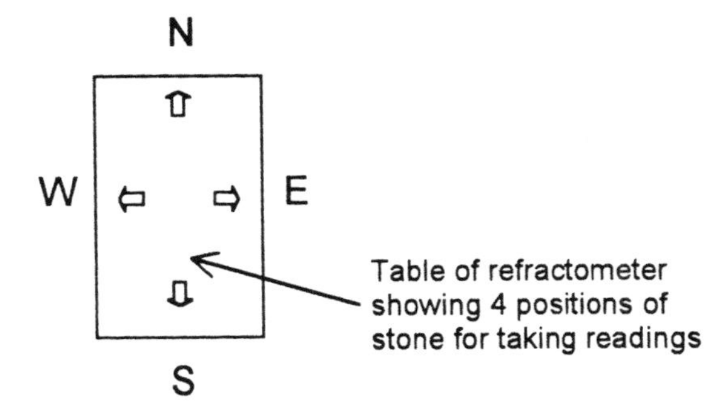

Figure 11.3 The north, east, south and west positions in which to place the stone in a refractometer

shadow edge is being observed. If glasses are worn it may be useful to remove them to get close enough to the eyepiece to see clearly the edges of the shadow edge.

7. Place the stone carefully to a north, east, west and south position as shown in Figure 11.3, noting any changes in the refractive index readings as the stone is moved through 360 degrees.

If a Dialdex refractometer is being used then the readings are taken from the indications on the drum on the side of the instrument but if a standard refractometer is being used looking down the eyepiece on to the scale of the instrument will reveal the shadow edge. If the shadow edge remains the same during the full rotation of the stone on the table of the refractometer the stone is singly refractive.

Should there be a difference in the refractive index readings during the exercise the stone is doubly refractive.

A note should be made of the different readings in the various positions N, E, S & W for the stone and the amount of double refraction (and the one subtracted from the other). A polarizing filter should always be used to establish whether double refraction is present, and any movement of the shadow edges should be looked for whilst rotating the polarizing disc through 360 degrees. An excellent stone to use for this exercise is peridot because it has such a large amount of double refraction and usually provides good clear readings on the refractometer scale.

It is useful to select a series of gemstones from the ones that are singly refractive and test these. Then doubly refractive gemstones should be tested for their readings and the amount of double refraction noted.

The important refractive index reading for each gemstone is given in Appendix 1, p. 275.

Polariscope

The polariscope (Figure 11.4) is a useful instrument for the detection of single or double refraction or the presence of false double refraction (this false double refraction being called anomalous).

When the filters in the figure are set in a certain given position (called crossed polars), and an unknown gemstone is examined between them, the resultant pattern produced can aid the identification of the gemstone. In order to reach this position the following exercise should be carried out. With the light turned on, the top filter is revolved through 360 degrees and the light intensity changes from light to dark during

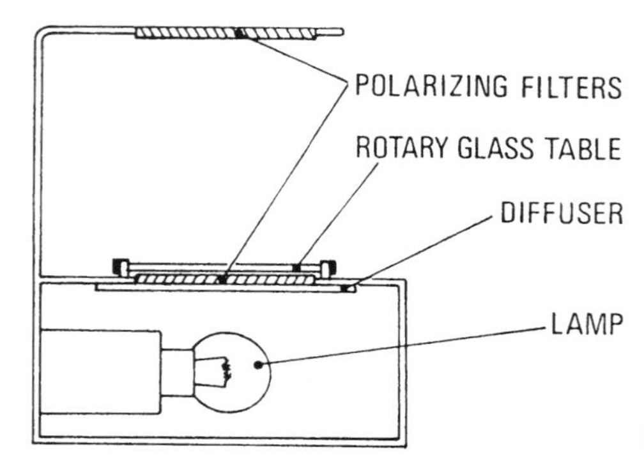

Figure 11.4 A polariscope

the rotation. This means that the filters are parallel in one position and crossed in the other. In the former it would be seen as light when viewed from the top and dark when the filters are in a crossed position. This sounds very complicated but is easily understood in practice.

When the gemstone is examined through the top filter with the filters crossed there are several possible outcomes. If the stone is singly refractive there will not be a clear, distinctive change between the light and the dark areas in the stone as seen with double refractive gemstones, but it will remain dark through the full rotation of 360 degrees. If the stone under test shows a pattern of light/dark four times in 360 degrees rotation, then this indicates the stone is doubly refractive.

This instrument is not suitable for stones that are very dark in colour or opaque. There is in fact only a small number of gemstones where this instrument can be a positive aid to their identification: these include garnet, paste and synthetic spinel. These, under examination, display a patchy light pattern rather than a distinctive change from one light intensity to another. A gemmologist wishing to detect the characteristic of a doubly refractive coloured gemstone known as pleochroism can use the polariscope but it is not advisable to venture into this area until having acquired the basic skills of operating the other instruments.

Spectroscope

This gemmological tool (Figure 11.5) is sometimes considered the most difficult of all the instruments used for the identification of gemstones. The trouble is usually because of the type of light source being used rather than any fault of the instrument. The problem can readily be overcome by the use of a fibre optic light source, preferably with a variable intensity control.

The absorption spectra seen in coloured gemstones is due to the nature and presence of the different chemical elements found in them. A well-known example is the chromic oxide found in ruby, which results in the material's red coloration. If this were not present the gemstone would be colourless in its purest form. A ruby, when examined through the spectroscope, will display a particular pattern similar to a barcode which is the result of this element being present.

This photograph shows the doublet in the deep red as a bright emission line. The remainder of the red and orange is freely transmitted. Note the weaker and narrower absorption of the yellow and green, which produces the more orange-red colour in this direction. The 'blue window' with the two fine lines is a little wider than seen under

Figure 11.5 A range of spectroscopes

normal conditions. The violet is totally absorbed. This photograph has been taken in the direction of the extraordinary ray.

The extraordinary ray is the ray in a double refractive gemstone belonging to either the hexagonal, tetragonal or trigonal, crystal system. In these systems the velocity varies with direction in which the ray passes through the gemstone.

The gemstone peridot is coloured by the presence of iron, which gives it an identifiable pattern barcode spectrum reading. This, along with the correct refractive index readings for peridot, would assist in the conclusive identification of the gemstone from the other similarly green gemstones. The photograph shows a medium olive green coloured peridot with three evenly spaced lines, which are diagnostic for peridot in the blue at 496 nm, 473 nm and 453 nm. These can only be detected with care in subdued light surroundings, preferably using an eyecup to exclude extraneous light to the eye. The use of a polarizing filter makes the lines become stronger and therefore clearer.

There are two types of spectroscope on the market: the prism and the diffraction grating. Both have their own features and benefits for the user. One of the most popular and easily portable is the O.P.L.O., made by the gem instrument company manufacturers Genesis of Epsom. This model has a fixed slit and therefore requires no adjustment by the user. When used with a good fibre optic light source the clarity of the results seen can be remarkable. The diffraction grating spectroscope has an adjustable slit which allows the user to control the amount of light entering the instrument and thereby clarifying the resultant spectra seen.

It is clear from the photographs of the typical spectra found in pink YAG (yttrium aluminium garnet) and CZ (cubic zirconia) (Plates 1 and 2; produced by John S. Harris F.G.A. of Carlisle as part of his educational spectra photographs) that they contain various thick and thin lines and areas of colour absorption. As already indicated, their presence and position in the spectrum are an indication of trace elements in the gemstone. Natural zircon has a typical spectra that is characteristic and when seen together with the doubling of the back facets when viewed through the table facet with a lens, the stone can conclusively be said to be zircon (Plate 3). This photograph shows the bar coding for a natural zircon. The lines in the deep red and the two dominant lines in the red at 660 nm and 653 nm are prominent. There are extra lines close together in the orange red at 621 nm and 615 nm which are characteristic of the photograph having been taken in line with the extraordinary ray. The other lines in the yellow, green and blue may vary in intensity and position compared with those visible when the photograph is taken in line with the ordinary ray. The strength and number of lines in this spectrum are very pronounced and the student will seldom see a more striking absorption spectrum than that in zircon. However, many zircons will present a similar pattern and one or both of the dominant lines in the red are usually seen even in the weakest absorption spectrum of high type zircons, including the heat-treated stones.

Most of the textbooks on gemstone testing contain illustrations on the various spectra seen in the coloured gemstones and it is recommended these are consulted to develop techniques of taking spectrums using the different light sources available.

Microscope

No chapter on gemmological instruments would be complete without reference to this most valuable tool, an example of which is illustrated in Figure 11.6. There is not the room in a textbook like this to neither give the reader an in-depth account of the optical arrangement of the microscope nor cover the process of operating it. The trade

Figure 11.6 Monocular microscope with light source

textbooks dealing with the subject of testing gemstones cover these areas in a step-by-step manner. The microscope enables us to examine the inclusions within gem materials which assist in their identification. Plates 4–8 illustrate the typical inclusions found in some of the common gem materials.

Plate 4 shows the natural colour banding, sometimes called colour zoning, which is found in citrines. It can often be seen under low-power magnification.

In Plate 5, the doubling of the back facets is clearly seen under low magnification. This is due to the very high amount of double refraction seen within natural zircons. This particular gemstone is above the refractive index readings of the Dialdex refractometer. Careful use of the 10× lens would display this characteristic of the stone when examined under a bright light source. Plate 6 shows the colour zoning found in natural sapphire. Note the angle at which these colour bands meet. In natural sapphire it would be at 60 or 120 degrees. This is due to the crystal system to which sapphire belongs – trigonal. Plate 7 shows the inclusions found in synthetic sapphire: curved line and colour zoning. The cracks are called ‘chatter marks’ and they are often found in synthetic sapphire due to fast polishing of the gemstone. These curved growth lines would not be seen in natural sapphire but the growth in the natural would be straight lines, as shown in Plate 6. Plate 8 shows typical pattern of the inclusions found in a synthetic emerald produced by the flux fusion method.

The detection of natural emeralds from their synthetic counterparts is extremely difficult and requires careful use of the microscope and an up-to-date knowledge of the production of modern synthetics. There are some natural emeralds that possess typical inclusions which can be used as a diagnostic feature of the material. Caution should be exercised when examining any gem material, whether natural or synthetic, under the microscope as this needs to be coupled with the basic skills of firstly identifying the inclusions, and secondly understanding the nature of the inclusions and balancing this against a knowledge of the development of modern synthetics. The inclusions shown in Plates 4–8 are representative of the inclusions that can be found in these gemstones but they are not necessarily always present. A gemstone that is free from inclusions gives rise to the need for further research and caution before a conclusion can be drawn as to its identity.

Chapter 12

Glass and china

Glass

Raw materials

Glass-makers can produce many different kinds of glass by varying the constituents or by altering the proportions of the ingredients melted to produce the metal. They can make common glass, that is, soda lime glass, by fusing a mixture of sand, soda and lime. They can make blue glass by adding cobalt. If gold is added to the mix they can make glass of a rich ruby-red colour.

The metal which all the makers of fine glass in the UK produce today is known as lead crystal. This metal, which was developed by George Ravenscroft in the seventeenth century, is particularly well suited to the production of cut glass because it is both relatively soft and brilliant and also has an attractive colour.

The essential ingredients of lead crystal are silica sand, potash and red lead. Other chemicals are added, however, to facilitate the fusion of the material, improve its appearance and working qualities and ensure against discoloration.

Modern lead crystal is a highly sophisticated material, the result of centuries of development and experience. It is vital to obtain the right materials and it is just as important to ensure that these materials are pure and uncontaminated. It is expensive to mix and fuse a batch of glass and to produce glasswares from the resulting metal. It would be disastrous if the results of all this labour had to be scrapped because the products did not conform to the high standards a modern glass factory has to maintain. So the ingredients are carefully checked and the sand to be used is subjected to a series of purification processes.

Mixing the ingredients

The proportions of the materials used are also very critical and they are weighed out very accurately. A typical batch of 203.8 kilograms would contain 103 kilograms of processed sand, 59.4 kilograms of red lead, 32.4 kilograms of potassium carbonate, 7.7 kilograms of potassium nitrate, 0.7 per cent of borax and 0.6 per cent of arsenic, plus that tiny amount of decolorizer. English lead crystal contains 30 per cent of red lead, so conforming to the British Standard. On the Continent the lead content is normally only 24 per cent.

Figure 12.1 Refractory pots (by courtesy of Waterford Crystal Ltd)

Making the pots

The making of the pots in which the constituents of glass are fused is a skilled and exacting craft. Their production represents a considerable overhead to add to the costs of buying in and processing the ingredients from which the glass is made. The clay used to make the pots at the Stevens and Williams factory came from Rommerode in West Germany. This clay, called 'grosselmerode', produced pots capable of standing up to the considerable heat stresses imposed upon them in the furnace.

These refractory pots are built up laboriously by hand in batches of seventeen, and a batch will take between six and seven weeks to complete. The pots are made in the shape of an old-fashioned bee-skip with a mouth at the top to allow filling (Figure 12.1). When the pot has been completed it has to be left to dry out completely in a draught-free area, and this takes a further six months.

A glass furnace usually holds ten or twelve pots. They can only be used between twenty-seven to thirty times before they have to be discarded and replaced with new ones. This is because the chemical reaction between the clay and the metal during firing causes pitting. In addition the sand content in the clay gradually fuses to form a glass of a very inferior quality which could contaminate the melt.

The pots are filled every two days and one of them is changed every week, traditionally on a Friday afternoon. This rotation of pots ensures that there are always an adequate number in use to supply the glass-makers with metal. A broken pot is a major disaster in a glasshouse because it results in skilled craftsmen being left without any work to do.

Filling the pot and firing

The pots are filled by a team of men known as 'teasers'. They shovel in the mixed ingredients and cullet, which is broken glass. They add some of one and then some of the other, until a 50/50 mixture has been built up. The pot is now stoppered to avoid contamination and left for eight hours. By this time the level of the contents will have fallen, so the pot has to be topped up with further cullet. It is stoppered again and stays in the furnace for twenty-four hours at a maintained temperature of 1350 degrees centigrade. The stopper is again removed and the temperature is now reduced to 1150 degrees, and this temperature is maintained for a further sixteen hours. By this time the metal will have acquired the desired hot toffee consistency and is ready to be worked.

Inside a glasshouse

The making of glasswares in a glasshouse has been likened to a strange ballet carried out in the half-gloom. The workers are in continual motion, gathering the metal glowing orange from the furnace, revolving the irons to produce the centrifugal force necessary to keep the metal evenly distributed, their faces illuminated by the glow from the furnace and from the molten metal. Nobody who visits a glasshouse can fail to be fascinated by this strangely ritualistic performance. Nor can they fail to be impressed by the skill these craftsmen bring to their job, and by the fact that fine glasswares are still individually produced by hand, as they have been for many centuries. Old engravings cut centuries ago, show craftsmen employing the same simple tools and techniques that are used today. The vocabulary of the glass house, too, contains words that date back to the time when foreign craftsmen were brought over from the Continent to make glasswares here. The glass-making industry is a craft industry, steeped in tradition.

The chair

The glass-makers work in teams, known as 'chairs'. The work is, in fact, carried out in the proximity of a crudely made seat with extended arms, also confusingly known as the 'chair'. The number of members of the team varies, depending on the type of ware the chair is producing, but a wine chair, which produces stemwares, consists of four craftsmen. These are the servitor, the foot-blower, the bit-gatherer and the taker-in. The servitor, more often known as the gaffer, is the man in charge and sits in the chair. In a caterhole shop, where large articles such as jugs and vases are made, the man in charge is known as a 'workman'. He has a number of servitors under him who do the preliminary blowing and forming.

Tools and techniques

The tools the members of the chair use are simple ones. The marver consists either of a block of carbon steel or a hollowed-out length of wood. This piece of equipment is as old as the craft itself, and gets its name from the fact that it was originally a block of marble. The metal gathered from the furnace is rolled, or 'marvered', on the marver to ensure that it is evenly distributed on the blowing iron.

The irons consist of the hollow blowing iron about 6 feet (2 metres) in length and the solid pontil or 'punty' iron. The gaffer uses various tools to trim and shape the molten glass – shears, pincers and lengths of board. There are also the moulds. Nowadays only special pieces are free-blown, most standard lines are blown in cast-iron moulds. These moulds are usually mounted on a device known as a 'mechanical boy', because it replaces the boy whose job it was to open and close the non-mechanical hinged moulds. The mechanical boy allows the foot-blowers to operate the mould themselves by depressing a pedal.

Figure 12.2 Gathering the metal from the pot

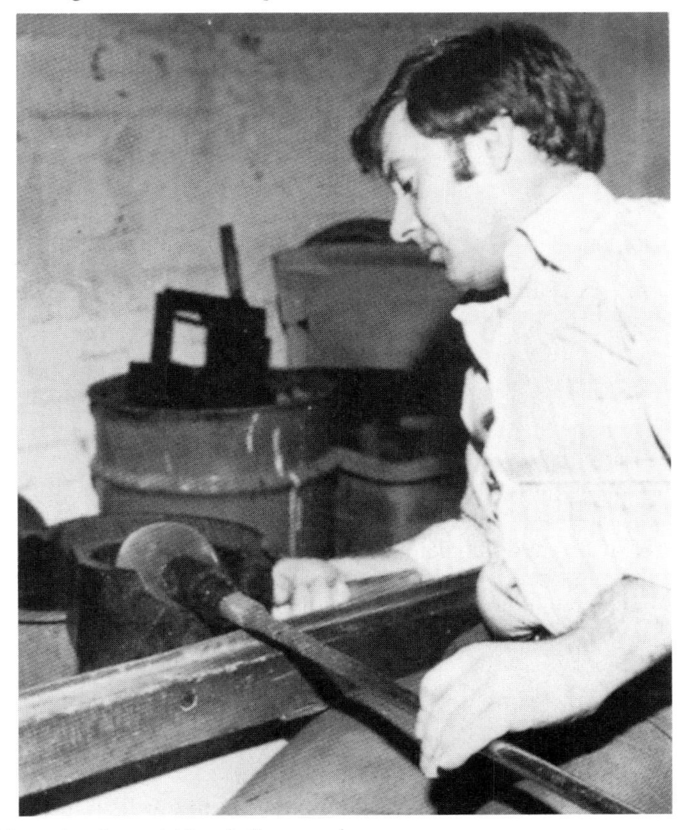

Figure 12.3 Marvering the metal in a hollow wooden marver

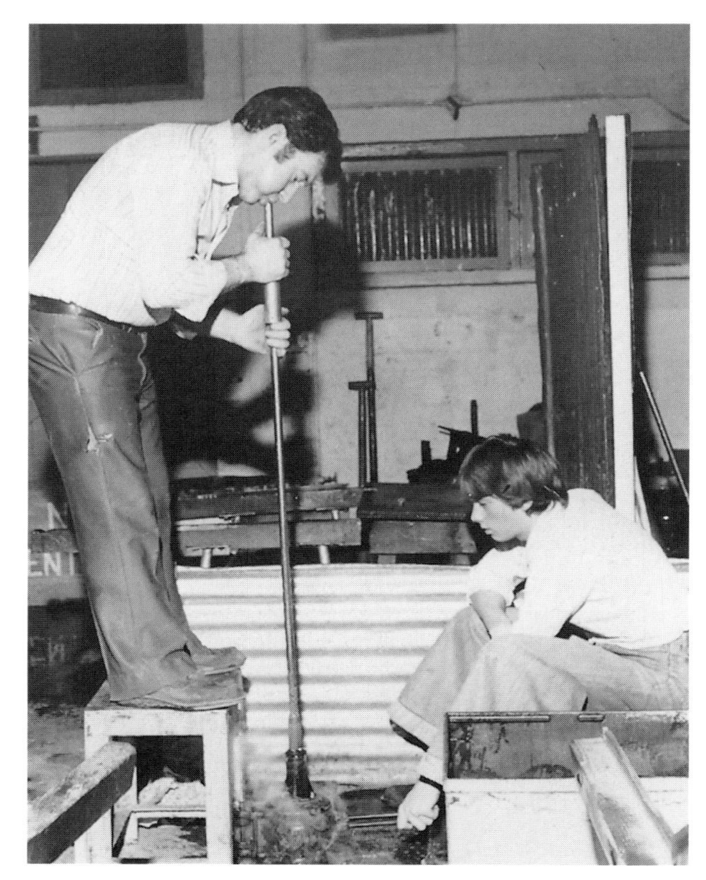

Figure 12.4 Blowing the bowl of a wine glass in a rnould. This foot-blower is not using a mechanical boy, and an apprentice has to open and close the mould for him

Making a wine-glass bowl

The first step in making a stemware is the gathering. by the foot-blower, of the molten metal from the pot on the end of the blowing iron (Figure 12.2). This may seem simple enough, but it takes experience to judge how much metal to take and know how to gather it.

The foot-blower now takes a gather of metal to the marver and marvers it (Figure 12.3). Still rotating the iron, the foot-blower now moves on to the mechanical boy, mounts it, inserts the metal in the mould and blows it until it assumes the shape of the mould (Figure 12.4).

Stem and foot

The foot-blower now carries the sphere of glass to the gaffer sitting in the chair. The gaffer places the blowing iron across the arms of the chair and continues to rotate it. Meanwhile the bit-gatherer is collecting a gather of metal from the furnace on the purity iron. This is carried to the chair and the gaffer up-ends the blowing iron,

Figure 12.5 The bit-gatherer passing the punty iron with the gather on the end of it to the gaffer, seated in the chair

reaches above his head, takes the punty iron from the bit-gatherer and allows the metal from it to flow onto the base of the blown sphere of glass (Figure 12.5). The blowing iron is then placed across the arms of the chair and rotated to form the stem, using a long pair of tweezers with wooden inserts (Figure 12.6). It seems incredible that these stems can be shaped to the right pattern within very close limits by eye alone. This operation is then repeated, but this time the gather which has flowed onto the end of the stem is shaped into a foot between two hinged boards, known as 'foot-boards'.

Cracking-off

There is now a sphere of metal on the end of the blowing iron, and the top of this sphere must be removed to leave a hollow bowl. This is the job of the taker-in. To remove the unwanted part, known as the 'moil', the taker-in scores a line round the globe of glass where the rim of the wine glass will eventually be. This is done with either a diamond or a hardened steel tool. The glass is then passed through a gas flame. Due to the different rate of expansion in the scored areas when heat is applied, the glass fractures at this point. In most factories today this operation is carried out in

Figure 12.6 The gaffer shaping the stem of a wine glass

an automated machine. The glasses are placed on revolving chucks which bring them in contact with the diamond tool and they are then automatically transported through a flame (Figure 12.7). A further automated machine then polishes the edges.

Previously, the edges were polished by a process called 'edge-melting'. The glass was mounted on a turntable and a flame directed on the rim, heating the metal to the point where it flowed and the sharp edge became softened. When the taker-in has cracked off the moil the glass is then carried as quickly as possible to the lehr. The moil is thrown into a bin known as the 'shrower' to be later ground up to make cullet. The glasswares travel slowly through the lehr over a period of between two and four hours. As they travel on, the temperature reduces until it reaches room temperature. Larger articles take longer to anneal. A large and heavy vase might spend as long as twenty-four hours in an annealing furnace.

Acceptance and rejection

Not every glass that passes through the lehr emerges unscathed. If a glass has been worked on over too long a period it may be already too late to redress the effects of the internal strains when it goes into the lehr. It will therefore crack on the conveyor belt and the fragments will be removed by the 'taker-off'. An acceptable level of breakages in the lehr would be about 2 per cent. The taker-off inspects every ware individually, rejecting faulty blanks at this stage before they pass to the cutting shop. There are a number of common faults to look for. These include stones, cords, blisters, speckles, bad shearing, mould marks, crizzle, surface imperfections, and uneven distribution.

Figure 12.7 Automated cracking-off

The rate of rejection is high, and this is, of course, reflected in the price of fine glasswares. About 30 per cent of all wares produced will be rejected either when they come out of the lehr or when they have been inspected after they have been cut. If a factory makes plain glasswares, or glasswares with plain bowls, the rejection rate can be as high as 50 per cent.

Of the rejected glasswares coming out of the lehr some will be consigned to the shrower, the bin which holds the glass that will be converted in to cullet. Wares which exhibit such serious faults as crizzling would be completely scrapped because to re-use them would be to perpetuate the problems. Some faults can be removed in the course of cutting, and the taker-out will decide what can be salvaged in this way.

Decorating glass

Glass can be decorated in a hundred and one different ways. This decoration may be integral, that is, incorporated in the glass in the process of forming the metal into a ware. Effects such as latticinio and trailed decoration used to create decorative stems are included in this category. Then there are the various ways in which decoration can be added to glass after a blank has been produced. Either the glass may be cut away, or decoration such as gilding or enamelling may be applied to it (Figure 12.8).

Glass is, of course, an inherently beautiful material, and there are those who consider that any form of decoration is extraneous, that the line and form which the designer dictates are sufficiently beautiful. Indeed, when looking at some of the graceful products of the leading Scandinavian glasshouses it is not difficult to understand why some people feel that this is glassware at its best. Similarly, eighteenth-century wine glasses, with the decoration limited to a teardrop or an air-twist in the stem, are

Figure 12.8 Painting flowers in enamel on the opaque white surface of cased glass in a Bohemian glass factory. Afterwards the enamel is fired

eminently satisfying to look at and pleasant to use. But whether or not to cut glass is a form of sacrilege, as Ruskin believed, the fact remains that the majority of glasswares sold in the UK today are decorated. By far the most popular form of decoration is cutting.

However, the glass-maker in the glasshouse can achieve many attractive effects in the course of working the metal, and much can be achieved by varying the constituents of that metal. Although these techniques are not widely used by the large British glasshouses in the production of table glass, they are still employed by smaller firms and designer-craftsmen in the creation of decorative glasswares in the UK and on the Continent.

Early glasswares were invariably coloured because the glassmakers did not know how to make colourless glass. There is a world of difference, however, between the accidently coloured glasswares of the past and the coloured glasswares produced during the past two centuries or so. Today it is possible to produce a wide range of coloured glasses of predictable hues. Blue glass is produced by the addition of a small quantity, about 0.05 per cent, of either cobalt oxide or copper oxide to a batch. Opaque white glass is produced by using a calcium antimony compound. A fine ruby red glass can be produced by adding gold or copper.

Balusters and twists

The baluster stem is probably the most satisfying example of integral decoration ever produced by a glass-maker. It is functional as well as decorative, and its subtle curves exploit the beauty of the material. It was borrowed from the baroque balustrade and was first applied to glass in the seventeenth century. It is seen at its best in the wine glasses made in England during the eighteenth century, a perfect foil to the simple undecorated bowls. It is still very much with us today. Probably half of all the stemware patterns produced in the major glasshouses have baluster stems, though these are now, more often than not, faceted. The knops which make up the stems are created freehand by the gaffer.

Eighteenth-century balusters often had a teardrop enclosed in them. These tears were induced by pegging the metal, pushing a metal spike into the gather and then covering the hole thus created with further metal. As the metal is drawn out and shaped, the air bubble so created assumes the characteristic form. It may well have been pure chance that originally directed the gaffer's attention to the possibility of further drawing out the bubble and twisting it to create what amounted to a spiral tube running down inside the length of the stem. These so called 'air-twists' were first produced in the 1730s, and by the 1760s the technique was extensively used. Air-twists are not common in modern stemwares, but they do exist. For many years Stuart produced an air-twist pattern appropriately called Ariel. The Czech glass-makers also produced a pattern with air-twist stems.

The earliest air-twists were produced by elongating a single air-bubble, but the designs were soon elaborated by pegging a number of holes in the metal, covering these, and then twisting the metal as it was drawn out to produce the stem. Today Stuart's gaffers had a special tool with four spikes which did this pegging in a single operation.

The next development was to produce opaque and coloured twists, though these never seem to have enjoyed the same popularity as the air-twist. These were produced by a quite different technique, a number of short opaque, white or coloured glass rods were incorporated in the gather before this was twisted and drawn out to create the stem.

Glass-makers have evolved many other ways of decorating glass in the course of working it. These include dipping the molten glass in water and then reheating it to produce an ice-like effect and incorporating threads of opaque glass in clear glass to create lace patterns. However, these types of decoration have only a limited appeal as novelties.

Cutting and engraving

By far the most popular form of decoration over the past two hundred years has been that produced by cutting the surface of lead crystal with diamonds or abrasive wheels. The most familiar of these forms of cut decoration is called glass-cutting. The first operation to be carried out on blanks is to mark them up with vertical and horizontal guide lines which are painted on the blank (Figure 12.9). These lines are no more than guidelines within which the cutter will work. The fact that there is some latitude allows the cutter to remove minor flaws in the blank in the course of making cuts. The more complicated the pattern cut on a ware, the easier it is to eliminate these blemishes.

This explains the seeming contradiction that the simpler the design of a cut-glass ware the higher priced it is likely to be. If the cutting is restricted to a few simple facets round the base of a stemware bowl then a perfect blank has to be used, and perhaps only 40 to 50 per cent of blanks coming out of the lehr would be acceptable. If a very elaborate design is cut on to the blank, then perhaps 75 to 80 per cent of the blanks could be used. So it pays a glasshouse to concentrate its bulk production on the more elaborate patterns in its range.

The glass-cutter sits at an electrically powered lathe, which turns a carborundum or sandstone grinding wheel. The ware is held above the wheel and, looking through it, the cutter follows the guidelines, first making the diagonal facets dictated by the design. Then the necessary straight cuts are made to fill out the pattern. Water pours

Figure 12.9 Painting the horizontal guidelines on a stemware blank

continually over the wheel from a small pipe placed above it which washes away the glass dust produced by the cutting and keeps the work cool. If the glass is not kept cool then 'firing' may occur, resulting in uneven cutting or, even worse, in the surface of the glass flaking off.

Dressing the wheels

The cutting is carried out in two stages. First comes the roughing, which is done over a carborundum wheel (Figure 12.10). These are shaped or 'dressed' by the rougher to the profile required for the particular type of cut to be produced.

After being used to cut two or three pieces the wheels have to be re-roughed, or 'tinned-up' as it is called, to re-open the grain so that they will cut efficiently. Then, after cutting some twenty wares, the wheels will have to be dressed all over again. The life of one of these wheels is only about two weeks, after which it will have worn down to the point where it can no longer be used. In a factory employing a hundred cutters the replacement of the carborundum wheels is a major expense.

The rough and the smooth

When the rough cutting has been completed the facets that have been created will have a rough texture. They will be white in colour and will lack sharp definition. The next process is to smooth these facets on brown sandstone wheels. These sharpen up the facets and smooth the surfaces to a sleek transparent grey colour and remove any of the carborundum that may have been left on the glass during the roughing process.

Though the cuts are now smooth, they still have to be polished. First they are

Figure 12.10 Rough-cutting the glass on a carborundum wheel (by courtesy of Stevens and Williams (Royal Brierley))

'wooded', that is, polished with pumice applied to a wheel made from poplar wood. They are then acid-polished. This is done by first washing them thoroughly and then dipping the glass in a bath of hydrofluoric and sulphuric acids.

Final inspection

When the wares have been thoroughly cleaned, they are ready for inspection. The inspector will be on the lookout for any faults in the blank which have not been eliminated in the cutting. Some of these could still be rectified, but faults which cannot be remedied condemn the wares to be sold off as seconds. The inspector will also be checking that the cutting is even, smooth and sharp, and that there is no chipping. None of the cuts must go too deep, piercing right through the glass. Any glassware which fails to live up to the high standards set by the firm is either rated a second or scrapped altogether.

Intaglio-cutting

A typical intaglio-cut glassware resembles the cutting on a seal stone. Compared with the mathematical cutting of cut glass wares, intaglio cutting is far more naturalistic. The curved forms of stylized leaves and flowers are represented in a realistic way. The intaglio-cutter also works with a sandstone wheel turned by an electric lathe, but the wheels are smaller than those used by the cutters of cut glass. The cutter also works under the wheel, not over it and this allows the freedom to create the curved cuts which are characteristic of intaglio decoration.

The designs which the intaglio-cutter produces may be acid-polished like cut glass, but often they are left with the frosted finish which they have when they come from the wheel. Sometimes part of the design will be polished while other parts of it are left frosted by stopping them off before the ware goes into the acid bath.

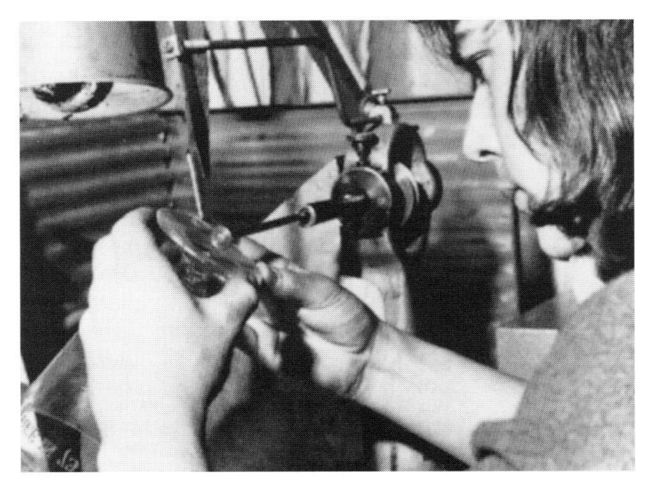

Figure 12.11 Copper-wheel engraving (by courtesy of Wedgwood)

Engraving

Another method of decorating glass by cutting away the metal is engraving. There are two distinct types. The metal can be scribed with a diamond. This is a technique usually employed by individual artists in their own studios working on blanks supplied by the glasshouses. The pieces they produce are invariably one-offs and are, of course, extremely expensive.

The other technique of engraving glass is known as wheel-engraving. This is carried out under a copper wheel of small diameter which is charged with a mixture of oil to prevent 'firing' and emery dust, which acts as the cutting agent (Figure 12.11). Though wheel-engraving is less laborious than diamond engraving, it nevertheless employs many hours of a highly skilled craftsman's time. The technique is therefore generally only used for commissioned pieces.

Sand-blasting

Because of the high cost of either intaglio-cutting or engraving it is now common practice to decorate popular crested waves by sand-blasting. The result closely resembles engraving, but the practised eye can readily detect which technique has been used to add these embellishments. Nevertheless, sand-blasting does produce a very acceptable result at a price which makes such pieces available to a far wider cross-section of the public.

Price justification

It is not hard to understand why a suite of cut crystal stemwares now costs many hundreds of pounds. The cost of the raw materials, particularly the red lead and potassium carbonate, is not inconsiderable. Then there is the expense of ensuring that the raw materials are uncontaminated, of mixing them in exact proportions and of ensuring the safety of the operatives. The cost of labour in the glasshouse and in the cutting department is also very high, because not only is glass-making very labour-intensive, but the craftsmen are highly skilled and have to be paid accordingly.

China

Clays, bodies and glazes

Pottery is made from decomposed granite, which has become transformed into what we know as clays. These clays contain varying quantities of the white mineral kaolinite, consisting of aluminium oxide, silica dioxide and water. Silica dioxide is the chemical name for quartz, one of the main constituents of granite. Quartz is the mineral which is fused to make glass. It can be appreciated, therefore, that pottery and glass are, like second cousins, loosely related.

Clay is a plastic material, which means that it can be readily shaped or moulded. This plasticity results from the fact that the kaolinite particles of which clay is composed are plate-like. These platelets slide over each other, lubricated by the water present in the material.

It was the existence of very large deposits of pure clay in many parts of China which led the Chinese to produce the translucent white pottery called porcelain at a very early date. One of the main constituents is white clay. Today the majority of fine quality tablewares and figurines are made from white clays with a high kaolin content.

What is pottery?

It would perhaps be useful at this point to define what we mean by 'pottery', for the word is often misunderstood and its meaning limited to low-fired earthenwares. It is, in fact, one of two words which are used to describe all those wares which are produced by driving out the water from clay by heat. The other word used to describe all these diverse products of the potter's kiln is 'ceramics'. These two generic words cover the whole spectrum, from the most delicate of bone china to sanitary wares and plant pots.

The clay plays a vital role in determining the type of pottery that is produced from it. The red clays found in The Potteries in the West Midlands lent themselves to the production of earthenwares and stonewares. The clays of Devon and Dorset proved suitable for the production of jasper wares, and so on. But it is not only the clay which a potter uses that dictates the nature of the body from which plates and cups, bowls and vases are fashioned. Any additions to the clay and the temperature of firing also determines the type of pottery produced. Though there are many sub-types of pottery, all the wares which can be produced by dehydrating clay can be divided under two main headings. There are, broadly speaking, earthenwares and stonewares. The essential difference between these results from the temperature at which they are fired.

Earthenware

A furnace at a temperature of between 800 and 1000 degrees Centigrade is hot enough to drive out all the water present. This includes not only free water but also the water chemically bound into the kaolinite. The resulting material, called by the descriptive name earthenware, is not only relatively fragile but is also porous. Earthenwares are usually covered with a glassy skin, known as a glaze, to render them waterproof. An unglazed form of earthenware is called terracotta, which simply means cooked earth.

Stoneware

To produce stoneware, clay is placed in a kiln at a temperature of 1250 degrees Centigrade or higher, with the aim of forcing the clay to fuse to form a vitreous mass. Even this temperature would not be high enough to fuse pure kaolinite. To induce fusion the potter adds powdered feldspathic stone to the clay and this acts as a flux. Stonewares are impervious to water and are much harder and much stronger than earthenwares. They probably got their name because, in their unglazed form, they have a texture like stone, and also ring like stone when they are tapped. Only relatively few stonewares, jasper for example, and some studio-produced wares, are left unglazed. When a glaze is added to a stoneware this is done to improve its appearance and to make it easier to clean, not in order to waterproof it.

These two main types of pottery can be subdivided. Earthenwares which have a salt glaze are variously called faience, Delftware and maijolica. The main types of stoneware are commercially described as stoneware, jasper, porcelain and bone china. Earthenwares are mainly differentiated on the basis of their decoration. The different types of stoneware result from different ingredients in their bodies which produce quite distinct physical characteristics, and from their being fused at different temperatures.

Porcelain

The Chinese made their porcelain from two materials – clay with a high kaolinite content and white China stone, which they called *petuntse*. This consists largely of feldspar, a silicate of aluminium in which soda and potash are also present. China stone is fusible whereas the kaolinite clay is infusable. In the kiln the China stone melts to form a cement that binds together the infusible particles of the clay.

Royal Worcester, which is the only pottery in the UK to produce hard-paste porcelains today, make their body from 50 per cent china clay, 25 per cent feldspathic stone and 25 per cent quartz. This is also the formula which Royal Copenhagen have adopted. The quartz is added to make the melt more viscose, that is, less fluid, thus preventing the body becoming too soft during firing. The purpose is to prevent distortion. In the furnace the quartz fuses somewhat sluggishly over a wide temperature range as the pots that are being fired travel through the furnace.

Bone China

Bone china was an English innovation, probably introduced at the Bow factory, though some authorities attribute it to Chelsea and others to Spode. It dates from the middle years of the eighteenth century. Today all the leading English factories making fine china produce bone china. The body consists of 25 per cent china clay, 25 per cent Cornish stone and 50 per cent calcined bone ash made from cattle bones imported from South America. Cornish stone is, to all intents and purposes, the same as the petuntse used by the Chinese potters. The bone ash has a number of functions. Not only does it make the behaviour of the body during firing much more predictable, but it also increases the mechanical strength of the fired body, and produces a body which is pure white in colour.

It might be assumed that hard-paste porcelain, because it is fired at very high temperatures, would be stronger than bone china, but in fact bone china is the mechanically stronger of the two. Porcelain is also not so white or translucent as bone china.

Glazing

The term 'glaze' derives from the word 'glass', and the act of glazing consists of covering a pottery body with a glass-like outer skin. While the glazes added to earthenwares are both decorative and functional, rendering them impervious to water, the glazes applied to stonewares are almost entirely decorative. An unglazed jasper jug functions just as well as a stoneware jug that has been glazed. In the case of ordinary stonewares and earthenwares a coloured glaze is usually used, and this is very often the only decoration. On porcelains and bone china wares it is usual to apply a white glaze, and this contributes to their translucency.

The glazes used for porcelains and for bone china wares do not just lie over the body, adhered to it but separate from it as are those on earthenware and stoneware. They combine with the body, and fuse into it. This is because the glazes used for these wares are chemically very similar to the bodies. The reason bone china is so highly translucent is because of the close relationship between body and glaze. The body consists of two types of crystal in its fused form. These have well-matched refractive indices which are similar to the refractive index of the glaze. Therefore, rays of light passing through the material are not dispersed but are bent at a consistent angle, as they are by water or glass.

Preparing the ingredients

The first step in the process of making pottery, whether tablewares, decorative wares or figurines produced in bone china, porcelain, stoneware or earthenware, is to prepare and mix the ingredients. In the case of bone china these consist of calcined cattle bones, china clay brought from Cornwall or France, and feldspathic stone from Cornwall. The bones, which amount to 50 per cent of the constituents of bone china, provide the refractory element which resists the action of the heat in the furnace. The china clay in bone china gives the resulting body plasticity, making it easier for the potter to shape it. It also provides stability in the furnace, forming, as it were, the skeleton of a ware during the firing. The feldspathic stone serves as a flux. This means that it helps the clay body to flow when it is heated and eventually to vitrify.

Once the bone and the feldspathic stone have been ground to a fine powder, they are placed in a slip tank, where the powdered material is thoroughly mixed with clay and water to form a paste. The paste is then passed over a series of electromagnets that remove any iron that is present. It is important to remove every trace of iron because even in very small quantities it could colour the body.

Next, the paste goes into a filter press where excess water is squeezed out of it to produce square or disc-like slabs of the material. These are then left for a period to cure.

After storing the body, it is passed through a 'pugging machine'. This simultaneously further refines the body and removes any air which could result in bubbles being incorporated in the pots that will be made from it. The body comes out of the pugging machine in the form of a continuous cylinder which is cut with a wire into suitable lengths as it emerges. To make flatwares such as plates and dishes, and in some factories to produce cups as well, this body is used in the form in which it comes out of

Figure 12.12 Semi-automatic plate-making at the Minton factory

the pugging machine. However, in a modern factory many hollowares and figurines are cast in moulds. To make these the clay must be in a state in which it will pour readily. To create this smooth, creamy casting material, known as 'slip', water is added to the body together with a small quantity of soda.

Making plates

In a modern factory plates are made either on a semi-automatic or on a fully automatic machine which employs the same basic principals as the semi automatic technique. On the semi-automatic machine, known as a jigger, the potter sits at a wheel on which a mould has been mounted which will dictate the form of the top of the plate. An appropriately sized lump of clay is thrown down into this mould, and as this revolves and spreads out over the mould the potter brings down a metal arm, a profiling tool, which shapes the upper side of the clay into the form of the bottom of the plate. The plate is thrown in the same way that a craft potter throws a dish, but it is done upside-down. The mould and the profiling tool ensure that the plate conforms to the required shape (Figure 12.12).

Making hollowares

The production of hollowares by casting in a mould begins with the making of a model, which has to be over-sized because the body shrinks in the kiln.

The models are carved from clay with sculptor's tools, and it is from these models that the moulds are produced. Most wares are assembled from a number of castings, so first the model is inspected and the decision taken as to how best to cast it. It is then cut up into the desired component parts. The body of a tea-pot, for instance, will be cast separately, as will the handle and the spout, while a separate mould will be produced to cast the lid. Figurines may involve the production of nine or ten separate castings. The various pieces then have plaster of Paris poured round them, and when this has set the resulting block is cut open and the model removed. A sprue hole is

Figure 12.13 Pouring the slip into a mould to produce part of a figurine at the Royal Doulton factory (by courtesy of Royal Doulton)

created, the mould is reassembled and the slip is poured in through the sprue (Figure 12.13). The slip is left in the mould for about twenty minutes, by which time a layer of it will have set round the walls of the mould. The residual slip is then poured out.

Next, the moulds are broken open. The cast components of the object are then assembled. When these components have dried to a leather hardness they are, as it were, glued together with slip. Then the assembled pot is left to dry.

Sponge, fettle and tow

When the wares from the plate department and the casting shop have dried out, a series of operations are carried out on them known as 'sponge', 'fettle' and 'tow'. Any bumps and lumps which mar the smooth surface of flatwares, hollowares and figurines are removed using damp sponges and small scalpels called fettling tools. The junctions between the assembled components of the hollowares are also smoothed. A tow, a mop usually of flax fibre, is used to smooth flatwares and to round off any sharp edges. After these operations all the pieces are individually inspected. Because every operation involves cost, any sub-standard ware is either sent back for rectification or scrapped before any further work is carried out on it.

Biscuit-firing

Bone china is fired for twenty-four hours at a temperature of 1250 degrees Centigrade in the biscuit furnace. During the firing the wares are supported on 'bats' or 'saggers' made from a heat-resistant material. Hard-paste porcelain is fired for a similar period at 900 degrees Centigrade. Bone china wares emerge from the kiln with a matt biscuity surface, and during the firing they will have shrunk by one-eighth their original size. The matt surfaces of the wares are then brushed to remove any coarse material from them and the wares are individually inspected once again. The wares are then loaded into a vibrating machine along with wood chips which smooth them to the desired degree. Too high a polish is not desirable, because this would result in the glaze not adhering to the biscuit surface.

Applying the glaze

In British factories under-glaze decoration is very rarely used these days, so the next operation is to apply the glaze. The glaze may either be applied by dipping or spraying.

Next the feet of the vessels are cleaned of glaze so that they will not stick to the pins or pegs that support them in the zircon frames in which they are placed before going into the glazing, or glost, kiln. The wares stay in this kiln for a period varying from five and a half hours to eight hours. The glaze on bone china is fired at 1100 degrees Centigrade and that on porcelain at 1400 degrees.

When the glazed pottery comes out of the glost kiln the marks left by the supporting pins or pegs are ground out of good-quality pottery, though cheaper pots often show these peg marks. A well-finished foot is a sign of a good-quality pot.

Decorating pottery

Pottery is often already decorated during the course of manufacture. For example, some blanks will have patterns impressed on them by the mould. Plates may be fluted simply by the plate-maker placing a suitable mould on the wheel. Similarly, if flutes are carved on the model of, say, a teacup and a plaster of Paris mould made from this, the mould will impart the imprint of these flutes onto the slip which is subsequently poured into it. Even such delicate decoration as half-pearls in the form of a circlet round the rim of a plate or a saucer can be applied to the blanks from the moulds.

In some instances further decoration may be added to a blank before the biscuit-firing. At the Royal Copenhagen factory skilled craftswomen model tiny flowers by hand, which will be used to embellish some of the more elaborate pieces of the Flora Danica pattern. Another type of decoration which can be applied before a ware is fired is piercing. This is also a feature of patterns produced by Royal Copenhagen, not only Flora Danica but also the lace versions of Blue Fluted. The holes in the body, which produce a lace-like effect, are created by pushing a metal tool through the clay and then trimming the edges with a fettling tool. Needless to say, both these hand operations add considerably to the cost of a piece.

Decoration and price

Whether or not a piece of bone china or porcelain has been decorated in the course of making, it will usually have further decoration applied to it in the painting shop. This decoration may consist of no more than a gold or enamelled band round the rim, but many designs call for much more elaborate decoration. This can vary from polychromatic motifs applied from a lithographic sheet to the most elaborate paste gold work. The price of tablewares, or, for that matter, decorative wares, is very largely dictated by the amount and type of decoration applied to them. A factory with a reputation to

Figure 12.14 Making the cameo decoration for jasperware at the Wedgwood factory. Body is pressed into a 'pitcher' mould made from fired pottery. Then, using a spatula, the intricately shaped ornament is eased out of the mould (by courtesy of the Wedgwood Group)

uphold will make only one quality of blank, the very best it is possible to make. So all the bone china blanks made at, say, the Minton factory will have cost the firm the same amount of money to make. But when those plates have been decorated the difference in price can be dramatic. A Minton plate decorated with a simple band of colour might retail at £15 or £20. The identical blank could, however, have been first decorated with acid gold and then with elaborate gold paste work, and the most decorative of plates in this firm's gold paste range sell for over £2000 each. Royal Copenhagen make two designs of blue under-glazed tableware, Blue Fluted and Blue Flowers. Both are painted by hand, but it takes a skilled decorator twice as long to paint a piece of Blue Fluted as it takes to paint the piece in the Blue Flowers design. This fact is reflected in the prices. Blue Fluted is almost exactly twice as expensive as Blue Flowers.

Hand-painted pottery

In the past all fine pottery was painted by hand. Metallic oxides were mixed with ground-up silica and the mixture was then fused to produce a frit, which was then ground to a fine powder and suspended in turpentine. This enamel was then painted on the pottery with a paint brush, and then the enamel was fired. There are still a large number of painters decorating pottery at the leading factories both in the UK and abroad. At Royal Copenhagen, for instance, they employ no fewer than six hundred painters, and a similar number work at the Meissen factory. Some reproduce set patterns, and they become so skilled at doing this that they can repeatedly reproduce a design with great precision. Some special pieces are decorated freehand, the painter is given a degree of freedom in interpreting the pattern (Figure 12.15). Invariably both bone china and porcelain figurines are hand painted.

Figure 12.15 Decorating a piece of Flora Danica over the glaze with enamels. In the background is a replica of the original wash-drawing from the eighteenth-century book (by courtesy of Royal Copenhagen)

Under-glaze painting

Although most bone china is painted over the glaze, some European porcelain is still painted under it. This has the advantage that the fused enamel is protected by the glass-hard glaze. However, the painter's palette is limited when working under the glaze, because many enamels would be discoloured by being subjected to the high temperatures of the glost furnace. Only certain metallic oxides will withstand this heat. A notable example is cobalt oxide, used to create the rich blue for which the Worcester Pottery, among others, was famous in the eighteenth century. This is the same blue which is used to decorate the Royal Copenhagen patterns, Blue Fluted and Blue Flowers. Other colours which can be used for under-glaze painting are green, produced by making a frit with copper oxide, and a rather rusty red shade, resulting from the use of an iron oxide frit. Towards the end of the eighteenth century, and increasingly during the nineteenth, the public favoured pottery decorated with a whole palette of colours which allowed floral designs to be rendered in a realistic way and popular paintings to be translated into decoration for pottery. And ever since then the English factories have concentrated on polychrome decoration over the glaze. The enamels can then be fired at around 840 degrees Centigrade, temperatures which they can happily withstand.

Transfer printing

The technique of transfer printing, which the Irish engraver John Brooks invented about 1750 when he was working for the short-lived Battersea Enamel Works, is still

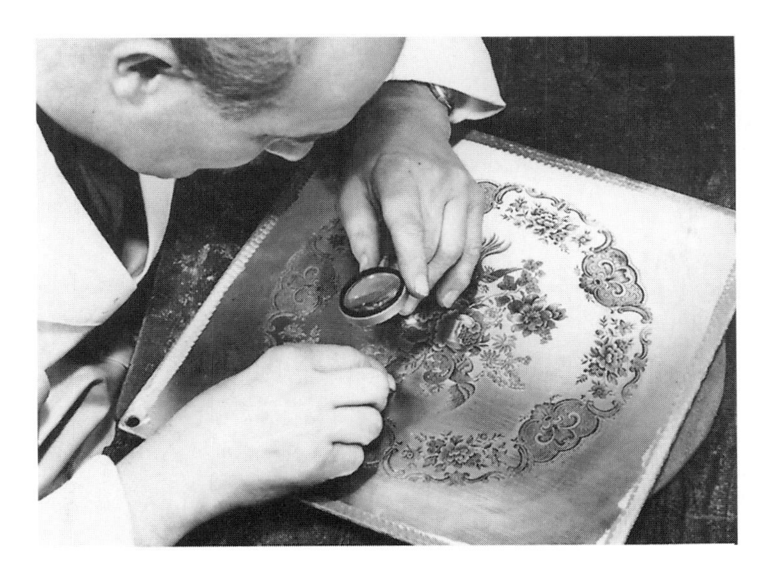

Figure 12.16 Engraving a copper plate. This technique has now been largely superseded by lithographic printing

used today, though only for special pieces and short runs. The process consists of first producing a copper plate, engraved with the required design (Figure 12.16). This plate is then 'inked' with a metallic oxide in suspension. A piece of tissue paper is then laid over the inked plate and copper plate and paper are cranked through a roller. The pressure applied to the roller imprints the design from the copper onto the tissue. The tissue paper is applied to the pottery, and after the ink has dried on the surface of the blank, the tissue paper is carefully and laboriously sponged off. What is now left on the pot is a monochrome outline, which is then fired. Such well-known designs as Willow Pattern were produced until quite recently by transfer printing.

Lithographic printing

Though similar in some ways to transfer printing, this recently introduced technique has a number of advantages over it. The most important of these is that it makes it possible to reproduce a polychromatic design in one operation with a minimum of labour. The lithographic sheet is also much easier to handle than the tissue paper used in transfer printing, and there is no need to spend time getting rid of the tissue by laborious sponging.

The production of the sheets used to transfer this decoration to the pottery is though, very expensive. These sheets are either produced by a lithographic process or by silk screen printing. If the silk screen technique, which tends to produce more muted shades, is used, many operations are required to obtain subtle colours. The silk screen process consists of transferring the colours to a piece of silk. Individual transfers are used to stop off areas where particular colours are not required. Because of the cost of producing the sheets, lithographic printing is only applicable when large numbers of pieces are going to be decorated with the same design, so spreading the initial outlay. This is why transfer printing is still employed for short runs.

Figure 12.17 Affixing a lithographic transfer to the ware (by courtesy of Royal Doulton)

Lithographic prints consist of a sheet of plastic film on which the design has been printed. The designs are printed in enamels and are protected by a backing sheet. All an operative has to do is to soak off this backing sheet and then apply the lithographic sheet to the cup or plate to be decorated. Even this simple operation does, however, call for some skill and experience, because the design has to be laid down in precisely the right position to ensure that all the pieces in a service match (Figure 12.17).

Once the transfer has been applied to the ware it is left to dry. Then the decorated ware goes into the kiln, which fires the enamel colours and burns off the plastic film. A piece of pottery decorated in this way resembles a piece that has been hand-painted, but with experience it is not difficult to recognize a printed design from a painted one. A sure indication is, of course, price. If a piece that is decorated with an elaborate polychromatic pattern is modestly priced, the decoration must have been applied by printing.

Gold-enrichments

Gold-enrichment is frequently used on tablewares and decorative pieces of pottery. At its simplest, this consists of a band of gold round the rim of a plate, a saucer or a cup. This, whether it is a coloured band or a gold one, may be applied to the pieces with a brush, or if the pieces are produced in large numbers it may be applied automatically as the pieces revolve on a turntable (Figure 12.18) Gold may also be used to enrich handles, lids, etc. and is put on by hand with a brush. Gold in powder form is suspended in a suitable oil, and when the wares go into the kiln and are fired at between 850 and 900 degrees Centigrade, the oil is driven off and the gold fused onto the surface of the pottery. Either one or two coats of gold may be applied and fired, depending on the quality which a particular factory aims to achieve. The difference between a single coat and two coats is very noticeable. A single coat tends to have a rather brassy look, while two coats produce a deep, rich golden colour. On good-

Figure 12.18 This machine automatically applies a gold band to the edge of plates

quality wares this is further enhanced by hand-burnishing. This method of giving a metal a high polish is well known in the silver trade, and it consists of rubbing the metal with a piece of agate mounted on a stick (Figure 12.19).

Acid gold

This is a technique invented in 1863 by James Leigh Hughes, and the patent rights were later acquired by the Minton factory. The technique remains very much a Minton

Figure 12.19 Hand-burnishing the gold with agate

speciality, but a number of other factories, Worcester and Rosenthal among them, also employ it. It makes the gold band applied to pottery more interesting by creating a three-dimensional effect. The first step is to etch a design into the surface of the pottery. This is done by affixing a transfer print or a lithographic sheet of the required design to the ware. This sheet consists of a black and white design. The black areas are printed with a compound that is resistant to acid. When this strip has been applied, the rest of the ware is covered with Brunswick black applied with a brush. Brunswick black is also resistant to acid. The ware is now immersed in a tank of hydrofluoric acid which bites into the glaze in those areas which are not protected by the acid resist. The ware is kept in the acid for a period long enough to allow the acid to eat into the glaze to a depth of 0.5 millimetre (2 thousandths of an inch). The acid is then washed off in cleansing tanks and the ware is dried. Then the Brunswick black and the lithographic resist is dissolved away in a solvent bath and the ware is again thoroughly washed and dried. The white surface of the ware now has a band of lace-like intaglio decoration running round it. Gold is hand-painted over this decoration and fired. A second coat of gold is applied and this is fired in its turn and then burnished. The result is a gold-enrichment that also has a texture which imparts an additional richness to it. This is an expensive process, not so much because of the cost of the gold but because of the amount of exacting hand-work involved.

Gold-paste decoration

This is the ultimate in rich decoration, and it is labour-intensive and time-consuming (Figure 12.20). Every piece to be decorated in this way is first very carefully inspected to ensure that it is perfect, because there would be no point in using a less than perfect blank. The technique is really a form of that style of decoration known as *pâte-sur-pâte.* Like so many pottery techniques, *pâte-sur-pâte* was first used by the Chinese potters, but the Sèvres factory in France adopted it in 1851 and it was brought over to the Minton factory later in the century. Subsequently Minton developed their gold-paste form of decoration from it and it has remained the speciality of this factory ever since.

The role of the inspector

The importance of the work of the inspectors who check the wares at every stage in their production in leading factories in the UK and on the Continent cannot be emphasized too strongly. It is clearly in the interest of these factories to ensure that expensive additional operations are not carried out on faulty blanks. Thus the inspectors who check these blanks after they come out of the biscuit kiln, and again when they emerge from the glost furnace, are making an important contribution to the economics of production.

It is standard practice in every major factory producing bone china or porcelain wares for every piece to be inspected before it is despatched. The inspectors responsible will be looking for a variety of common shortcomings, such as black spots caused by foreign matter having got into the body or the glaze, unevenness or surface marks, discoloration, blemishes in the decoration, marks below the gold on rims, incorrectly placed printed decoration, etc. A ware that shows any of these defects will either be sent back to the factory to be rectified, where this is possible, designated as seconds or at worst scrapped altogether. Just how exacting the final inspection is can be judged by looking through a delivery of seconds from any factory with a high rep-

Figure 12.20 Minton's Shrewsbury Green pattern, decorated with elaborate paste gold (by courtesy of Royal Doulton Tableware Ltd)

utation for quality. On some of the pieces, it is true, the faults may be apparent, but on many the reason for their rejection will be almost imperceptible. Indeed, it is often the blue pencil mark put on the piece by the inspector that directs attention to some tiny blemish.

Glossaries

Gemmology – glass – jewellery – pottery – silverware – horology

These glossaries are not intended to be exhaustive but contain terms which are commonly met in descriptions of goods

Gemmology

AMORPHOUS Unlike a crystal an amorphous material does not have a definite structure.

ASTERISM The star-like phenomenon seen in various gems when they are cabochon-cut. Such stones are called star-stones, e.g. star-sapphire, star-ruby, star-rose quartz.

BAGUETTE A simple rectangular cut. Literally translated it means a 'rod'.

BAROQUE PEARL A pearl that is irregularly shaped.

BIREFRINGENCE Another term for double refraction.

BLISTER PEARL A pearly accretion attached to the shell, sometimes cut away to produce a half-pearl.

BOULE A mass of synthetic gem material in the form of a carrot, produced by the Verneuil process.

BRILLIANT-CUT The most popular cut for diamonds. A total of 58 facets are cut on the stone including the table and the culet. The culet is nowadays optional.

BUTTON PEARLS Round-topped pearls with a flat base.

CABOCHON To polish a stone into the form of a dome. All star stones are 'cut' cabochon.

CAMEO A carved bas-relief. Cameos are carved both on gem materials such as the chalcedonies and on mollusc shells. The latter are properly called 'shell cameos'.

CARAT WEIGHT The standard (metric carat) used for weighing gems. The metric carat equals 200 milligrams (one-fifth of a gram).

CAT'S-EYE An optical effect visible in cabochon gemstones; it appears like a ray of light crossing the dome of a cabochon. The best known cat's eye is the chrysoberyl cat's-eye. The effect, caused by a series of short tubes or needles, is known as chatoyancy.

CHALCEDONY A micro-crystalline variety of quartz. Agate is a chalcedony, where the colour is distributed in bands, and onyx is black and white banded chalcedony. Most onyx is now produced by staining agate.

CHATOYANCY A band, or streak of light, which moves across the domed surface of a cabochon-cut stone. The phenomenon is known as cat's-eye effect and is due to the reflection of light from cavities or fibres within a stone. It is observed in various gem species.

CLARITY The term used to describe the degree of freedom from inclusions of a gemstone.

CLEAVAGE Certain gems cleave in definite directions, e.g. diamond has a perfect octahedral cleavage.

CONCHIOLIN Organic material (dark in colour) secreted by molluscs which produce pearls. It is also known as conchine.

CRYSTAL A solid which has an atomic structure arranged in an orderly manner. There are seven crystal systems into which gems are divided.

DISPERSION A term for the 'fire' seen in transparent gems. It is due to the ability of faceted gems to separate the colours of the spectrum when light is refracted and reflected within them.

DOUBLE REFRACTION The splitting of a ray of light into two rays when it enters certain gems belonging to crystal systems other than the cubic. In crystals of the cubic system and in amorphous substances light is refracted but continues as a single ray.

DOUBLET A stone composed of two or more pieces of genuine or man-made gem material cemented together. Sometimes pigment is included in the cement layer with the object of using inferior material to simulate an expensive stone.

ESSENCE D'ORIENT Name given to a fish-scale preparation used in the production of imitation pearls.

FABULITE A common trade name used for strontium titanate.

FACET A flat polished area on a cut gemstone, hence a faceted stone.

FEATHER A name given to inclusions in gems (especially sapphires) which are like feathers in appearance. They are, in fact, layers of crystalline and liquid inclusions.

FIRE-MARKS Hair-like cracks on facets of gems caused by overheating during cutting and polishing.

FOUR C'S An expression evolved to describe the four factors which influence the value of a diamond – colour, clarity, cut and carat weight.

FLUX FUSION A method of producing synthetic stones.

FRESHWATER PEARLS Pearls found in mussels which live in rivers.

GOODS A term commonly used by stone dealers to describe the stones they buy and sell.

GRAIN The pearl grain is one-fourth of a carat.

HARDNESS The usual scale of hardness used in the jewellery trade is the one introduced by the scientist F. Mohs. The scale is:

Diamond	10	Apatite	5
Corundum	9	Fluorspar	4
Topaz	8	Calcite	3
Quartz	7	Gypsum	2
Feldspar	6	Talc	1

The scale is arbitrary and the difference between the hardness of diamond (10) and sapphire (9) is greater than that between sapphire and talc (1). There are also intermediary figures, e.g. emerald is rated about 7½, chrysoberyl is rated 8½.

HEAT TREATMENT Many stones have their colour altered by heat treatment. Well-known examples are the production of white zircon, a popular diamond simulant, from brown zircon. Most commercial citrine is heat treated amethyst, and most blue aquamarine is the result of heat treating green aquamarine. Most tanzanites are brown when found; the blue coloration results from heat treatment. Sapphires are heat treated to improve their colour.

INCLUSIONS Most gems contain inclusions which may be minerals, liquid or gas-filled libella (bubbles). Diamond often contains inclusions of carbon, garnet, ilmenite and other minerals.

INTAGLIO A seal stone into which a monogram or device has been incised is described as intaglio cut.

LAP A revolving circular disc which can be loaded with abrasive, against which gemstones are held to produce facets. A lapidary is a cutter of coloured stones.

LEVERIDGE GAUGE The most accurate type of gauge for estimating the carat weight of a gemstone by measurement.

LOUPE A small magnifying glass.

LUSTRE This is the surface brilliance of a stone. It depends upon the amount of light reflected from the surface, which is governed by the degree of polish and the hardness. There are various types of lustre. These include adamantine, as in diamond;

vitreous or glassy, as in ruby and sapphire; resinous, as in amber; waxy, as in turquoise; pearly, as in pearl.

MAKE A term describing the cut of a diamond. A stone will be less valuable if it suffers from faults of make.

MARQUISÉ Boat shaped. This is a fashionable variation on the brilliant-cut for diamonds.

MATRIX The rock which contains a mineral. For instance, ruby is commonly found in a matrix of zoisite, and the matrix containing the ruby is often slabbed and polished or carved into bowls.

MIÊLEÉ Small stones less than ¼ carat in weight. A term usually applied to diamonds.

MINERAL An inorganic substance, hence gem minerals, as distinguished from organic gems such as pearls and jet.

MIXED-CUT A mixture of brilliant-cut above the girdle and step-cut below the girdle used for coloured stones.

MOTHER-OF-PEARL Material obtained from the inside part of the shell of certain molluscs.

NACRE Substance consisting mainly of carbonate of lime, which is secreted by certain molluscs. A pearl is composed of nacre.

NATURAL As found in nature. A natural stone rather than a man-made synthetic.

ONYX MARBLE (correctly called MARBLE). Banded calcite. Not a true onyx, which is a form of crypto-crystalline quartz.

PASTE The glass used for imitation stones.

PAVILION The part of a cut gemstone below the girdle, hence pavilion facets.

PIPE A mass of rock that has solidified in the neck of a volcano, for example the famous diamond pipes of South Africa consisting of 'blue ground'.

PLAY OF COLOUR An effect seen in opal and labradorite, due to interference or scattering of light from thin platelets or spheres near the surface of a stone.

REFRACTION When a ray of light entering a transparent or translucent stone is bent from its normal path. The refractive index of a stone is a measure of the degree to which light is bent when entering or leaving it. Refraction, and internal reflection, of light in a properly proportioned stone accounts for its 'fire' *(see* DISPERSION).

RHINESTONE A name sometimes given to glass imitations of diamond.

ROSE-CUT An early cut in which the base of the stone is polished flat and the top has triangular facets cut on it and is pyramidal in profile.

SINE OF AN ANGLE The ratio in a right-angled triangle between the side opposite the angle under consideration and the hypotenuse. It is an important definition for ascertaining refractive indices of gems.

SOUDÉ STONES French term for soldered. It is usually applied to composite stones (e.g. beryl- or quartz-crown and base with coloured layer between) imitating emerald. Another version is produced from synthetic white spinel.

SPECIFIC GRAVITY The ratio of the density of a substance to water. The measurement of specific gravity of gemstones is one of the methods used to identify them.

SPECTROGRAPHIC ANALYSIS Analysis of the spectrum produced by a stone in order to identify it.

STAR A rayed effect caused by light reflected from inclusions in a gemstone.

STEP-CUT A cut in which the facets are parallel to the girdle.

STRIAE Another name for growth lines (usually straight) seen in natural gems, and the usually curved growth lines in Verneuil synthetic gems.

SYNTHETIC A man-made stone. Some synthetics like synthetic corundum closely resemble natural stones in their chemical composition and physical properties.

Others, like yttrium aluminate and cubic zirconia, have no counterpart in nature and should be called synthetic simulants.

TABLE The large facet on the top of a stone, parallel with the girdle. The facets cut above the girdle of a stone are also described as table facets.

TRANSPARENCY Measure of light penetration in substances. There are transparent stones, semi- or partly transparent, translucent stones through which only some light passes and opaque stones through which no light passes.

WHITE A quite inaccurate adjective used to describe diamonds of high quality, for example 'blue-white'. Such stones are not white but almost colourless, tinted with blue or more often yellow.

Glass

ACID-POLISHING After glass is cut the facets have to be polished. This is done by acid-etching in a bath containing hydrofluoric and sulphuric acid.

AIR-TWIST A form of stem decoration common in the second half of the eighteenth century and still used today. It is produced by inducing a bubble in the metal by pegging it and then drawing out the metal and twisting it.

ANNEALING Taking out the stresses in glasswares by passing them through a lehr.

BLISTERS Air bubbles which result from gases created during the melting of the constituents of the metal not being driven off.

BLOWING The expansion of metal taken from the furnace by the craftsman blowing through his hollow blowing iron. Glass can either be free-blown, blown in a mould by a foot-blower or blown mechanically in an automated machine.

CASED GLASS Glass consisting of a second gather of a different colour surrounding the initial gather. Cameo glass is produced by carving the outer layer of cased glass.

CHAIR The group of men who make glasswares. For example, a wine chair consists of four men. The leader of the group is called the gaffer. A chair is also the crude wooden seat with extended arms on which the gaffer sits.

COLOURED GLASS Glass may be coloured by the addition of metallic oxides. Cobalt oxide produces blue glass, copper oxide green glass, etc.

CORDS Lines, or streaks, in the glass resulting from poor mixing of the constituents or from the joining of metals of different temperatures.

CRACKING-OFF Removing the moil by scribing the ware with a diamond tool and then passing it through a gas flame. Nowadays the process has been automated.

CRISTALLO A type of glass first produced in Venice.

CRIZZLE A disease of glass resulting in internal cracking, the formation of a sour-smelling liquid and eventual disintegration.

CULLET Broken glass from the shrower which is put in the pot and melted with the other constituents. It assists fusing, but is also included for economic reasons. Makers of fine glasswares use only their own cullet.

CUT GLASS Glass that has been faceted by grinding, first on a carborundum wheel and then on a sandstone wheel.

EDGE-MELTING Melting the rim of a glass in a gas flame, after the moil has been cracked-off, to round the sharp edge. Nowadays the edge is usually ground in automated machinery.

ENAMELLING Decorating glasswares with enamels consisting of a frit of metallic oxides and glass. After this has been painted on the glass it is fired.

ENGRAVED GLASS Glass can be engraved in different ways, with a diamond tool or under a small copper wheel charged with carborundum powder and oil.

FIRING The process of fusing the enamel or gold decoration on a glassware.

FLINT GLASS Glass made from ground-down flint pebbles instead of silica sand. Flint is an impure form of quartz and therefore largely consists of silica dioxide.

GAFFER The leader of the chair who forms the stems and feet of stemwares.

GATHER The metal which the foot-blower or the bit gatherer takes from the pot on his blowing iron or punty iron.

GLORY HOLE The small furnace located in the caterhole shop where larger wares are produced. These wares may have to be reheated a number of times as the workman forms them and they are placed in the glory hole for this purpose.

HOBNAIL-CUTTING A style of cutting that produces small circular mounds or points. It is a style popular in Ireland and nowadays is associated with the Waterford factory.

HYDROFLUORIC ACID An acid which readily attacks silicates and which is therefore used both for etching and polishing glasswares.

INTAGLIO-CUTTING A method of cutting carried out below a small sandstone wheel, giving far more freedom to the cutter than cutting over a larger wheel. Intaglio-cutting is often left frosted, but is sometimes polished.

LEAD CRYSTAL Glass containing red lead. In the UK there is a British Standard for lead crystal which stipulates that it must contain 30 per cent lead. On the Continent most glasshouses produce a lead glass containing 24 per cent lead.

LEHR The tunnel oven through which wares pass on a moving belt to remove the stresses in the glass. The wares are first raised to working temperature then slowly cooled.

MARVER A steel sheet, or recessed block of wood, on which the foot-blower rolls the metal on his blowing-iron to ensure it is true. Also the act of marvering.

METAL The molten glass taken from the pot in the furnace.

MOIL The end of the sphere attached to the blowing iron after blowing that is cracked-off before the ware goes into the lehr.

POT The ceramic vessel in which the constituents of glass are melted in the furnace.

PUNTY IRON Sometimes called a pontil iron. This is the solid iron rod on which the gathers are taken from the pot to form the stem and foot of a wine glass.

SAND-BLASTING Engraving glass by directing a jet of sand onto the surface.

SEEDS Small bubbles in glass due to the gases formed during the fusion of the metal not being completely driven off.

SERVITOR Another name for the gaffer who occupies the chair in a wine chair.

SHROWER The bin into which broken vessels from the lehr and the moils from the blowing irons are put to be later used as cullet.

SIDEN A ware which is out of true.

SODA LIME GLASS Glass whose constituents are soda and lime and silica sand. Soda lime glass contains no lead.

TEARDROP An air bubble induced in the metal used to form the stem of a stemware by pegging.

WOB FOOT A ware whose foot is out of true.

WORKMAN The gaffer in a caterhole shop.

Jewellery

AIGRETTE A gem-set head ornament in the form of a plume, worn to one side of the hair.

AJOUR Any setting which enables the pavilion facets of a gem to be seen.

ALBERT A slender chain of precious metal for men's evening wear. Women wore them as guards or muff chains. It was named after the Prince Consort to Queen Victoria.

ALLOY A metal compound resulting from melting two or more metals together. For example, carat golds are alloys of gold, silver, copper, etc.

ALLUVIAL A metal or gemstone which has been transported, usually by water, from the matrix in which it was formed and is found in an alluvial deposit.

AMULET An ornament worn to protect the wearer from malign influences. The Ancient Egyptians invoked the protection of their gods by depicting them on the jewels they wore.

ARABESQUES Flowing leaf and scroll work, often in low relief.

BANDEAU A narrow head ornament, usually set with gems.

BASSE-TAILLE A form of enamelling in which the surface of metal is recessed to varying depths to receive enamel.

BEADING Round grains or beads formed by a special tool to secure a gem in its setting. Often called millegrain.

BEZEL A thin piece of metal inside the edge of a box. A name given to the surround of metal which secures the crystal to a clock or watch case. The name is also given to the rub-over settings of some rings, and to certain facets in brilliant-cut diamonds.

BOX SETTING A closed form of setting in which a gem is enclosed in a 'box'. The edges of the metal box are pressed on to the girdle of the stone.

BRAZING Another term for hard soldering.

BROACH A tapered tool used for shaping, trimming or enlarging holes.

CAGE-BACK SETTING A gem setting which is cage-like. It is often used for cluster rings.

CALIBRE-CUT Stone cut to a special shape to fit a design.

CAMEO A carved gem, or shell, in which the carved design stands out against a darker or lighter background.

CANNETILLE WORK Gold, sometimes silver, wire-work used to decorate nineteenth century jewellery.

CARTOUCHE An ornament in the form of a paper scroll, often used as a surround for a coat of arms.

CHAINS Curb or cable: solid or hollow, circular or oval links slightly twisted. Fetter: long links. Fetter and trace: long links with smaller oval links. Fetter and three: long links separated by three short links. Prince of Wales: U-shaped links soldered in overlapping style. Trace: equal-sized oval links.

CHAMPLEVÉ Recesses gouged out of metal and subsequently filled with enamel.

CHANNEL SETTING A metal setting holding stones on two sides only.

CHATELAINE Long chains worn by women for carrying keys, seals, watches, sewing cases and other small objects. A large hook behind the top of the ornament was inserted over the waistband.

CHENIER A name for a joint tube.

CIRE PERDU Lost-wax process of casting.

CLIP A piece of jewellery similar to a brooch, but attached by means of a spring clip instead of with a pin.

CLOISONNÉ Strip is soldered on to metal to create divisions, or cloisons, into which are placed stones or enamel.

COLLET The setting of a stone.

CROWN Upper part of a cut gem.

DAMASCENE An art of decorating a metal by inlaying it usually with a more precious one. It is used on sword blades.

DIAMANTÉ The description widely used for white paste used in imitation jewellery.

DIE The tool used to form a piece of metal in a stamp. Hence die-sinking, which means to make a die for die-stamping.

DUCTILITY The property of a metal that allows it to be drawn out to form wire.

ELECTRUM A natural alloy of gold and silver. Many ancient wares were made from electrum.

ENGINE-TURNING A machine-engraving process for all-over decoration of a metal surface. Popular forms are 'barley', 'foxhead' and 'straight line'. At some periods it was fashionable to lay transparent enamel over engine-turned surfaces.

ENGRAVING Decorating metal by cutting lines into the surface with a sharp graver.

FILIGREE Ornamentation of fine twisted wire.

FINDINGS Semi-manufactured small parts of jewellery such as settings, catches, galleries, wires and snaps.

FLANGE A reinforcing rim inside a bezel which acts as a bearing for a gem.

FOIL A thin piece of metallic substance placed at the back of an imitation or natural gem to emphasize its brilliance, or to impart colour to a stone.

GALLERY A decorative border, usually perforated.

GILDING METAL A base metal alloy, mainly of copper (80 to 90 per cent) and zinc (10 to 20 per cent).

GILT The application of a thin layer of gold to a base metal, i.e. mercurial gilding. Nowadays, gilding is invariably carried out electro-chemically in a plating bath.

GIRANDOLE EARRINGS Triple pendant earrings.

GOLD FILLED An American term used to describe rolled-gold.

GOLD PLATING The covering of base metal with a skin of gold. Today this is done electrolytically in a plating bath.

GRANULATION Individual grains of gold used to decorate jewellery.

GUARD An old name for a long chain.

GUARD RING A ring worn to prevent loss of a more valuable ring.

GYPSY RINGS AND GYPSY SETTING A wide gold ring usually with single diamonds embedded in it. Hence 'gypsy setting'. The effect of being deeply set is often increased by an engraved star whose points spread out from the stone.

INTAGLIO The opposite of cameo. A carved design hollowed out of a gem or decorative mineral.

JUMP-RING A wire ring which can be opened to allow another ring to be attached. Commonly used for charms and chains.

LEMEL Sweepings from a jewellery workshop that are melted and re-refined.

LUNATE Crescent-shaped.

MALLEABILITY The physical property of metal that allows it to be stretched by hammering or rolling. Gold is the most malleable of all metals.

MARCASITE A disulphide of iron. Not so common in nature as pyrite, another disulphide of iron which is usually called 'marcasite' in the jewellery trade.

MARQUISE RING Pointed oval shape.

METAL CORE A term formerly used to describe a form of rolled gold. Not acceptable to assay offices.
MILANESE Small links of chain interwoven to form a mesh ribbon.
MIZPAH RINGS Letters MIZPAH were engraved, or left in relief, on broad gold rings. The word has a biblical reference.
MOUNT The metal part of a piece of jewellery before the stones are set into it. Hence, the mounter is the man who produces the mount into which the setter inserts the stones.
NIELLO WORK Incisions cut into metal which are filled with a soft dark-grey compound composed of lead, silver and copper with borax and sulphur. In modern niello the ground surrounding the design is filled in rather than the design itself. In cheaper goods the niello is merely painted on as a background and recesses are not cut into the metal.
PARURE A matching suite of jewellery.
PAVÉ A number of stones set in to a mount to produce a paved effect.
PENDANT An ornament that hangs from a chain or a pin.
PICKLE SOLUTION Acid diluted with water used for cleaning purposes such as removing the coating of oxides formed during annealing.
PINCHBECK An alloy of copper (83 parts) and zinc (17 parts) invented by Christopher Pinchbeck, a London watchmaker (1670–1732). His metal, which was gold coloured, was used for watch cases, jewellery and trinkets.
PIQUÉ D'OR A design formed by driving small gold rods into tortoiseshell or ivory.
PLIQUE-A-JOUR A rare open form of enamelling. Enamel is laid on a temporary metal ground with the different colours separated by wires. After firing, the enamel is separated from the background to produce a stained glass window effect.
REFINING The processes of removing impurities from a precious metal.
RIVIÈRE Gem-set necklaces, with each gem in its own collet graduating in size from a central gem.
ROLLED GOLD Gold is simulated by bonding a block of gold to a thicker block of copper and rolling out the bimetallic block to form sheet from which jewellery and watch cases are made.
SAUTOIR Long ropes of pearls ending with two long tassels of pearls.
SETTER The craftsman who secures stones in a mount.
SHANK The part of a ring that encircles the finger and to which the settings for the stones are attached.
SHOULDER The upper part of a ring shank near the setting which is often decorated.
SLEEPER RING A thin gold ring used to keep the hole open after an ear has been pierced for earrings.
SNAKE CHAIN Often referred to as Brazilian chain. Small metal cups joined to form a flexible chain.
STRASS A name applied to paste or glass imitation stones. Derived from its inventor Josef Strass.
TOUCHSTONE A block of basalt, black quartz or ceramic material. Used for testing metals by the comparison of streaks produced by rubbing the metal on the stone with those produced by needles of known quality.
TRANSLUCENT Passing light imperfectly. For example, translucent enamel.

China

ACID GOLD A process of producing intaglio gold decoration by etching the glaze with hydrofluoric acid and then covering the incised area with gold.

BAT A slab of body placed on the jigger to make a plate or on the jolly to produce a cup.

BISCUIT Pottery which has a matt surface after firing. Hence a biscuit kiln and biscuit firing.

BODY The prepared ingredients from which articles of pottery are made.

BONE The calcined cattle bones which are ground down and used to produce bone china.

BONE CHINA A type of pottery made from 50 per cent bone ash, 25 per cent china clay and 25 per cent Cornish stone.

CASTING Pouring slip into a plaster of Paris mould to produce hollowares or figurines.

CERAMIC A generic term applied to all forms of fired clay. This is derived from the Greek *keramikos,* meaning pottery.

CLAY Decomposed granite consisting of aluminium oxide, silica dioxide and water. When the water is driven off by heat this plastic material becomes permanently hard.

COBALT BLUE The oxide of cobalt which, when mixed with a powdered glass, produces the blue enamel used mainly for under-glaze decoration.

CORNISH STONE A feldspathic stone found in Cornwall which is similar to the China stone used to produce porcelain.

CRAZING Fine cracks in the glaze caused by shrinkage during the cooling of the glaze. It can be caused by placing wares in too hot an oven.

DELFTWARE A tin-glazed earthenware made in Delft, Holland. The term ‘English Delftware’ is used to describe similar wares made in England.

DRESDEN CHINA A name applied to Meissen porcelain, but also used for other wares made in Dresden and not infrequently to figurines in the Meissen style made elsewhere.

EARTHENWARE Low-fired clay which remains porous unless covered with a glaze. Terracotta is an unglazed earthenware. Faience is a tin-glazed earthenware.

ENAMEL Metallic oxides mixed with ground glass and suspended in an oil.

FAIENCE A tin-glazed form of earthenware sometimes known as Majolica or Delftware.

FELDSPAR A rock largely composed of aluminium silicate which acts as a temper for kaolinite particles. It is used in bodies and glazes.

FETTLE To trim excess material from a piece of pottery before firing. The fettling tool is a small-bladed scalpel.

FIRE-SPECKLING A defect in pottery caused by alien black material such as carbon in the body or in the glaze.

FLUTED A ware decorated with grooves imparted by a mould.

GILDING Applying gold which is subsequently fired.

GLAZE A glass-like substance applied to a body after it has been fired. Glazes are used to make earthenwares waterproof, to decorate stonewares and to improve the appearance and produce the translucency of bone china and porcelain.

GLOST FURNACE The furnace in which a glaze is fused onto the body of pottery wares.

HYDROFLUORIC ACID The only acid which will attack silica products such as pottery and glass. It is used in the acid gold process.

IMARI A type of Japanese pottery widely copied in Europe.

IN-GLAZE DECORATION Decoration applied to a glazed ware. The ware is then fired at a temperature which causes the glaze to melt and absorb the colour. Colours are restricted to those enamels which will withstand the high temperature of the kiln.

JIGGER The semi-automatic machine used for making a plate. It consists of a powered turntable on which a mould is placed and a profiling tool that shapes the base of the plate.

JOLLY A semi-automatic machine similar to the jigger, which is used for making cups.

KAOLINITE Aluminium silicate. This is the major constituent of kaolin, the white clay used to produce porcelain and bone china.

KILN An alternative name for furnace, e.g. biscuit kiln, glost kiln.

LITHOGRAPHIC PRINTING The application of enamel decoration printed on a plastic sheet either by the lithographic or the silk screen process. This decoration is usually polychromatic. Subsequent firing fixes the enamels and burns off the plastic sheet.

MAJOLICA (OR MAOLICA) Named after the island now known as Majorca, this was originally a tin-glazed earthenware made in Italy. The word was later used to describe similar earthenwares made in England.

MARKS Over the centuries potters have marked their wares in various ways, using painted, incised or impressed marks. A number of guides to antique pottery marks are available.

MOULD The hollow plaster of Paris block into which slip is poured to produce hollowares and figurines.

OVER-GLAZE DECORATION Enamels painted or printed onto the ware after it has been glazed.

PIERCED DECORATION Decoration produced by cutting holes in the body before it is placed in the biscuit furnace for firing.

PITCHER MOULD A recessed mould into which body is pressed to produce a casting. The cameo decoration on jasper wares is cast in pitcher moulds.

PORCELAIN A translucent body evolved by the Chinese. The secret of porcelain was discovered at Meissen in the early eighteenth century. It consists of a clay largely composed of kaolinite and feldspathic stone, nowadays usually with the addition of quartz.

POTTER'S WHEEL The horizontal wheel on which a potter throws his pots. Sometimes operated by a foot-treadle but nowadays usually mechanically powered.

POTTERY A generic word to cover all articles made from clay, including porcelain, bone china, stoneware and earthenware.

PRINTED DECORATION Decoration produced by applying an enamel, printed on tissue paper from a copper plate.

PUGGING The process of kneading the body in a purpose-built machine to ensure a smooth mixture with the correct water content.

RELIEF DECORATION Decoration that stands out, in relief, from the background.

RESIST A protective material. It is used in the acid gold process to restrict the etching effect of the acid to the required areas.

SAGGER A protective casing of fire clay used to contain wares during firing in the kiln.

SLIP Body in a liquid state used for pouring into moulds to cast wares. It is also used to decorate earthenwares and stonewares when it is applied after the wares have been formed.

SPONGE To smooth the surface of pottery with a damp sponge before firing.

STONEWARE A ware which is part-way between an earthenware and a porcelain. It consists of a mixture of clay and a fusible stone and is fired at a high temperature, which results in its being impervious to water.

TERRACOTTA Unglazed earthenware of an orange-brown colour, used mainly for garden ornaments.

THROWING The process of fashioning a ware on the potter's wheel.

TOW To polish the surface of pottery with a fibre mop, usually made of flax but sometimes of hemp or jute.

TRANSLUCENT Allowing the passage of diffused light. Bone china is the most translucent form of pottery.

UNDER-GLAZE DECORATION Enamel applied to the biscuit body before the glaze is applied and fired. A technique developed by the Chinese potters and widely used in Europe in the eighteenth century. Colours are restricted to those enamels which will withstand the temperatures of the glost kiln.

VITREOUS Glassy. Glazes give pottery a vitreous surface.

Silverware

ACANTHUS ORNAMENT A leaf-like decoration derived from classical architecture. It is often seen on eighteenth century silver.

ANNEALING Keeping metal in a malleable condition by heating and cooling after work hardening has taken place.

ANTHEMION An ornament introduced by the Adam brothers into England after its discovery on the walls of Herculanium. It consists of a stylization of honeysuckle flowers.

APPLIQUÉ Applied ornament.

BALUSTER From the balusters of a classical balustrade. It is used to mean a complex curved form such as is found on the stem of an early eighteenth century candlestick, i.e. baluster candlesticks.

BAROQUE ORNAMENT Ornament inspired by Renaissance architecture. Used on 16th, 18th and 19th century silverwares.

BAYONET JOINT Consists of two projecting lugs fixed to the top half of a cylindrical part, and engaged in two notches in the lower half. It is used for casters, cruets, etc.

BEAKERS Normally straight-sided drinking vessels. Some have slightly everted lips.

BLACKJACKS Waxed and blackened leather drinking vessels, somewhat like riding boots, with a silver mounted rim and handle. Most date from the seventeenth century. They vary in height from about 7 to 15 in., and hold between a quart and a gallon. Large blackjacks may originally have been used for carrying beer or ale from the cellars.

BLEEDING BOWLS Small shallow silver bowls used by surgeons between about 1670 and 1720. They have bulging sides, a flat everted rim, and a single handle.

BRITANNIA METAL An alloy used for hollowares consisting of approximately 93 per cent tin, 5 per cent antimony and 2 per cent copper. Wares made of Britannia metal are usually sold with an electroplated deposit of silver on them. Britannia metal is also known in the trade as 'white metal', and sometimes as 'Vickers metal', after James Vickers who is supposed to have first used it to produce tea-pots, spoons, etc. in about 1780.

BUFF A cloth or leather mop used for polishing, hence 'to buff' a piece of silverware.

BURNISH To impart a bright finish with a hardened steel tool.

CANDELABRUM A candlestick with two or more branches.

CARTOUCHE An area of plain silver, surrounded by ornamental designs, intended to have a coat of arms engraved or chased on it.

CASTER A vessel for holding and scattering pepper, salt, or sugar. The first casters made their appearance after the restoration of Charles II. They usually consist of a body with a separate pierced and domed cover through which the contents can be cast upon food. The method of joining the cover to the body is mostly by means of a bayonet fixture or of a snap-in rim. Screw tops are not common. In the eighteenth century casters were often made in sets of three – two small and one larger. Other types of caster are the spice dredger said, probably erroneously, to have been used in the kitchen. This is fitted with a loop handle on the side, a style still sometimes used for small modern pepper casters. There is also the muffiner, a small caster, often octagonal in form, for spreading cinnamon on hot buttered muffins.

CHASING Decoration applied with a hammer and punch. Repoussé-chased means raised decoration to which the detail has been applied by the chaser. Cast-chasing is chased up casting.

CHOCOLATE POTS First made after the introduction of drinking chocolate from the West Indies at the end of the seventeenth century. They are rare, and since coffee- and chocolate-pots conformed to much the same shape, distinction between them is often very difficult. Many chocolate-pots were provided with a small additional lid in the cover through which the stirrer (molinet) could be inserted.

COASTER A wine bottle holder. The first coasters date from the late eighteenth century when it was found convenient to place wine bottles in a stand (usually of silver with a wooden base) to protect the table top. The base is usually covered with baize. Pairs were not uncommon, and sometimes two coasters were mounted on a stand on wheels, when they were known as wine wagons or double coasters.

CREAMERS A vessel made to hold cream.

CUT-CARD WORK A form of applied ornamentation consisting of shapes cut from thin sheets of silver and applied to silver articles.

DAMASCENING A method of decoration produced by inlaying with gold or silver.

DISH RINGS Usually of Irish make though the earliest known example was made in England. They served as holders for dishes, and were probably first intended to protect the table from the heat of the dish. Most are spool-shaped with intricately pierced motifs decorating the sides.

EPERGNE A centrepiece for a table. In medieval and Tudor days the importance of a guest accorded him a position above or below the salt which was placed before the host. The use of the great salt died out with dining in the great hall, but the idea of a centrepiece was retained in some households. Towards the end of the eighteenth century the large centrepiece was reintroduced from France. Many of these centrepieces incorporated candelabra and fruit bowls, some even incorporated casters and cruet bottles.

FINIAL The ornament topping a piece of silverware. For example, urn finials were placed on the top of Adam silverwares.

FLUTING Vertical channelling used on silverwares.

GADROON A lobed border which looks like a series of hemispheres, sometimes elongated. It was used on silver of the William and Mary period and at the beginning of the nineteenth century.

HOLLOWARE Dishes, cups, etc., are described as hollowares. Spoons and forks are described as flatwares and knives as cutlery, though the term cutlery is frequently applied also to spoons and forks.

INLAY The cutting away of a metal surface and then infilling usually with a metal or other material of a different colour.

LOVING CUP An urn-shaped cup with two handles.

MAZER A wooden bowl with a silver mount round the rim. They were made in the fifteenth or sixteenth century.

MONTEITHS Bowls with a notched rim, which is usually removable. They are called, supposedly, after the Scotsman who invented them. The rims were usually elaborately ornamented with scrolls and cherub heads. They were made in the late seventeenth and early eighteenth centuries and were probably used to rinse or cool wine glasses.

OLD SHEFFIELD PLATE Plate made from bi-metallic sheet produced by bonding a block of silver to a block of copper and rolling it down.

PARCEL-GILT An old term meaning 'partly gilt'.

PICKLING SOLUTION Various acids (nitric, hydrochloric, sulphuric) in water, used for removing films of oxide and sulphide from the surface of metals.

PIERCING Decoration that is produced by cutting through the metal, originally with a hammer and punch and later with a piercing saw or with press-tools.

PLACE SETTING The cutlery and flatware necessary to set one place at a table. Nowadays, some manufacturers offer their cutlery, and flatware boxed as place settings.

PLANISHING Evening out the hammer marks on a hand-raised silver vessel with a planishing hammer.

PLATE A term correctly used to describe silverwares, but 'silver plate' is a term frequently used, incorrectly, to describe silver-plated wares.

PORRINGER Small two-handled bowls introduced about 1650. Some were provided with covers which could be reversed to form a small dish.

QUAICH A shallow bowl with straight handles, of Scottish origin.

REED A pattern often seen on spoons and forks consisting of narrow convex ridges.

REPOUSSÉ WORK Relief decoration hammered up from the back. Detail was added by chasing – hence repoussé-chased.

SCONCE A wall ornament to which candle holders were attached.

SNUFFER Scissor-like implement used for extinguishing candles.

SPINNING Producing hollowares by forcing metal over a chuck in a lathe.

STANDING SALT A tall salt used as a centrepiece in the sixteenth and seventeenth centuries.

STERLING A silver alloy which contains 925 parts per 1000 of silver.

STIRRUP CUPS These are sometimes still handed to the huntsman while he is on horseback. They have no foot, and are often shaped as a horse's head, fox's head or hound's mask.

TAZZA Italian word meaning 'cup'. Often applied incorrectly to a flat dish supported on central stem. The word 'cup' is preferable to tazza.

VINAIGRETTE A small box for containing aromatic vinegar soaked in a sponge.

WAITER A small tray, often footed.

WINE FUNNEL Wine funnels and wine funnel stands were made in silver towards the end of the eighteenth century. The funnel is tapered and bent at the end to direct the stream against the side of the vessel into which the wine was poured, so avoiding undue agitation of the liquid.

Horology

ARBOR An axle in a watch or clock.

AUTOMATIC A self-winding watch.

BALANCE The controller used in watches and clocks. Through the escapement it controls the power and activates the gears to move the hands. The balance is impulsed by the escapement.

BALANCE SPRING The spring which causes the recoil of the balance wheel. Often called the hairspring.

BALANCE STAFF The axle or arbor of the balance.

BANJO CLOCK A clock in the form of a banjo, a style very popular with American makers in the nineteenth century.

BARREL A brass drum holding the mainspring of a watch or a spring-driven clock.

BEAT A normal watch beats 18 000 times an hour, or 5 times a second. The 36 000 train watches beat 36 000 times an hour.

BEZEL The metal rim holding the crystal of a watch.

BOB The lump of metal on the end of a pendulum.

BRIDGE A metal arch supporting the arbor of a train-wheel.

CENTRE-SECONDS A seconds hand attached to the pinion at the centre of the dial of a watch or clock. Sometimes called 'sweep-seconds'.

CHAPTER RING The ring calibrated for hours and minutes which allows the time to be read off.

CHAPTERS The hour marks on a chapter ring.

CHATELAINE An ornamental chain from which a watch was hung. It was originally the chain worn by the chatelaine of a castle from which her keys were suspended.

CHRONOGRAPH A combined timekeeper and timer. The timing mechanism can be stopped and started without interfering with the watch mechanism.

CHRONOMETER A name which horologists would like to reserve for timekeepers fitted with a detent escapement. Nowadays, it is commonly applied to any watch certificated by an observatory or testing bureau.

COCK Like a bridge, but supported only at one end. A balance cock supports the balance staff.

CROWN The serrated button used for winding and setting the hands of a watch.

CRYSTAL Commonly used to describe a mineral watch 'glass'.

CURB PINS Two pins placed on either side of the outer end of a hairspring. They can be moved by moving the index, and in so doing the length of the hairspring is altered and the timekeeping of the watch can be corrected.

EBAUCHE All the parts of a watch movement other than the escapement, balance, jewels and springs. Ebauches S.A. is the giant Swiss company with many factories, that produces the ebauches for the Swiss assembly plants.

ELECTRONIC CLOCK A term loosely used for any clock containing a transistor or a chip. A true electronic clock has electronic circuits instead of mechanical gearing. The solid-state clock is a true electronic clock.

END STONE A flat disc of synthetic corundum placed behind a jewelled bearing. Sometimes called a cap jewel, it takes the end thrust of a pivot and keeps out dust. The end stones on the balance pivots are spring loaded to provide shock protection.

ESCAPE WHEEL The wheel which the pallets of the escapement stop and release.

ESCAPEMENT The mechanism which controls a watch or clock. In a jewelled-lever watch, the escapement consists of a lever with a square-sectioned synthetic corundum pallet in each arm of the fork, and a wheel with L-shaped teeth on which these pallets act.

FUNCTION A term used to describe the various readouts available on an electronic watch. An hour readout is one function, a minute readout another and so on. A five-function watch might display the hours, minutes, seconds, day of the month and month.

FUSÉE A cone-shaped component with grooves cut in it. A length of gut or chain attached to the mainspring barrel is led round the grooves to equalize the power output of the mainspring. It is still used for mechanical marine chronometers.

GRANDFATHER CLOCK Another name for a long-case clock.

GRANDMOTHER CLOCK A smaller version of the grandfather clock.

HAIRSPRING Another name for the balance spring.

HERTZ A measure of oscillation, i.e. times per second. Quartz crystal oscillations are expressed in hertz. Oscillations of a 'megahertz' quartz crystal are measured in millions of hertz per second.

IMPULSE PIN The ruby pin fixed to the fork of the escapement; it impulses the balance.

INDEX The regulator which moves the curb pins that alter the length of the hairsprings.

ISOCHRONISM Occupying equal time. A pendulum or a balance is described as isochronous if the time of its swing does not vary.

JEWEL A watch or clock bearing made of a hard gemstone. Nowadays, jewels are made from synthetic corundum. The jewel acts as an oil reservoir and also reduces wear.

JEWELLED LEVER A watch with a jewelled-lever escapement.

KEYLESS MECHANISM The gearing by which a manually wound watch is wound.

LCD Liquid-crystal display. A continuous readout system applied to solid-state watches. The liquid crystals arranged as segments of a figure '8' become visible as a result of electronic impulses.

LED Light-emitting diodes. A type of digital readout. The diodes light up as a result of electronic impulses to display numerals composed of dots or lines. Such displays were produced by the wearer of an LED watch pressing a button, or activating the readout in some other way, necessary to conserve the limited battery power.

LIGHT CLOCK A clock fitted with photo-electric cells. The light energy absorbed by these cells is translated into electrical energy and this is used to wind the clock.

LIGNE A measure of the size of a watch movement. It is a measure of diameter. A ligne is the equivalent of just under 2.256 millimetres.

LUG The ears on a watch case which support the lug pins to which the strap is attached.

MAINSPRING The coiled spring which provides the power to drive a watch or clock mechanism.

OUT OF BEAT A watch in which the tick is alternately loud and soft because the action is not symmetrical.

PALLET The part of the escapement that detains and releases the escape wheel. In a jewelled-lever movement the pallets are 'jewelled' with synthetic ruby. In a pin-lever the pallets consist of steel pins.

PENDULUM A swinging weight that provides a clock with an isochronous time standard.

PERPETUAL CALENDAR A calendar watch or clock which automatically corrects itself for months of different length and for leap years.

PINION An arbor on which teeth have been cut which mesh with the teeth of the wheels of a watch or clock train.

PIN-PALLET A watch escapement which has steel pins instead of jewelled pallets attached to the forks. It is sometimes called a pin-lever.

PIVOT The end of an arbor or staff. The part which runs in the bearing.

PLATE A brass disc, or plate, with holes in it which forms the frame of a watch or clock.

PLATFORM ESCAPEMENT An escapement mounted on its own platform. Commonly used for small clock movements.

POISE It is necessary to poise a balance wheel by removing metal from parts of the rim or the screws, so that it will run absolutely flat. An out-of-poise balance results in serious positional errors.

POSITIONAL ERRORS The timekeeping of a mechanical watch varies as a result of its being placed with its dial up or its dial down, its winding crown up or its winding crown down. These variations are called positional errors. The main reason for the accurate timekeeping of the 36 000 train is that the faster the beat of a watch the smaller the positional errors.

PUSH PIECE The push buttons which start, stop and return to zero the timing hands on a chronograph or timer.

RATE The amount by which a watch or clock loses or gains.

REGULATING Adjusting a watch or clock to improve its rate.

REPEATER A clock or watch with a mechanism which could be activated to make the striking mechanism repeat the hour and quarters of its last strike. This device was used when a lamp or candle had to be lit before the clock, or watch, dial could be read at night.

ROSCOPF The name once used on the Continent for pin-pallet watches. Roscopf made cheap watches in Switzerland in the middle of the nineteenth century.

ROTOR The eccentric weight in an automatic watch which swings when the wearer moves his arm, and winds the watch through a series of gears. Also part of an electric clock.

SELF-WINDING WATCH Today, usually called an automatic. A watch with a pivoted weight which swings when the wearer moves, and winds the mainspring through a series of gears.

SET-HANDS The mechanism which makes it possible to move the hands of the watch without interfering with its movement.

SHOCK ABSORBER A method of protecting the balance staff from damage by fitting springs behind the end stones to absorb shocks.

SKELETON WATCH OR CLOCK A watch or clock with pierced plates which allow the action of the movement to be seen.

SOLID STATE Without moving parts. An LCD watch is an example of a solid-state watch. A quartz watch with hands is not solid-state, because it has a gear train to impulse the hands.

SPLIT SECONDS TIMER A timer with two seconds hands. One hand can be stopped by pressing a push piece. When the push piece is pressed again this split seconds hand will catch up with the other one. It allows the performance of two athletes to be timed simultaneously.

STOP WATCH A timer which has no timekeeping mechanism. It is used for sport and industry.

SYNCHRONOUS CLOCK A clock driven by a motor synchronized to the mains supply.

TIME-ELAPSE WATCH A watch with a moveable bezel which allows the wearer to see how much time has elapsed since the watch was set. It is a feature of divers' watches and the watches made specially for rally drivers.

TIMER A stop watch used for timing sport and industrial operations.
TRAIN A series of gear wheels and pinions which translate the beat of the escapement into hours, minutes and seconds, and drive the hands.
WATCH TIMER An electronic instrument for testing the rate of watches.
WHEEL A gear wheel in the train of a watch or clock.
WINDING BUTTON The serrated button used for winding and setting the hands of a watch. Often called a crown.
WORLD TIME CLOCK OR WATCH A watch or clock on which it is possible to read off the time in the principal cities of the world simultaneously.

Appendices

Appendix 1

The properties of gemstones and recommended trade names

(by courtesy of the National Association of Goldsmiths)

The nomenclature is not intended to be exhaustive and some of the rare gemstones and minerals are not tabulated

Species	*Colour*	*Recommended trade name*
Amber – Hardness 3–2½. S.G. 1.04–1.10. R.I. 1.539–1.545		
Amber	All colours	Amber, with or without appropriate colour description
Andalusite – Hardness 7–7½. S.G. 3.1–3.2. R.I. 1.63–1.64		
Andalusite	Green, red, brown	Andalusite, with or without appropriate colour description
Beryl – Hardness 7½. S.G. 2.67–2.80. R.I. (Double) 1.57–1.58		
Emerald	Bright green[1]	Emerald
Aquamarine	Pale blue, pale greenish-blue	Aquamarine
Beryl	White	White beryl or Goshenite
	Green[2]	Green beryl
	Golden, yellow	Golden or yellow beryl
	Pink	Pink beryl, Morganite

[1] Colour due to chromium.
[2] Colour not due to chromium.

Species	*Colour*	*Recommended trade name*
Chrysoberyl – Hardness 8½. S.G. 3.70–3.74. R.I. (Double) 1.74–1.75		
Chrysoberyl	Yellow, yellowish-green, yellowish-brown, brown	Chrysoberyl
Chatoyant Chrysoberyl	Translucent yellow to greenish or brownish – showing chatoyancy	Chrysoberyl cat's-eye
Alexandrite	Green to greenish-brown by daylight, red to reddish-brown by artificial (tungsten) light	Alexandrite
Coral – Hardness 4. S.G. 2.68		
Coral	Red, pink, white, sometimes black	Coral, with or without appropriate colour description
Corundum – Hardness 9. S.G. 3.94–4.01. R.I. (Double) 1.76–1.77		
Ruby	Red	Ruby
	Red, with star effect	Star-ruby
Sapphire	Blue	Sapphire
	Blue, grey, etc., with star effect	Star-sapphire
	All colours other than the above	Yellow sapphire, green s., pink s., mauve s., etc.
Diamond – Hardness 10. S.G. 3.52. R.I. (Single) 2.42		
Diamond	White, yellowish-white, yellow, brown, green, pink, red, mauve, blue, black	Diamond
Emerald – See Beryl		

Species	*Colour*	*Recommended trade name*
Feldspar – Hardness 6. S.G. 2.55–2.71. R.I. (Double) 1.52–1.53		
Orthoclase	White	Adularia
	Yellow	Orthoclase
Moonstone	Whitish with bluish shimmer of light	Moonstone
Microcline, Amazonite	Opaque green	Amazonite or Amazon stone
Oligoclase and Orthoclase S.G. 2.66	Whitish-red-brown – flecked with golden particles	Sunstone Aventurine feldspar
Labradorite S.G. 2.71 R.I. 1.55–1.56	Ashen grey with bluish or reddish or yellowish or green gleams	Labradorite

Species	*Colour*	*Recommended trade name*
Fluorite, Fluorspar – Hardness 4. S.G. 3.18. R.I. 1.43		
Fluorite, Fluorspar	Green, yellow, red, blue, violet, etc.	Fluorspar, Fluorite, with or without appropriate colour description
	Banded blue and other colours	Fluorspar or Blue John

Species	*Colour*	*Recommended trade name*
Garnet – Hardness 7½–6½. S.G. 3.3–4.3. R.I (Single) 1.7–1.9		
Garnet	All colours	Garnet
Almandine Hard. 7½. S.G. 3.8–4.2 R.I. 1.76–1.81	Violet-red	Almandine or almandine garnet
Pyrope Hard. 7½–7. S.G. 3.6–3.8 R.I. 1.74–1.76	Red to crimson	Pyrope or pyrope garnet
	Pale violet	Rhodolite or rhodolite garnet
Spessartite Hard. 7. S.G. 3.9 –4.2 R.I. 1.80–1.81	Brownish-red, orange-red	Spessartite or spessartite garnet (or spessartine if preferred)

continued

Species	*Colour*	*Recommended trade name*
Garnet *continued*		
Grossularite Hard. 7 S.G. 3.5–3.7 R.I. 1.74–1.79	Pale green and other colours (translucent)	Grossularite, grossular garnet or massive grossular garnet
	Orange – yellowish-red, Orange – reddish-brown	Hessonite or hessonite garnet
Andradite Hard. 7–6½. S.G. 3.85 R.I. 1.89	Yellow	Andradite or andradite garnet
	Green, yellowish-green	Andradite or Demantoid or demantoid garnet
	Black	Malanite garnet (R.I. 1.8–2.0)
Uvarovite Hard. 7. S.G. 3.42–3.5 R.I. 1.84	Emerald-green	Uvarovite or uvarovite garnet

Iolite – Hardness 7½–7. S.G. 2.6. RI. (Double) 1.54–1.55

Iolite Cordierite	Blue and dingy brown	Iolite or Cordierite

Jade (Pyroxene) –Hardness 7–6½. S.G. 3.2–3.5. R.I. (Double) 1.66–1.68

Jadeite	Green, whitish with emerald green flecks, mauve, brown, orange, opaque to translucent	Jadeite, or Jade
Chloro-melanite	Dark green or nearly black, with white flecks, opaque to translucent	Chloromelanite, or Jade

Jade (Amphibole) – Hardness 6½–6. S.G. 2.9–3.1. R.I. (Double) 1.61–1.63

Nephrite	Green, white, single coloured and flecked, opaque to translucent	Nephrite

Species	*Colour*	*Recommended trade name*
Lapis-lazuli – Hardness 6–5½. S.G. 2.45–2.95. R.I. 1.5		
Lapis-lazuli	Blue (opaque) often with brassy specks of pyrite; opaque whitish-light blue	Lapis-lazuli

Malachite –Hardness 5½. S.G. 2.3–2.4. R.I. 1.50		
Malachite	Green veined, banded	Malachite

Marcasite – See Pyrite and marcasite

Opal – Hardness 6½–5. S.G. 2.0–2.1. R.I. (Single) 1.44–1.46		
Opal	Milky with rainbow-like play of colours	Opal or white opal
	The same on dark background	Opal or black opal
	Transparent straw-coloured or colourless, iridescent	Opal or water opal
Fire opal	Fiery red to browny-red	Opal or fire opal
Matrix opal	Flecks of opal in matrix	Opal matrix

Pearl – Hardness 3½. S.G. 2.71		
Pearl	All colours	Pearl, with or without appropriate colour description

Pearl (Conch) – Hardness 3½. S.G. 2.84		
Conch pearl	Pink, white (no pearly lustre)	Conch pearl, with or without appropriate colour description

Species	*Colour*	*Recommended trade name*
Peridot, Olivine – Hardness 7–6½. S.G. 3.3–3.4 R.I. (Double) 1.65–1.69		
Olivine	Yellowish-green, olive green, brown	Peridot

Species	*Colour*	*Recommended trade name*
Pyrite and marcasite – Hardness 6½–6. S.G. 4.9–5.1		
Pyrite Marcasite	Brassy-grey with metallic sheen	Pyrite or Marcasite

Species	*Colour*	*Recommended trade name*
Quartz – Hardness 7. S.G. 2.65. R.I. (Double) 1.54–1.55		
Rock crystal	Colourless	Quartz or rock crystal
Amethyst (natural colour)	Light or dark violet	Amethyst
Amethyst (heat-treated)	Yellowish, brownish-yellow	Citrine or golden or yellow quartz
	Reddish, reddish-brown, reddish-yellow	Quartz with or without appropriate colour description
	Green	Green quartz
Citrine	Yellow, brownish-yellow	Citrine or golden or yellow quartz
Smoky quartz	Smoky or brownish-yellow to black – when brownish-yellow to brown or smokey brown	Smoky quartz Cairngorm, or brown quartz
Rose-quartz	Milky rose-pink	Rose-quartz

Species	*Colour*	*Recommended trade name*
Quartz (with inclusions)		
Prase	Leek-green	Prase
Chatoyant quartz	Whitish-grey, greyish-green, greenish-yellow, blue, with shimmering streaks of light	Quartz cat's-eye
Crocidolite (pseudo-morph)	Yellowish-brown, brownish-golden yellow with shimmering streaks of light	Tiger's-eye

continued

Species	Colour	Recommended trade name
Quartz (with inclusion) *continued*		
Crocidolite (pseudo morph)	Like tiger's-eye but greyish-blue	Falcon's-eye
Aventurine quartz	Yellowish-browny-red, yellow, brown, red, or green, with small flakes of mica	Aventurine quartz

Quartz (Cryptocrystalline) Chalcedony group

Species	Colour	Recommended trade name
Chalcedony (translucent)	Grey to bluish	Chalcedony
Chrysoprase	Apple-green and light green	Chrysoprase
Cornelian	Red in various shades	Cornelian
Heliotrope	Dark green with red spots	Bloodstone Heliotrope
Jasper	Whitish, yellow, red, green, brown, etc.	Jasper
Plasma	Leek-green	Plasma
Agate Hard. 7. S.G. 2.59–2.67 R.I. 1.53–1.54	Banded in various colours, white, yellow, grey, red, brown, blue, black, etc.	Agate, onyx, sardonyx, etc., as appropriate
	Milky with green or rust-coloured moss-like inclusions	Moss-agate

Ruby and Sapphire – See Corundum

Serpentine – Hardness 2½–5. S.G. 2.5–2.7. R.I. 1.57

Species	Colour	Recommended trade name
Serpentine	Translucent green	Bowenite
	Emerald-green with black spots	Williamsite
	Green, grey-green, whitish- and reddish-brown rock	Serpentine

Species	*Colour*	*Recommended trade name*
Sphene – Hardness 5–5½. S.G. 3.45–3.56. RI. 1.9–2.05		
Sphene	Yellow, green, brown and grey	Sphene, with or without appropriate colour description

Spinel – Hardness 8. S.G. 3.58–3.65. R.I. (Single) 1.72		
Spinel	All colours	Spinel; or red s., pink s., orange s., etc., respectively

Spodumene – Hardness 7–6½. S.G. 3.18. R.I. (Double) 1.65–1.68		
Spodumene	Yellowish-green, brownish-green, pale yellow	Spodumene
Hiddenite	Bright green[1]	Hiddenite
Kunzite	Rose-pink, lilac, violet	Kunzite

[1] Colour due to chromium.

Topaz – Hardness 8. S.G. 3.50–3.56. R.I. (Double) 1.61–1.63		
Topaz	All colours	Topaz; or white t., pink t., blue t., etc., respectively

Tourmaline – Hardness 7½–7. S.G. 3.0–3.20. R.I. (Double) 1.62–1.64		
Tourmaline	All colours	Tourmaline, or red t., green t., parti-coloured t., etc., respectively

Turquoise – Hardness 6–5. S.G. 2.6–2.9. R.I. (Double) 1.60–1.65		
Turquoise	Sky blue, blue, bluish-green, greenish	Turquoise
Turquoise matrix	Flecks of turquoise in matrix	Turquoise matrix

Species	*Colour*	*Recommended trade name*
Zircon – Hardness 7½. S.G. 3.95–4.70. R.I. (Double) 1.79–1.99		
Zircon	All colours	Zircon, or blue z., red z., etc., respectively

Zoisite – Hardness 6. S.G. 3.35–3.38. R.I. 1.703–1.696		
Zoisite	Crystals usually brown but changed to blue by heat treatment. Also found in massive green form	Zoisite

Appendix 2

Exemptions from hallmarking

This appendix lists the articles exempted from compulsory hallmarking under Part II of Schedule I of the Hallmarking Act 1973, as amended by the Hallmarking (Small Silver Articles) (Exemption) Order 1975, and the Hallmarking (Exempted Articles) (Amendment) Order 1975.

Exempted articles

1. An article which is intended for despatch to a destination outside the United Kingdom.
2. An article which is outside the United Kingdom, or which is in course of consignment from outside the United Kingdom to an assay office in the United Kingdom.
3. Any coin which is, or was formerly at any time, current coin of the United Kingdom or any other territory.
4. Any article which has been used, or is intended to be used, for medical, dental, veterinary, scientific or industrial purposes.
5. Any battered article fit only to be remanufactured.
6. Any article of gold or silver thread.
7. Any raw material (including any bar, plate, sheet, foil, rod, wire, strip or tube) or bullion.
8. Any manufactured article which is not substantially complete, and which is intended for further manufacture.
9. Any article which is wholly or mainly of platinum, and which was manufactured before January 1, 1975.

Articles exempt if of minimum fineness

10. Any article which
    (*a*) is wholly or mainly of gold or of silver or of gold and silver assaying in all its gold parts not less than 375 parts per thousand and in all its silver parts not less than 800 parts per thousand; and
    (*b*) was manufactured before the year 1900 and has not since the beginning of the year 1900 been the subject of any alteration which would be an improper alteration if the article had previously borne approved hallmarks.
11. Any musical instrument, where the description is applied to the mouthpiece, and the mouthpiece is of minimum fineness.

12. Any article containing only one precious metal, being a metal of minimum fineness and of a weight less than that specified in the following table:

Gold	1 gram
Silver	7.78 grams
Platinum	0.5 gram

13. Any article which is wholly of one or more precious metals of minimum fineness and which is so small or thin that it cannot be hallmarked.
14. Any article which is of minimum fineness and which is imported temporarily (whether as a trade sample, or as intended for exhibition or otherwise) and for the time being remains under the control of the Commissioners of Customs and Excise.

14A. Any article, any precious metal in which is of minimum fineness, and which either
 (*a*) contains gold and platinum but not silver, and the weight of the gold parts of which exceeds 50 per cent of the total weight of the precious metals in the article, that total weight being less than 1 gram; or
 (*b*) contains silver and either gold or platinum or both gold and platinum, and the weight of the silver parts of which exceeds 50 per cent of the total weight of the precious metals in the article, that total weight being less than 7.78 grams.

14B. Where an article described in paragraph 12 or paragraph 14A above contains materials other than precious metals the article shall not be taken as falling within that paragraph unless it complies with Part III of Schedule 2 to this Act.

Existing exemptions

15. The following articles of gold, if manufactured before January 1, 1975, and (except in the case of articles mentioned in sub-paragraph (*d*) below) of minimum fineness:
 (*a*) rings, except wedding rings, pencil cases, lockets, watch chains and thimbles,
 (*b*) articles consisting entirely of filigree work,
 (*c*) articles so heavily engraved or set with stones that it is impossible to mark them without damage,
 (*d*) jewellers works, that is the actual setting only in which stones or other jewels are set, and jointed sleeper earrings.
16. Subject to the exceptions below, the following articles of silver, if manufactured before January 1, 1975 and (except in the case of articles mentioned in (*e*) below) of minimum fineness:
 (*a*) lockets, watch chains and stamped medals,
 (*b*) mounts the weight of which is less than 15.55 grams,
 (*c*) articles consisting entirely of filigree work,
 (*d*) silver articles the weight of which is less than 7.78 grams,
 (*e*) jewellers works, that is the actual setting only in which stones or other jewels are set.
17. Articles of gold or silver manufactured before January 1, 1975, other than articles mentioned in paragraphs 15 or 16 above, and being of such descriptions as, under any enactment in force immediately before the passing of this Act, to be specifically exempt from hallmarking.

The only change to the exemptions from the Hallmarking laws introduced on January 1, 1999 was to change the date from 1900 to 1920 for articles manufactured which have not been subject to any alteration if the article had previously borne approved hallmarks

'Minimum fineness' means for gold the standard of 375 parts per thousand, for silver the standard of 800 parts per thousand and for platinum the standard of 850 parts per thousand..

Appendix 3

The lever escapement

(*Text and illustrations by courtesy of the British Horological Institute*)

The parts (Figure A.1)

The lever escapement consists of the following parts:
Escape wheel, usually with 15 teeth.
Pallet assembly. Pallet staff 9; lever 7, with notch and horns 12; pallet stones – entrance 10, exit 8. The pallet stones or jewels are generally of synthetic sapphire or ruby. They are assembled firmly in the slots of the pallet frame and are generally secured by shellac to make adjustment easy. The lever terminates in a notch, flanked by two horns, forming part of the safety action (see Figure A.1). *Banking pins* 6, which limit the movement of the lever.

Balance assembly. Balance spring 2; large or impulse roller 4; small or safety roller 5; impulse pin 14. The large roller carries the impulse pin, or ruby pin as it is sometimes called, which operates on the notch of the lever and unlocks the escapement, and also receives the impulse.

The small roller with the guard pin or dart forms the main part of the safety action to prevent overbanking. A crescent-shaped piece, known as the passing hollow, is cut out of its circumference.

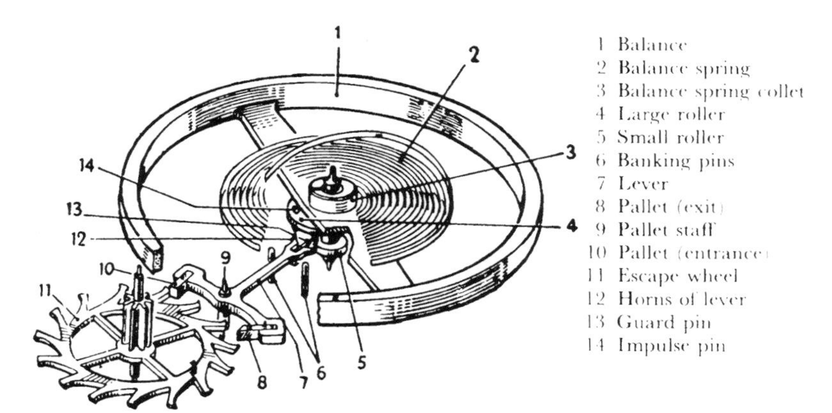

Figure A.1 Parts of lever escapement. This is a straight-line arrangement. The escape wheel has club teeth

Action of the escapement (Figure A.2)

The balance rotates in the direction of the arrow. The locking corner of tooth B of the escape wheel rests against the locking face of the entrance pallet. The inpulse pin is clear of the lever notch, the lever resting against the left-hand banking pin. The angle at which the pallet stone is set in relation to the locking face of the wheel tooth is such as to draw the stone into the wheel and hold it there, so that a knock will not cause the lever to move; thus the guard pin is free of the roller.

Unlocking (Figure A.3). The balance, having completed its swing to the left, is brought to a stop before it starts turning in the opposite direction as a result of the action of the balance spring. The impulse pin enters the notch of the lever and strikes against its right inside face. The pallets are thus caused to turn on the pallet staff and the entrance pallet stone is withdrawn from the shaded tooth B. The instant the locking corner of the entrance pallet stone reaches the locking corner of the tooth, unlocking is complete and the wheel starts to turn.

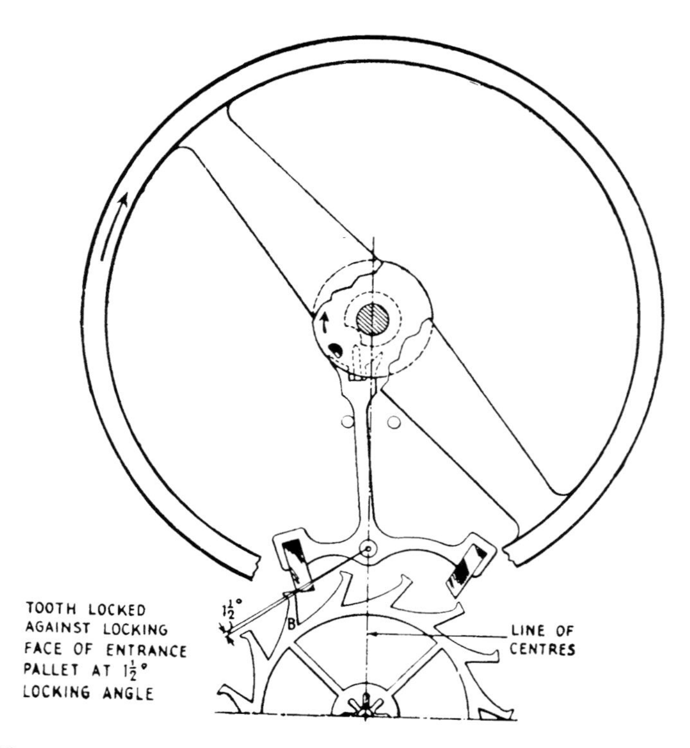

Figure A.2 The escapement when locked

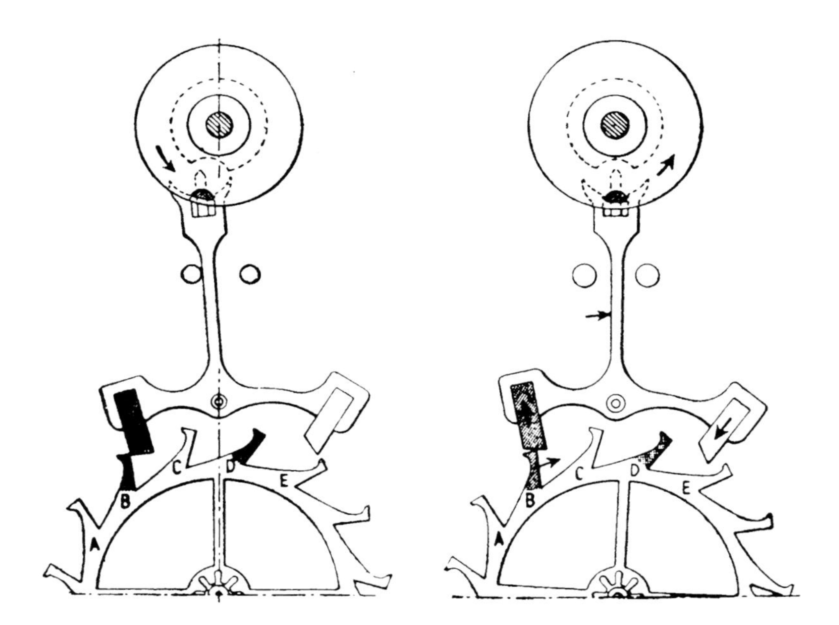

Figure A.3 Unlocking commences

Figure A.4 Impulse – first phase

Impulse (Figure A.4) – first phase. The impulse is in two phases. As the entrance pallet stone continues to move away from the shaded tooth B, and the lever is forced over the right by the action of the impulse pin in the lever notch, the locking corner or toe of tooth B forces its way along the impulse plane of the stone, thereby pushing the stone out of the way. This action causes the left-hand inner face of the lever notch to press on the impulse pin, and gives impulse to the balance. In unlocking, the impulse pin moves the lever and withdraws the pallet stone from the wheel tooth. In the impulse stage this is reversed and the impulse pin is assisted on its way by the lever notch.

Impulse (Figure A.5) – second phase. When the locking corner of the shaded tooth B reaches the discharge corner of the stone, the impulse plane of the tooth takes over and continues the impulse by pressing upon the discharge corner of the pallet stone, causing the pallets to continue turning still further. The instant the discharge corner of the tooth reaches the discharge corner of the stone impulse is complete, and the useful work of this particular 'escape' is finished.

Locking (Figure A.6). The escape wheel now moves freely until the shaded tooth D drops on the exit pallet stone. The impulse pin leaves the lever notch. The lever is close to the right-hand banking pin, but not yet quite in contact with it. The balance continues to rotate in the direction of the arrow. This is the supplementary arc. At the same time that the entrance pallet was progressively moved clear of tooth B during unlocking and impulse, the exit pallet was similarly moved into position to intercept and lock tooth D. Drop brings the locking corner of tooth D on to the locking surface of the exit pallet stone and locking at drop takes place. This does not complete the locking action, because the stone has still to be drawn further into the escape wheel, and the lever brought firmly against the banking pin.

Run to the banking (Figure A.7). As soon as locking at drop has taken place, the run to the banking starts. This movement is caused by the draw action due to the angle at which the pallet stone is set in relation to the locking face of the wheel tooth. It continues until the lever is held against the banking pin and it permits the wheel to turn slightly further after locking at drop. It also ensures that the guard pin will be held clear of the circumference of the safety roller. Locking at drop, plus run to the banking, give the total lock.

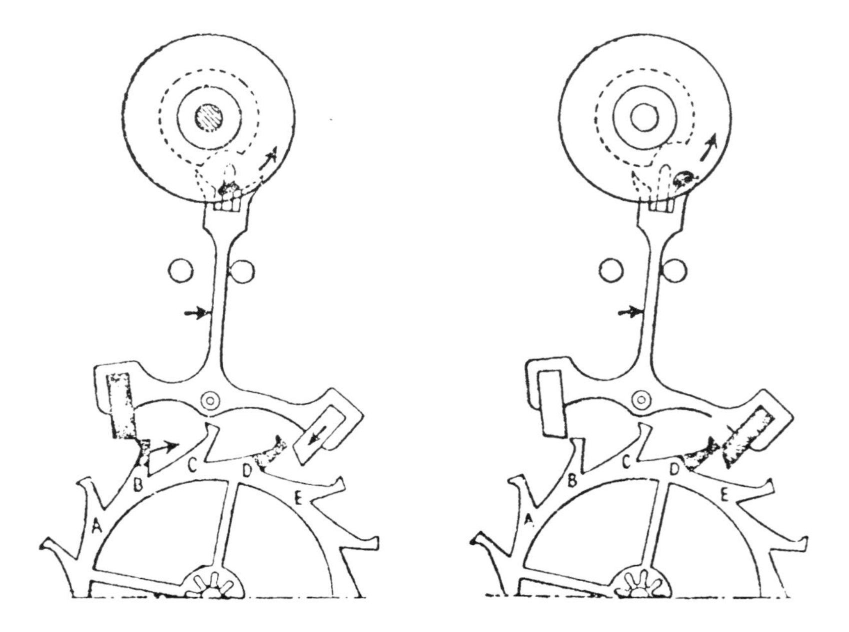

Figure A.5 Impulse – second phase

Figure A.6 Locking at drop

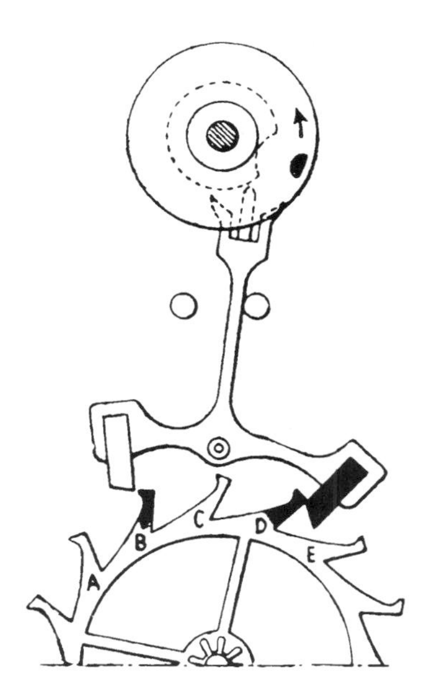

Figure A.7 Run to the banking

Draw provides safety against random operation of the safety action should the watch receive a knock. If the watch should receive such a severe jar as to cause the lever to be moved away from the banking pin, the guard pin or dart makes contact with the safety roller. This prevents the escapement from being unlocked until the guard pin can enter the passing hollow. This can only happen when the impulse pin is about to enter the lever notch to unlock the escapement at the correct instant. The horns of the notch take over the safety action during the short time when the guard pin is entering the passing hollow and until the impulse pin is safely in the notch.

When the balance makes its next return swing the same action takes place in the opposite direction. Tooth D is unlocked and gives impulse to the exit pallet stone. Tooth A is then locked on the entrance pallet stone – and so the action is repeated.

The balance is absolutely free or detached throughout its swing, except for the comparatively short time the impulse pin is actually in the lever notch, while unlocking and impulse are taking place. The free part of the swing is called the free, or supplementary, arc.

Appendix 4

Touchstone testing in the workshop

The most effective way of testing metals in the retail jeweller's workshop is to use the touchstone. This has served jewellers well for many centuries, and allows them not only to identify one metal from another, but with practice will enable them to determine the carat quality of unhallmarked gold. Testing sets that eliminate the need to use a touchstone and needles have become available. Gradually, because they are more convenient and easier to use, they are tending to replace the touchstone.

A touchstone is a slab of basaltic rock, which is harder than most metals, and fine-textured. Good touchstones are extremely difficult to find these days and the jeweller may have to make do with a 'stone' made of ceramic material.

Besides a touchstone a set of touch needles of known carat quality are required. Nitric acid and hydrochloric acid are also required. For the preliminary tests a dilute nitric acid is used, consisting of one part of distilled water and one part of concentrated acid. Nitric acid has no effect on gold, but it attacks such metals as copper and silver, with which it is usually alloyed. The principle of the touchstone test is to make a streak on it with the metal to be tested and then drip a few drops of the acid on to the streak and study the reaction. To ascertain the carat quality of the gold the reaction of the test streak is compared with a streak made with a touch needle of known carat quality.

To ensure that the sample is not rolled gold or gold plated a scratch should be made in an inconspicuous place with a small file. This will remove the outer layer of gold. The initial test is then made by placing a drop of the nitric acid on the streak. If the streak is yellow, the indications are:

(*a*) Streak dissolved indicates a copper alloy, for example the base metal beneath the gold layer of rolled gold or gold plate.
(*b*) Streak darkened indicates 9 carat gold.
(*c*) Streak unchanged indicates 14 carat gold, or a higher carat.

If the streak is white, the indications are:

(*a*) Streak dissolved indicates either a silver, nickel or tin alloy.
(*b*) Streak darkened indicates a 9 carat white gold.
(*c*) Streak unchanged indicates a higher than 9 carat white gold, a platinum or palladium jewellery alloy, a stainless steel, an aluminium alloy or a lead alloy.

The final test is made by adding a drop of hydrochloric acid to the drop previously placed on the streak. The indications are now:

(*a*) A white precipitate in the liquid indicates a silver alloy.
(*b*) A darkened streak, which dissolves, indicates a 9 carat gold.
(*c*) A white streak, unchanged after the initial test, but which becomes coated with a white film, indicates a lead alloy.
(*d*) A white streak unchanged after the initial test, but which dissolves on the addition of hydrochloric acid, indicates an aluminium alloy.
(*e*) A streak unchanged after the first test, but which is very slowly dissolved on the addition of hydrochloric acid, indicates either a higher gold than 9 carat, a palladium jewellery alloy of 950 fineness, or a stainless steel.
(*f*) A streak which is unchanged on the addition of hydrochloric acid to the first drop indicates a platinum jewellery alloy of 950 fineness.

To complete the simple touchstone test, for discrimination among the white alloys, a slip of white blotting-paper is used to absorb the liquid from the touchstone. A faint brown stain indicates a palladium–white gold. A brown stain indicates a palladium jewellery alloy. A faint yellow-green stain indicates a nickel–white gold or stainless steel, where white streaks were not dissolved during the initial test with nitric acid. (See indications white (*c*), and final (*e*).) To distinguish between nickel–white gold and stainless steel the touchstone is cleaned and dried, a new streak is made, the touchstone is warmed, and one drop of hydrochloric acid is added. By this treatment stainless steel is dissolved, but nickel–white gold is not.

An effective confirmatory test for white alloys may be made in the following manner: a streak is made on the touchstone, a little finely-powdered potassium iodate (less than one-tenth of a grain) is sprinkled on the streak, and one or two drops of concentrated hydrochloric acid are added. After one or two minutes the touchstone is washed under a very gentle stream of water and the streak is observed. A 950 fine platinum jewellery alloy streak remains unchanged, a 950 fine palladium jewellery alloy streak shows a red colour, nickel–white gold or stainless steel is dissolved and a palladium–white gold gives a paler colour than a palladium jewellery alloy.

The simple touchstone test does not give any reliable indication of the fineness of a silver alloy; to ascertain this an additional test is necessary, using a solution prepared by dissolving 30 grains of silver nitrate in one fluid ounce of distilled water and adding a drop of concentrated nitric acid. A drop of the solution is placed on a fresh streak. If the streak is unchanged, the indication is sterling silver (925 fineness), or Britannia silver (958.4 fineness). A brown stain on the streak indicates a fineness below 900, and the depth of the colour intensifies as the proportion of base metal increases until, with a white base metal, a black stain is obtained.

A simple distinction between sterling or Britannia silver and nickel silver (nickel–copper–zinc alloy) can be made with a solution of one part by weight of chromic acid, six parts by weight of concentrated nitric acid, and two parts by weight of water. When a drop of this solution is applied to the streak, a red coloration indicates silver, whereas a green coloration indicates nickel silver.

If it is necessary to ascertain the fineness of a yellow gold alloy, a series of test acids is required:

For gold alloys below 9 carat – One fl. oz. of distilled water is mixed with 1 fl. oz. of concentrated nitric acid, a few drops at a time.

For gold alloys from 9 to 14 carat – One fl. oz. of No. 1 acid (above) is mixed with 3 drops of a saturated solution of common salt in distilled water.

For gold alloys from 14 to 18 carat – One fl. oz. of distilled water is mixed with 2 fl. oz. of concentrated nitric acid and 10 drops of a saturated solution of common salt in distilled water.

For gold alloys above 18 carat – One fl. oz. of No. 3 acid (above) is mixed with 1 grain of potassium iodide.

In addition, a set of test needles of alloys of known carats is necessary so that comparison streaks can be made near to the streak of the alloy under examination. Thus, the reaction of the alloy under test to the appropriate acid can be closely compared with the reaction of a known alloy.

Experience and good judgement are required for reliable evaluation of gold alloys.

An indication of the fineness of a platinum alloy may be obtained by using a test acid made up of 1 fl. oz. of concentrated hydrochloric acid, one-third fl. oz. concentrated nitric acid, one-sixth fl. oz. distilled water, and one-third oz. of ammonium chloride crystals. This test acid must be allowed to stand until it has a yellow colour; it is then ready for use. The streak of a platinum alloy, of fineness below 950, is attacked in proportion to the relative amount of base metal present in the alloy.

The indications of the touchstone tests may be found to depend somewhat on the conditions of test, e.g., on the type of touchstone used, the temperature of the reagents, and the experience of the operator. Reliability in the identification of alloys may nonetheless be ensured providing touch needles of the standard alloys are available, so that comparison tests may be made.

If a touchstone is not available, spot tests may be made by the application of the reagents to a small area which has been filed at an inconspicuous place on the article. Owing to the somewhat limited area available, in comparison with a touchstone test, it may be advantageous to watch the course of the spot test through a magnifying glass, and finally to note the extent to which the metal has been stained or etched by the reagents.

All the chemicals used in the touchstone tests are poisonous, and the acids are highly corrosive to the skin and clothing. Inhalation of acid fumes must be avoided.

Appendix 5

Some useful tables

Weights for precious stones and pearls

The metric carat of 200 milligrams is the unit used for precious stones other than pearls, which are generally reckoned by grains which are quarter-carats (i.e. 50 milligrams).

4 grains	=	1 metric carat
1 metric carat	=	200 milligrams
5 metric carats	=	1 gram

A useful comparison is:

141¾ metric carats = 1 oz. avoir.

Troy, avoirdupois and grams

	Troy grain	*Troy dwt.*	*Troy ounce*	*Avoir. ounce*	*Avoir. pound*	*Grams*	*Kilo-grams*
1 troy gram	1	0.04167	0.00208	0.00229	0.00014	0.0648	0.00006
1 troy dwt.	24	1	0.05	0.05486	0.00343	1.55518	0.00156
1 troy ounce	480	20	1	1.09714	0.06857	31.1035	0.03110
1 avoir. ounce	437.5	18.22917	0.91145	1	0.06250	28.34954	0.02835
1 avoir. pound	7000	291.66666	14.58333	16	1	453.59264	0.45359
1 gram	15.43235	0.64301	0.03215	0.03527	0.00220	1	0.001
1 kilogram	15.432349	643.0145	32.15073	35.27394	2.20462	1000	1

Some British coins as weights

Gold coins	*Standard weight* (grains)	*Remedy* (grains)
Sovereign £l	123.27447	0.20
Half-sovereigb – 10*s*.	61.63723	0.15
Silver (up to 1946)		

Oz. troy and grams equivalents

Oz. troy	*Grams*	*Grams*	*Oz. troy*
1	31.104	1	0.0321
2	62.207	2	0.0643
3	93.310	3	0.0964
4	124.414	4	0.1286
5	155.517	5	0.1607
6	186.621	6	0.1929
7	217.724	7	0.2250
8	248.828	8	0.2572
9	279.931	9	0.2893

Melting points of metals

Metal	°C	°F
Osmium	2700	4892
Iridium	2454	4449.2
Ruthenium	2450	4442
Rhodium	1955	3551
Chromium	1830	3326
Platinum	1773	3223.4
Palladium	1555	2831
Copper	1083	1981.4
Gold (fine)	1063	1945.4
22 ct. yellow	1003	1837.4
18 ct. yellow	905	1661
14 ct. yellow	838	1540.4
9 ct. yellow	830	1526
Silver (fine)	961	1761.8
Silver (sterling)	893	1639.4
Brass	940	1724
Aluminium	660	1220
Zinc	419	786.2
Tin	232	449.6

Specific gravity

Specific gravity is the ratio of the weight of a metal (or other body) to the weight of an equal volume of water at a standard temperature and pressure, the weight of water being taken as 1.

Osmium	22.5
Iridium	22.41
Platinum	21.4
Gold (fine)	19.36
22 ct. yellow	17.7
18 ct. yellow	15.58
14 ct. yellow	13.4
9 ct. yellow	11.3
Rhodium	12.44
Ruthenium	12.2
Palladium	12.00
Silver (fine)	10.53
Sterling	10.40
Copper	8.94
Brass	8.5
Tin	7.3
Chromium	7.14
Zinc	7.14
Aluminium	2.70

Index